T0177507

A Place in History

Photo of John Kendrew on the grounds of Peterhouse, Oxford University, in the early 1960s. Reproduced from an item in the Kendrew Archives, Weston Library, Oxford

A Place in History

The Biography of John C. Kendrew

PAUL M. WASSARMAN

OXFORD
UNIVERSITY PRESS

OXFORD
UNIVERSITY PRESS

Oxford University Press is a department of the University of Oxford. It furthers
the University's objective of excellence in research, scholarship, and education
by publishing worldwide. Oxford is a registered trade mark of Oxford University
Press in the UK and certain other countries.

Published in the United States of America by Oxford University Press
198 Madison Avenue, New York, NY 10016, United States of America.

Library of Congress Control Number: 2019036511
ISBN 978–0–19–973204–3

1 3 5 7 9 8 6 4 2

Printed by Sheridan Books, Inc., United States of America

For Eveline

A man's soul and reason are his own and he must
go whither they beckon.

Arthur Conan Doyle

It is the fate of every human being to be a unique
individual, to find his own path, to live his own life,
to die his own death.

Oliver Sacks

Contents

Figures and Tables

Figures

Tables

Preface

This is a biography of Sir John Cowdery Kendrew, from his birth in Oxford on Saturday, March 24, 1917, to his death in Cambridge on Saturday, August 23, 1997. In research begun in 1946, as a 29-year-old research student at the Cavendish Laboratory in Cambridge and continued independently into the early 1960s, John was the first to determine the three-dimensional structure of a protein at atomic resolution using X-ray crystallographic methods. Low- and high-resolution structures of myoglobin, a protein found in abundance in muscle tissue of all vertebrates, were solved in 1957 and 1959 and published in the journal *Nature* in 1958 and 1960, respectively.

John's tremendous achievements demonstrated at long last that proteins were not aggregates of small molecules but were true macromolecules, and their structures could be solved by X-ray crystallographic methods. For the first time, one could visualize predicted structural features of proteins, such as alpha-helices, in some detail. It was not until five years later, in 1965, that a high-resolution structure of a second protein, the enzyme lysozyme, was published and not until the late 1960s that several other high-resolution structures of proteins, including hemoglobin, became available. It should be remembered that John was *nulli secundus*, "second to none," having determined a high-resolution structure for myoglobin in 1959.

Having solved the three-dimensional structure of myoglobin, an accomplishment worthy of a share of the Nobel Prize in Chemistry in 1962, John began an odyssey away from the life of a bench-scientist to that of an organizer, administrator, and promoter of science on an international scale. He became a central figure in the evolution and spread of molecular biology in Europe and America during the 1950s, '60s, and '70s and left behind a rich legacy matched by few others, even other Nobel laureates.

In 1958, John founded the *Journal of Molecular Biology* and served as its editor-in-chief for nearly 30 years. The journal drew articles from the best investigators in a rapidly emerging field of research and is still published today, 60 years later. At about the same time John became a part-time consultant to the British Ministry of Defense, which led to other governmental appointments and eventually to a knighthood in 1974. From his early 40s into his late 70s, John took on numerous national and international leadership roles that allowed him to serve as a positive force for science, especially for European science.

Among other national and international activities, John doggedly pursued the goal of creating an international laboratory of molecular biology in Europe. This goal was finally realized with establishment of the European Molecular Biology Laboratory (EMBL) in Heidelberg, Germany, but only after more than a decade of John's tireless efforts and relentless diplomacy. Max Perutz pointed out that "Kendrew became the moving spirit in the founding of EMBL, which would never have come into being but for his drive and shrewd diplomacy."[1] John was appointed the first director-general of EMBL in 1974 and served in that capacity until 1982 when he became the 33rd president of St. John's College, Oxford. In 1987 John retired to The Old Guildhall, his home in Linton, and he died in Cambridge 10 years later in 1997, at the age of 80.

There is a widely held belief that scientists do not make particularly good subjects for biographies. Erwin Chargaff felt that many scientists "lead monotonous and uneventful lives"[2] and Peter Medawar concluded that "the lives of academics, considered as Lives, almost always make dull reading."[3] On the other hand, science writer and biographer Thomas Hager believes that "the lives of scientists and effects science has on our world are much more nuanced, complex, interesting . . . and far more important, often in unexpected ways . . . than most people realize."[4] John's life was neither monotonous nor uneventful, and his legacy continues to be recognized today by research scientists and major institutions.

In this biography, certain aspects of John's early life are described. We see him as an only child brought up from the age of four without a mother by a rather difficult father, as a schoolboy at Dragon School and Clifton College, as an undergraduate at Cambridge University, and as a scientific advisor in Britain, the Middle East, and South East Asia during World War 2 (WW2). It was not until he was a teenager that John reunited with his mother who was by then a well-known Italian art historian living in Florence. It was during the war that John worked in Ceylon/Sri Lanka with John Desmond Bernal, an eminent crystallographer who would significantly affect the direction of John's post-war career. Also during the war, John met and fell in love with Shoshana "Suzy" Ambache in Cairo and seriously considered marriage to her and conversion to Judaism. In 1948 he married Elizabeth Gorvin-Jarvie, a medical student and widow of John's close friend, John Jarvie, who was killed tragically in WW2. They divorced in 1956, and John remained a bachelor for the rest of his life, but had relationships with a number of women until his death in 1997.

These aspects of John's life, in particular, molded his personality, influenced his decision to become a scientist, and affected the course of his professional career and his personal relationships. Although science and technology always played principal roles in John's life, his love of the arts, including classical music, literature, painting, photography, architecture, and archeology were his passions and enriched his life. For John it can truly be said that *scientia sine ars nihil est*, "science without art is nothing."

Following John's death in August 1997, memorials were held in Cambridge and Oxford and obituaries appeared world-wide in newspapers and scientific journals (Appendix A.2). However, the only substantial published attempt at John's biography was written in 2001 by Ken Holmes for the *Biographical Memoirs of Fellows of the Royal Society*. This is a thoughtful summary of John's career and offers keen insight into his background and contributions to science and society. Ken and his wife Mary knew John for 40 years in Cambridge and Heidelberg and thought of him as a "much appreciated older brother for whom [they] had a deep affection."[5] Ken Holmes also hosted a long, in-depth interview with John in Cambridge on Wednesday, June 18, 1997, just two months before John died.

Why didn't John write his autobiography? According to some, John considered it, but concluded that too many autobiographies and biographies of scientists had already appeared, there was no need for another. When I asked a colleague of John's whether he would approve of my writing his biography, the answer was a definite no. "Why not," I asked. "Because John didn't do it himself" was the answer.[6] I strongly suspect that John's intention to keep his personal life private may have been an even stronger motive for not writing his autobiography. Apparently he requested that his personal diaries and letters be destroyed following his death. This was a decision similar to that of the great German naturalist and explorer, Alexander von Humboldt (1769–1859), who burned or destroyed many of his letters in order to keep his personal life very private. John was always very reluctant to speak or write about his private life, and he turned down several requests to do so. On the other hand, he left his very extensive archives to the Bodleian Library in Oxford, possibly anticipating that some day following his death someone would wish to write his biography.

Perhaps John would be surprised to learn that biographies or autobiographies of several of his colleagues and friends have been written since his death in 1997. For example, books on the lives of John Desmond Bernal (F. Aprahamian and B. Swann, 1999; A. Brown, 2007), Lawrence Bragg (G.K. Hunter, 2004; J. Jenkin, 2008), Sydney Brenner (S. Brenner and L. Wolpert, 2001; E.C. Friedberg, 2010), Francis Crick (M. Ridley, 2006; R. Olby, 2009), Dorothy Hodgkin (G. Ferry, 1998), Aaron Klug (K.C. Holmes, 2017), Max Perutz (M. Paterlini, 2006; G. Ferry, 2007), Fred Sanger (G.G. Brownlee, 2014), Maurice Wilkins (M. Wilkins, 2005), and Solly Zuckerman (J. Peyton, 2001). There is also a large amount of biographical material from Jim Watson dating as far back as 1968 in *The Double Helix: A Personal Account of the Discovery of the Structure of DNA* and as recent as 2002 in *Genes, Girls, and Gamow: After the Double Helix* in both of which John plays a significant role. Watson came to Cambridge to work with John as a postdoctoral fellow at the Cavendish in the early 1950s. John also is a prominent figure in *Max Perutz and the Secret of Life* (G. Ferry, 2007), *Designs for Life; Molecular Biology after World War II* (S. de Chadarevian, 2002), *A Nobel Fellow on Every Floor* (J. Finch, 2008), and *A Sense of Purpose: Recollections* (S. Eban, 2008).

In 1992, "A Little Ancient History" appeared in the journal *Protein Science* in which Richard Dickerson reminisced about his time as a postdoctoral fellow with John in the late 1950s working on the high-resolution structure of myoglobin at the Cavendish. In 2009 the *Journal of Molecular Biology* celebrated "50 Years of Protein Structure Determination" by publishing three semi-biographical articles by Richard Dickerson, Bror Strandberg, and Michael Rossmann, in each of which John plays a major role. Like Dickerson, Strandberg also worked with John in the late 1950s, and Rossmann was present at the Cavendish at the same time, working with Max Perutz on the structure of hemoglobin. In 2017, the *Journal of Molecular Biology* celebrated the 100th anniversary of John Kendrew's birth by publishing a special issue titled "Marking the Milestones in Structural Biology." A brief biography of John, titled "A Personal Perspective: My Four Encounters with John Kendrew," by this author appeared in the special issue.

Wherever possible John's own words from his notes, correspondence, interviews, speeches, and publications have been included in this biography since they reveal a great deal about him. Reading John's own words often provides considerable insight into his intellect, character traits, and personality.

It should be noted that this biography is intended for a general audience interested in John's life and legacy but possessing a limited background in science. The physics, mathematics, and technical jargon associated with X-ray diffraction and protein crystallography have been minimized here as much as possible. However, sources of information are provided in the Appendix (S.1) for those wishing to learn more about technical aspects of the subject.

This biography is bound to have some errors, omissions, and other faults for which the author takes sole responsibility. Despite the inevitable faults it is hoped that readers will gather enough information about John Kendrew's life and legacy to draw their own conclusions about the man and his contributions.

Acknowledgments

Sir John Kendrew was an obsessive collector and organizer throughout his life. He left his extensive archives to the Bodleian Library, Oxford, with some sections subject to restricted access at the present time. The documents were deposited over a two-year period, April 1987 to April 1989, and are located in the Department of Western Manuscripts at the New Bodleian/Weston Library. Appendix A.1 provides detailed descriptions of the contents.

The Kendrew archives (Appendix A.1) were organized by the late Mrs. Jeannine Alton, former executive director of the Contemporary Scientific Archives Center, who characterized the collection as reflecting

> on almost every page what are probably the best-known features of Kendrew's personality: on the one hand his methodical and analytical power, his meticulous not to say obsessive insistence on accuracy and comprehensive documentation, shown in his lifelong interest in record-keeping and the devising of recondite systems for information storage and retrieval; on the other hand, an aloofness or elusiveness of temperament which sets certain limits to personal relations. There are a steadiness and control, a detachment combined with seemingly tireless application which constitute a formidable intellectual armoury and which are present from the earliest records. (Appendix A.1, 248)

The contents of the Kendrew archives are described in two catalogues of more than 500 pages. There is a third supplementary catalogue of 43 pages, compiled by Adrian Nardone and Peter Harper, that organizes some of John's personal papers received in July 1998 from the executor of his final will, John A. Montgomery, following John's death. The archives are contained in about 400 boxes and include correspondence, photographs, research notes and notebooks, press clippings, lunch and dinner menus and seating arrangements, programs for theatrical and concert performances, and nearly 9,000 pages of schoolboy notes, notebooks, and essays. All the material is related to John's schooling, war-time service, research, government service, society and committee work, and so forth. The archives are a tremendous resource and I've relied very heavily on them, as well as on Google Web Search, Wikipedia (the web-based free encyclopedia), and various articles, books, and reviews most of which are listed under sources in the appendix (S.2).

It should be noted that the Kendrew archives provide very little information about John's relationship as a child and adult with his father, mother, and extended family, or as an adult with his wife and thereafter with female partners. Presumably his personal diaries contained such information, but the diaries, if they still exist, were not made available to me. Furthermore, I was not granted access to the St. John's College archives that possibly could have provided additional valuable information about John's five-year term as president of the college.

I am very grateful to many people connected with this biography. Many thanks to Colin Harris, past superintendent of the Special Collections Reading Rooms at the Weston Library, for his friendship and seeing to it that my time in Oxford was always productive and enjoyable; to the staff of the Special Collections Reading Rooms at the Weston Library for their assistance and patience; to Gay Sturt, Tom Gover, Neil Ingram, Charles Knighton, and Annette Faux for providing access to archives at the Dragon School, Oxford, Clifton College, Bristol, and Laboratory of Molecular Biology, Cambridge; to Mark Bretscher, Tony Crowther, Barbara Pearse, and Richard Henderson in Cambridge, George Brownlee, Susan Brownlee, Ross McKibbin, and Gay Sturt in Oxford, and Ken Holmes, Mary Holmes, and Frieda Lennert-Glöckner in Heidelberg for their advice and generous hospitality over several years; to Susan Black-Wilsmore in London for her openness, graciousness, and friendship; to Frances Bickerstaff who kindly opened her home, The Old Guildhall in Linton, to me; to Tony Crowther and Richard Henderson in Cambridge and Jim Bieker, Rob Krauss, Eveline Staedeli-Litscher, and Brinton Taylor-Parson in New York City who read various versions of the biography, corrected errors, and provided valuable suggestions on how to improve the text; to the Max Planck Institute in Heidelberg for accommodations at their guest house in 2010 and to St. John's and Lincoln Colleges in Oxford and Darwin College in Cambridge for accommodations and assistance during visits to England in 2009, 2010, 2013, 2014, 2016, and 2017.

I extend a very special thanks to Richard Henderson and Tony Crowther at the Laboratory of Molecular Biology, Cambridge, Ross McKibbin at St. John's, Oxford, and Gay Sturt at the Dragon School, Oxford. Their friendship, collegiality, and considerable efforts on my behalf are deeply appreciated.

I am very grateful to John A. Montgomery, executor of John Kendrew's final will, who granted me copyright permission on October 11, 2018, to reproduce in this biography material that is included in John's archives at the New Bodleian/Weston Library, Oxford. I also thank the New Bodleian/Weston Library for permission to reproduce the information about the Kendrew Archives presented in Appendix A.1.

There were no immediate members of John's family for me to interview. However, many individuals, some of whom were John's friends and/or colleagues, as well as a few of John's female partners, participated in interviews and provided essential material for the biography. Their names are listed below. Interviews carried out in person are indicated by a bold asterisk, while others were carried out by telephone, letter, and/or e-mail. For their contributions I thank the following individuals: *Waltraud Ackermann, the late *Suzy Ambache, *Frances Bickerstaff, *Susan Black-Wilsmore, *Tony Boyce, the late *Sydney Brenner, *Mark Bretscher, Peter Bretscher, *George Brownlee, Maurizio Brunori, the late *Brian Clark, *Susan Coombs-Hess, *Jackie Couling, *Tim Cox, *Hallard Croft, *Tony Crowther, Jacques Dubouchet, Claire Earnshaw-Held, *Mark Freedland, *Tom Gover, Andrew Greany, *Freddie Gutfreund, Gillian Harris-Adams, *William Hayes, *Richard Henderson, *Ken Holmes, *Mary Holmes, the late *Hugh Huxley, *Neil Ingram, Charles Knighton, *Frieda Lennert-Glöckner, *Kevin Leonard, *Chris Llewellyn-Smith, *Julia Marton-Lefèvre, *Iain Mattaj, *Ross McKibbin, *Marek Mlodzik, *John Montgomery, the late *Konrad Müller, Christiane Nüsslein-Volhard, *Marta Paterlini, Philip Pattenden, *Sophie Petersen, the late Lennart Philipson, the late Michael Rossmann, *Michael Sela, *Kai Simons, David Smith, Nick Strausfeld, Tony Stretton, Lubert Stryer, *Gay Sturt, *John Meurig Thomas, *John Tooze, *Ruth Toureau, *Jim Watson, the late *Edith Weisz, *Eric Wieschaus, and *Margaret Yee.

I thank Jeremy Lewis, Senior Editor, and Bronwyn Geyer, Editorial Assistant, at Oxford University Press for their assistance, enthusiasm, and patience over a very long time.

Finally, it is essential to point out that I have been assisted and advised continuously on every aspect of this biography by my wife, Eveline Staedeli-Litscher. Her numerous contributions to the biography, especially to chapter 3 on John Kendrew's family, were absolutely invaluable. Eveline edited several versions of the biography over many years and her advice, enthusiasm, and steadfast support enabled me to forge ahead during difficult periods. It is largely due to Eveline that I continued to work on and finally completed this biography for which she has my undying gratitude and love.

John Kendrew's Timeline

1917—Born in Oxford to Wilfrid G. Kendrew and Evelyn M.G. Sandberg

1921—Permanent separation of parents; mother emigrates to Italy

1924—Attended Dragon School, Oxford

1930—Attended Clifton College, Bristol

1932—Reunited with mother, Evelyn Sandberg-Vavalà

1936—Attended Trinity College, Cambridge; BA, 1939, MA, 1942

1939—Research Student, Department of Physical Chemistry, Cambridge

1940—Joined Air Ministry, World War 2

1946—Left Air Ministry as Wing Commander, RAF; Honorary commission

1946—Research Student, Cavendish Laboratory; PhD, 1949, ScD, 1962

1947—Fellow, Peterhouse, Cambridge

1948—Married to Mary Elizabeth Gorvin-Jarvie

1954—Reader, Davy Faraday Research Laboratory, Royal Institution, London (part-time)

1956—Divorced from Mary Elizabeth Kendrew

1958—Published structure of myoglobin at low-resolution in *Nature*

1958—Founder and Editor-in-Chief, *Journal of Molecular Biology*

1960—Elected Fellow of the Royal Society

1960—Published structure of myoglobin at high-resolution in *Nature*

1960—Appointment at Ministry of Defense (part-time)

1961—Deputy Chief Scientific Advisor, Ministry of Defense (part-time)

1961—Death of mother, Evelyn Sandberg-Vavalà, in Italy

1962—Deputy Chairman, Head Structural Studies, MRC LMB, Cambridge

1962—Death of father, Wilfrid George Kendrew, in Cambridge

1962—Nobel Prize in Chemistry shared with Max Perutz

1962—Meeting at CERN, Geneva, to organize an international laboratory

1963—Made Commander of the British Empire (CBE)

1964—Purchased The Old Guildhall, Linton

1965—Awarded Royal Medal of the Royal Society

1969—Secretary General EMBO and EMBC

1971—Project Leader EMBL

1972—Honorary Fellow, Trinity College, Cambridge

1974—Made Knight Bachelor of the British Empire (KBE)

1974—First Director-General, EMBL, Heidelberg

1975—Honorary Fellow, Peterhouse, Cambridge

1982—33rd President, St. John's College, Oxford

1987—Honorary Fellow, St. John's College, Oxford

1987—Retired to The Old Guildhall, Linton

1997—Honorary Doctor of Law, Cambridge University

1997—Died in Cambridge at age 80

John Kendrew's Service

John Kendrew served as an advisor, chairman, consultant, deputy chairman, director, director-general, fellow, governor, honorary fellow, member, president, project leader, secretary, secretary general, trustee, or vice president for many professional societies, institutions, and governmental agencies in the UK and abroad. Among these were the following:

British Biophysical Society (1959)
Council, Biophysical Society, United States (1959)
Medical Research Council, Laboratory of Molecular Biology (1962)
Ministry of Defense (1962)
British Broadcast Corporation Advisory Group (1964)
Weizmann Institute of Science, Israel (1964)
International Union of Pure and Applied Biophysics (1964)
Council for Scientific Policy (1964)
Clifton College, Bristol (1964)
Institute of Biology (1965)
Council, Royal Society (1965)
Board of Directors, Academic Press (1968)
Defense Scientific Advisory Council (1969)
European Molecular Biology Organization (1969)
European Molecular Biology Conference (1970)
European Molecular Biology Laboratory (1971, 1975)
Scientific Council, Laboratory Molecular Embryology, Naples (1973)
Laboratory of Molecular Immunology, Naples (1973)
Max Planck Institut für Kernphysik, Heidelberg (1974)
Basel Institute of Immunology (1975)
British Museum (1974)
British Association for the Advancement of Science (1974)
International Foundation for Science (1975)
International Council of Scientific Unions (1974)
Max Planck Gesellschaft, Munich (1976)
UK National Commission for UNESCO (1976)

United Nations University, Tokyo (1980)
Confederation Scientific Technological Organizations Development (1981)
St. John's College, Oxford University (1982)
Commonwealth Science Council (1983)
Weizmann Institute Foundation (1983)
Science and Engineering Research Council (1984)
Joint Research Center, European Economic Communities (1985)
European Bioinformatics Institute (1993)
Oxford International Biomedical Center (1993)

John Kendrew's Awards and Honors

John Kendrew's scientific achievements and his extensive dedicated service to many academic, government, private, and international entities were recognized during his lifetime by various honors and awards, as well as by election to membership in several prestigious learned societies. Among these were the following:

Fellow, Royal Society (1960)
Nobel Prize in Chemistry (1962)
Honorary Member, American Society of Biological Chemistry (1962)
Commander of the Order of the British Empire (1963)
Foreign Honorary Member, American Academy Arts and Sciences (1964)
Royal Medal of the Royal Society (1965)
Member, Leopoldina Academy (1965)
Member, Board of Directors, Academic Press (1968)
Honorary Fellow, Weizmann Institute of Science (1969)
Honorary Fellow, Trinity College, Cambridge (1972)
Foreign Member, National Academy of Sciences, United States (1972)
Knight Bachelor, British Peerage (1974)
Honorary Fellow, Peterhouse , Cambridge (1975)
Corresponding Member, Heidelberg Academy of Sciences (1978)
Foreign Member, Bulgarian Academy of Sciences (1980)
Honorary Member, Royal Irish Academy (1981)
Honorary Professor, University of Heidelberg (1982)
Honorary Fellow, St. John's College, Oxford (1987)
Distinguished Service Award, Elsevier (1987)
William Procter Prize for Scientific Achievement, Sigma Xi (1988)
Foreign Fellow, Indian National Science Academy (1989)
Honorary Doctor of Law, Cambridge University (1997)
Honorary Degrees (ScD) from the Universities of Keele (UK, 1968); Reading (UK, 1968); Stirling (UK, 1974); Pécs (Hungary, 1975); Exeter (UK, 1982); Buckingham (UK, 1983); Complutense (Spain, 1987); Cambridge (UK, 1997)

Named Lectures—Herbert Spencer (1965); Hans Sloane Memorial (1966); Crookshank (1967); Robbins (1968); John R. Bloor (1968); Proctor (1969); Fison Memorial (1971); European Science Foundation (1977); Monsignor de Brun (1979); Saha Memorial (1980)

Nobel Laureates Cited

Adrian, Edgar (1889–1977)/Physiology or Medicine, 1932
Anfinsen, Christian (1916–1995)/Chemistry, 1972
*Appleton, Edward (1892–1965)/Physics, 1947
*Aston, Francis (1877–1945)/Chemistry, 1922
Beadle, George (1903–1989)/Physiology or Medicine, 1958
*Blackett, Patrick (1897–1974)/Physics, 1948
*Bragg, Lawrence (1890–1971)/Physics, 1915
*Bragg, William (1862–1942)/Physics, 1915
*†Brenner, Sydney (1927–2019)/Physiology or Medicine, 2002
Butenandt, Adolph (1903–1995)/Chemistry, 1939
Calvin, Melvin (1911–1997)/Chemistry, 1961
*Chadwick, James (1891–1974)/Physics, 1935
Churchill, Winston (1874–1965)/Literature, 1953
*Cockcroft, John (1897–1967)/Physics, 1951
Cori, Carl (1896–1984)/Physiology or Medicine, 1947
*†Crick, Francis (1916–2004)/Physiology or Medicine, 1962
Dale, Henry (1875–1968)/Physiology or Medicine, 1936
Dubochet, Jacques (b. 1942)/Chemistry, 2017
Eigen, Manfred (1927–2019)/Chemistry, 1967
Eliot, T.S. (1888–1965)/Literature, 1948
Frank, Joachim (b. 1940)/Chemistry, 2017
Glaser, Donald (1926–2013)/Physics, 1960
†Henderson, Richard (b. 1945)/Chemistry, 2017
Hershey, Alfred (1908–1997)/Physiology or Medicine, 1969
Hicks, John (1904–1989)/Economics, 1972
Hinshelwood, Cyril (1897–1967)/Chemistry, 1956
*Hodgkin, Dorothy (1910–1994)/Chemistry, 1964
Hofstadter, Robert (1915–1990)/Physics, 1961
Hopkins, Frederick (1861–1947)/Physiology or Medicine, 1929
†Horvitz, Robert (b. 1947)/Physiology or Medicine, 2002
Huber, Robert (b. 1937)/Chemistry, 1988)
Huxley, Andrew (1917–2012)/Physiology or Medicine, 1963

Jerne, Niels (1911–1994)/Physiology or Medicine, 1984

*†Kendrew, John (1917–1997)/Chemistry, 1962

*†Klug, Aaron (1926–2018)/Chemistry, 1982

†Köhler, Georges (1946–1995)/Physiology or Medicine, 1984

Krebs, Hans (1900–1981)/Physiology or Medicine, 1953

Landau, Lev (1908–1968)/Physics, 1962

†Levitt, Michael (b. 1947)/Chemistry, 2013

Lewis, Edward (1918–2004)/Physiology or Medicine, 1995

Libby Willard (1908–1980)/Chemistry, 1960

Luria, Salvatore (1912–1991)/Physiology or Medicine, 1969

Medawar, Peter (1915–1987)/Physiology or Medicine, 1960

†Milstein, César (1927–2002)/Physiology or Medicine, 1984

Monod, Jacques (1910–1976)/Physiology or Medicine, 1965

Moore, Stanford (1913–1982)/Chemistry, 1972

Mössbauer, Rudolf (1929–2011)/Physics, 1961

*Mott, Nevill (1905–1996)/Physics, 1977

Norrish, Ronald (1897–1978)/Chemistry, 1967

Northrop, John (1891–1987)/Chemistry, 1946

Nüsslein-Volhard, Christiane (b. 1942)/Physiology or Medicine, 1995

Pauling, Linus (1901–1994)/Chemistry, 1954; Peace, 1962

*†Perutz, Max (1914–2002)/Chemistry, 1962

*Queloz, Didier (b.1966)/Physics, 2019

†Ramakrishnan, Venkatraman (b. 1952)/Chemistry, 2009

†Roberts, Richard (b. 1943)/Physiology or Medicine, 1993

Röntgen, Wilhelm (1845–1923)/Physics, 1901

*Rutherford, Ernest (1871–1937)/Chemistry, 1908

*Ryle, Martin (1918–1984)/Physics, 1974

†Sanger, Frederick (1918–2013)/Chemistry, 1958, 1980

Sharp, Phillip (b. 1944)/Physiology or Medicine, 1993

Siegbahn, Karl (1886–1978)/Physics, 1924

Stein, William (1911–1980)/Chemistry 1972

Steinbeck, John (1902–1968)/Literature, 1962

*Strutt, John (1842–1919)/Physics, 1904

†Sulston, John (1942–2018)/Physiology or Medicine, 2002

Sumner, James (1887–1955)/Chemistry, 1946

Theorell, Hugo (1903–1982)/Physiology or Medicine, 1955

*Thomson, Joseph (1856–1940)/Physics, 1906

Todd, Alexander (1907–1997)/Chemistry, 1957

Urey, Harold (1893–1981)/Chemistry, 1934

von Euler, Hans (1873–1964)/Chemistry, 1924

von Laue, Max (1879–1960)/Physics, 1914

†Walker, John (b. 1941)/Chemistry, 1997

*Walton, Ernest (1903–1995)/Physics, 1951

*Watson, James (b. 1928)/Physiology or Medicine, 1962

Wieschaus, Eric (b. 1947)/Physiology or Medicine, 1995

Wilkins, Maurice (1916–2004)/Physiology or Medicine, 1962

*Wilson, Charles (1869–1959)/Physics, 1927

†Winter, Gregory (b. 1951)/Chemistry, 2018

*Laureate associated with the Cavendish Laboratory in Cambridge, England.
† Laureate associated with the MRC Laboratory of Molecular Biology in Cambridge, England.

British Peers and Knights Cited*

Lord Edgar Adrian (Baron; 1889–1977)
Lord Albert Alexander (Earl; 1885–1965)
Sir Edward Appleton (1892–1965)
Sir Raymond Appleyard (1922–2017)
Lord Herbert H. Asquith (Earl; 1852–1928)
Lord Clement Attlee (Earl; 1883–1967)
Lord Herbert Austin (Baron; 1866–1941)
Lord Francis Bacon (Viscount; 1561–1626)
Lord Stanley Baldwin (Earl; 1867–1947)
Sir Roger Bannister (1929–2018)
Sir Joseph Barcroft (1872–1947)
Sir Misha Black (1910–1977)
Lord Patrick Blackett (Baron; 1897–1974)
Sir Hermann Bondi (1919–2005)
Sir William Henry Bragg (1862–1942)
Sir William Lawrence Bragg (1890–1971)
Sir Herbert Butterfield (1900–1979)
Lord Henry Cavendish (Duke; 1731–1810)
Lord William Cavendish (Duke; 1592–1676)
Sir James Chadwick (1891–1974)
Sir Winston Spencer Churchill (1874–1965)
Sir John Cockcroft (1897–1967)
Lord Frederick Dainton (Baron; 1914–1997)
Sir Henry Dale (1875–1968)
Sir Thomas Dunhill (1876–1957)
Sir Samuel Edwards (1928–2015)
Sir Richard Friend (b. 1953)
Lord David Lloyd George (Earl; 1863–1945)
Lord Douglas Hague (Earl; 1926–2015)
Lord Douglas Haig (Earl; 1861–1928)
Sir William Hardy (1864–1934)
Sir Charles Harington (1897–1972)

Sir Henry Harris (1925–2014)
Lord Denis Healey (Baron; 1917–2015)
Sir John Hicks (1904–1989)
Sir Harold Himsworth (1905–1993)
Sir Cyril Hinshelwood (1897–1967)
Lord Quintin Hogg (Baron; 1907–2001)
Sir Frederick Hopkins (1861–1947)
Sir Andrew Huxley (1917–2012)
Sir John Kendrew (1917–1997)
Sir Aaron Klug (1926–2018)
Sir Hans Krebs (1900–1981)
Sir Harrie Massey (1908–1983)
Sir Peter Medawar (1915–1987)
Sir Edward Mellanby (1884–1955)
Lord William Morris (Viscount; 1877–1963)
Lord Herbert Morrison (Baron; 1888–1965)
Sir Nevill Mott (1905–1996)
Lord Louis Mountbatten (Earl; 1900–1979)
Sir Isaac Newton (1642–1726/27)
Sir Hugh Pelham (b. 1954)
Lord David Phillips (Baron; 1924–1999)
Sir Alfred Pippard (1920–2008)
Sir Edward Playfair (1909–1999)
Sir John Pope-Hennessy (1913–1994)
Sir Venkatraman Ramakrishnan (b. 1952)
Sir Richard Roberts (b.1943)
Lord Nathaniel Rothschild (Baron; 1910–1990)
Lord Ernest Rutherford (Baron; 1871–1937)
Sir Martin Ryle (1918–1984)
Sir Michael Scholar (b. 1942)
Lord Charles Snow (Baron; 1905–1980)
Sir Richard Southern (1912–2001)
Sir Michael Stoker (1918–2013)
Lord John Strutt (Baron; 1842–1919)
Sir John Sulston (1942–2018)
Lord Michael Swann (Baron; 1920–1990)
Lady Margaret Thatcher (Baroness; 1925–2013)
Sir John Meurig Thomas (b. 1932)

Sir Joseph Thomson (1856–1940)
Lord William Thomson (Baron; 1824–1907)
Lord Edward Thorneycroft (Baron; 1909–1994)
Lord Alexander Todd (Baron; 1907–1997)
Sir John Walker (b. 1941)
Sir Robert Watson Watt (1892–1973)
Sir David Willcocks (1919–2015)
Sir Gregory Winter (b. 1951)
Sir Frank Young (1908–1988)
Lord Solly Zuckerman (Baron; 1904–1993)

* *These individuals are cited in the biography, but their knighthood (Sir) or peerage (Lord) is not provided in the text.*

Abbreviations

Å	Ångstrom (1 Å is equal to 0.1 nanometers or 1 x 10^{-10} meters; there are 254-million Å in 1 inch; 1 Å is approximately the length of a chemical bond)
ASV	Air to Surface Vessel
BBC	British Broadcasting Corporation
CERB	Centre Européen de Recherches Biologiques
CERN	Centre Européen de Recherche Nucléaire
DG	Director-General
DSIR	Department of Scientific and Industrial Research
EDSAC	Electronic Display Storage Automatic Calculator
EM	Electron Microscopy
EMBC	European Molecular Biology Conference
EMBL	European Molecular Biology Laboratory
EMBO	European Molecular Biology Organization
HEPP	High Energy Particle Physics
ICSU	International Council of Scientific Unions
IUPAB	International Union for Pure and Applied Biophysics
JIGSAW	Joint Inter-Service Group for the Study of All-Out Warfare
JMB	*Journal of Molecular Biology*
LMB	Laboratory of Molecular Biology
MRC	Medical Research Council
MoD	Ministry of Defense
OR	Operations Research
ORS	Operations Research Station
RAF	Royal Air Force
SAC	Scientific Advisory Committee
TRE	Telecommunications Research Establishment
UK	United Kingdom
WW1	World War 1
WW2	World War 2

1

Introduction to John Kendrew

> [I] did not want to continue churning out structures . . . [so] I moved
> into the politics of science.
>
> John Kendrew

Personal Recollections

Fresh from graduate studies in biochemistry at Brandeis University in Waltham, Massachusetts, I traveled with my wife and three children from Boston to Cambridge, England, in the summer of 1967. It was the first time any of us had been on an airplane or abroad. I was to begin research as a Helen Hay Whitney Foundation postdoctoral fellow in the Division of Structural Studies at the Medical Research Council (MRC) Laboratory of Molecular Biology (LMB) on Hills Road in Cambridge. Trained in enzymology and protein chemistry in the Graduate Department of Biochemistry at Brandeis, my PhD supervisor, Nathan O. Kaplan (1917–1986), advised me to get some experience in protein crystallography if I planned to continue my career as a protein chemist. We felt that the very best place to gain this experience in the late 1960s was at the LMB where X-ray crystallography and other experimental approaches to determining protein structure flourished.

My sponsor at the LMB was John C. Kendrew, deputy chairman of the LMB Governing Board and head of the Division of Structural Studies (fig. 1.1). Max Perutz (1914–2002) was chairman of the LMB Governing Board and, in addition to John and Max, the Division of Structural Studies senior staff included David Blow (1931–2004), Hugh Huxley (1924–2013), and Aaron Klug (1926–2018). John had solved the three-dimensional structure of the protein myoglobin by X-ray crystallographic methods in 1957 and shared the Nobel Prize in Chemistry with his colleague Max Perutz in 1962. Under John's sponsorship I was to work with his close colleague, Herman Watson (1933–1994), on the three-dimensional structure of the glycolytic

Fig. 1.1 Photo of a portrait of John Kendrew drawn by William Lawrence Bragg. Reproduced from an item in the Kendrew Archives, Weston Library, Oxford

enzyme glyceraldehyde-phosphate dehydrogenase, isolated and crystallized from lobster muscle. Herman, who joined John in the late 1950s to work on the high-resolution structure of myoglobin, had been on sabbatical leave in the Graduate Department of Biochemistry at Brandeis while I was a third-year graduate student there. I attended several of Herman's lectures on X-ray crystallography and protein structure, and we got to know each other quite well during his stay. Since I had worked with dehydrogenases as a graduate student, Herman suggested that I try to extend the X-ray crystallographic work on glyceraldehyde-phosphate dehydrogenase carried out at the LMB by Len Banaszak (b. 1934), a former postdoctoral fellow with John.

John was my official sponsor at the LMB. However, I must confess at the outset that we had very little contact during my time in Cambridge. In fact, shortly after arriving there I learned that he was rarely to be found at the LMB, such that meetings not involving him were often held in his so-called vacant office. John himself estimated that he passed through London airport once every 10 days or so. An American postdoctoral fellow with John in the early 1960s recalled that even then John was away from the LMB most of the time and another fellow in the late 1960s recalled seeing John in the corridor only once or twice. One of John's research students in the mid-to-late 1960s claimed that he rarely had a word with John, not even a good morning. From about 1962 onward John was rarely to be seen in the LMB's canteen, a place where nearly everyone else congregated on a regular basis for coffee, lunch, tea, and—far more important than the food and drink—for conversation. John was rarely if ever seen at seminars, parties, or other events during my years at the LMB. However, I do remember him making a surprise appearance one day when the British Broadcasting Company (BBC) arrived to produce a short film about his research on myoglobin. At the time, Herman Watson had to help him search for some sperm whale and sturgeon muscle in the freezer for the film. John did not know where the freezer was located.

As I understood it in 1967, John spent two to three days each week serving on government committees in London, such as the Defense Scientific Advisory Council and Council for Scientific Policy. In addition, he devoted considerable time serving as a fellow and director of studies at Peterhouse, editor-in-chief of the *Journal of Molecular Biology* (*JMB*), an officer of the European Molecular Biology Organization (EMBO) and its Conference (EMBC), and as a vigorous campaigner for establishment of an international molecular biology laboratory. Although John was head of the Division of Structural Studies at the time, apparently the day-to-day operation of the division was left in the hands of his colleagues, David Blow, Hugh Huxley, and Aaron Klug, and a Structural Studies Committee that met monthly. John's administrative assistant, Susan Coombs-Hess, managed his office matters at the LMB and John's remaining, but greatly diminished research interests, were overseen by Herman Watson, with whom I collaborated briefly until he moved to Bristol University in the summer of 1968.

Soon after my arrival at the LMB, John invited me to his office for what he called a chat. I don't remember much about our conversation, other than John doing most of the chatting. He was polite, formal, and slightly imposing, and I was very ill at ease. To this day I still think of John as being well over six feet tall, although now I know he was about 5 feet 10 inches in height. I do remember being taken aback when John told me that his salary from the MRC was only about £1,600 (about $3,850) more than I was earning with my Helen Hay Whitney fellowship. I learned much later that John's MRC salary in 1967 was about £5,000 (about $12,000); at the time the average annual income in England was about £1,400 (about $3,360). I was very surprised at the small difference in our salaries; after all John was the deputy chairman of the LMB, head of the Division of Structural Studies, and a Nobel laureate. I was even more surprised that he would tell me about it. John ended our visit by telling me how much he enjoyed talking to me and that we would get together to talk again soon. We did meet again in his office for a brief chat, but not until another year had passed. As I recall, our second meeting and our last in Cambridge, was much like the first. We never discussed my research at the LMB, science in general, or anything else of particular importance that I can remember. When I left Cambridge for the United States it was without a goodbye from John—he was away.

More than 20 years later, I was attending a meeting in Israel and had the opportunity to meet John, now Sir John, once again. He had been retired for several years and was living in Linton near Cambridge after leaving as director-general (DG) of EMBL, Heidelberg, and then as president of St. John's College, Oxford. At a cocktail party in Jerusalem, Aaron Klug told me that John was attending the party, had been quite ill, and the prognosis was not good. Aaron strongly suggested that since I was so grateful to John for sponsoring my postdoctoral time in Cambridge that I take the opportunity to thank him in person.

Reluctantly I searched for John and found him by himself outside on a balcony. I introduced myself—it was unclear whether or not he remembered me—and told him how much I appreciated his invitation to join the LMB, how very much I valued my time there, and that it was one of the most memorable periods of my professional life. To my great surprise John became very emotional, his eyes filled with tears, and he thanked me effusively for telling him how I felt. At the time I thought his reaction uncharacteristic, certainly unexpected, and our conversation was short. However, I am extremely grateful for having had our third, and what proved to be our final, chat. Our meeting in Jerusalem has remained a vivid memory during the intervening years. John died in Cambridge six years later at the age of 80.

Molecular Biology

Much of this biography deals with molecular biology, a discipline that began to flourish in the 1950s due in part to publication of the structure of deoxyribonucleic acid, DNA, in 1953 by James Watson and Francis Crick, founding of the Molecular Biology Research Unit at the Cavendish Laboratory by John and Max in 1956, John's three-dimensional structure of myoglobin in 1957, and founding of *JMB* by John and Academic Press in 1958.

John discussed the origin and meaning of the term molecular biology in *The Encyclopedia of Molecular Biology*, a comprehensive volume of nearly 1,200 pages he edited in 1994 while in retirement:

> The term molecular biology seems first to have been used by Warren Weaver in his 1938 report to the Rockefeller Foundation . . . Astbury used the term a year later, in 1939, and it became increasingly common as time went on. It was apparently first used in the name of an institution in 1956, for the Medical Research Council Laboratory of Molecular Biology at Cambridge . . . in that of a journal in 1959, the *Journal of Molecular Biology* . . . in that of an international organization in 1963, the European Molecular Biology Organization . . . and in that of an encyclopedia in the present volume.
>
> There have been many attempts to formulate an adequate definition of molecular biology, many of them of course reflecting the intellectual preoccupation of their authors. Thus Erwin Chargaff wrote of it as "the practice of biochemistry without a licence" a definition ironically intended, but which did indicate that its devotees in the early days were generally trained in other fields. Indeed few of the early molecular biologists were brought up as biologists; most were chemists or physicists . . . .
>
> Perhaps the most satisfactory brief definition is that of Jacques Monod: "What is new in molecular biology is the recognition that the essential properties of living things could be interpreted in terms of the structures of their molecules."
>
> At the beginning there were two schools of molecular biology. The first comprised those primarily interested in the *three-dimensional* structure, or *conformation,* of biologically important macromolecules, especially the proteins, and whose technique of choice was X-ray crystallography. Early practitioners were Astbury, Bernal, and Pauling. Perhaps the first experiment in molecular biology was Astbury's demonstration in 1929 that the X-ray pattern, and therefore the structure, of a human hair changed when it was stretched. Then, a little later, came those interested in biological *information* and its replication. These made up what may be called the *one-dimensional* school; early practitioners were Delbrück and Luria. The two schools, which at first had rather little to do with one another, began to have active connections in the early 1950s when it became clear that DNA was

the genetic material, and when one of those trained in the informational school—Watson—came to the Cambridge laboratory which was the center of the conformational school, and worked with Crick on the structure of DNA. Since that time the two schools have become intimately intertwined with one another and with other fields, and today the boundaries between biochemistry, genetics, molecular biology, and biophysics have become less and less well defined."

Laboratory of Molecular Biology

The MRC Unit for Research on the Molecular Structure of Biological Systems was founded in 1947 at the Cavendish Laboratory in Cambridge with only two members, Max Perutz and John Kendrew, the former director, the latter a research student. One year later the unit doubled in size to four members, Max, John, Francis Crick, and Hugh Huxley; the latter two also research students, Crick with Max and Huxley supervised by John, who was himself still a research student. The name of the unit was changed to the Molecular Biology Research Unit in 1956 and John, who received a PhD in 1949 at age 32, became its deputy director eight years later in 1957.

In 1962 members of the unit at the Cavendish, as well as other scientists from Cambridge and London, including Hugh Huxley, Aaron Klug, and Fred Sanger (1918–2013), moved into a newly constructed building, the MRC LMB, at the Addenbrooke's Hospital site on Hills Road, Cambridge. Max and John became chairman and deputy chairman, respectively, of the new laboratory. John was also appointed head of the Division of Structural Studies and the two other divisions, Molecular Genetics and Protein Chemistry, were headed by Francis Crick and Fred Sanger, respectively. A governing board for the LMB was established, which by 1967 consisted of six tenured staff members, John, Max, Sydney Brenner (1927–2019), Crick, Huxley, and Sanger (fig. 1.2). All of the board members were Fellows of the Royal Society, a learned society founded in 1660 making it the oldest scientific institution in the world. In the late 1960s an extension was added to the LMB building that nearly doubled the amount of space and permitted an expansion of the divisions and facilities with a significant increase in the number of research staff. Max stepped down as chairman of the LMB in 1979 and was replaced, as director, by Sydney Brenner (1979–1986; Nobel/Physiology or Medicine,

Fig. 1.2 Photo of the six-member Governing Board of the MRC LMB in 1967. Seated left-to-right are Hugh Huxley, Max Perutz, Fred Sanger, and Sydney Brenner. Standing left-to-right are John Kendrew and Francis Crick. Five of the six members were at the time or subsequently became Nobel laureates. Reproduced with permission of the MRC LMB, Cambridge

2002), followed by Aaron Klug (1986–1996; Nobel/Chemistry, 1982), Richard Henderson (1996–2006; Nobel/Chemistry, 2017), Hugh Pelham (2006–2018), and Jan Löwe (2018–present).

In 2013 members of the LMB moved into a brand new, world-class building on Francis Crick Avenue in Cambridge at the Cambridge Biomedical Campus (fig. 1.3). The building accommodates about 600 people, including 400 scientists, and consists of four divisions: Cell Biology, Neurobiology, Protein and Nucleic Acid Chemistry, and Structural Studies. The building cost more than £200 million (about $320 million) with much of it funded by the MRC using income from patents taken out on humanized antibodies

Fig. 1.3 Photo of the MRC LMB buildings in 1962 (top), 1968 (middle), and 2013 (bottom). From 1962 to 2013 the LMB was located at the Addenbrooke's Hospital Site on Hills Road, Cambridge. In 2013 the LMB moved to a new building on Francis Crick Avenue at the Cambridge Biomedical Campus. Reproduced with permission of the MRC LMB, Cambridge

created by Greg Winter (b. 1951; Nobel/Chemistry, 2018), head of the Division of Protein and Nucleic Acid Chemistry (1994–2006) and for a time deputy director of the LMB (2006–2011); he is currently master of Trinity College, Cambridge. The new LMB was opened by Queen Elizabeth II in May 2013 and for the five-year period 2012–2017 received nearly £170 million (about $270 million) in funding from the MRC. What Max and John started at the Cavendish in 1947 had grown roughly 200-fold or more over the intervening 65 years and produced more than a dozen Nobel laureates and nearly 50 fellows of the Royal Society.

Secret of Life

In the 1950s, two remarkable scientific discoveries were made by members of the Molecular Biology Research Unit at the Cavendish. First, Crick (Nobel/Physiology or Medicine, 1962), an Englishman, and Watson (Nobel/ Physiology or Medicine, 1962), an American, determined the three-dimensional structure of DNA and recognized its enormous biological/ genetic implications. Second, John and Max solved the three-dimensional structure of the proteins myoglobin at high-resolution and hemoglobin at low-resolution and thereby provided a way forward for other protein crystallographers. In both instances it was a first and for their discoveries all four scientists were awarded Nobel Prizes in 1962: Francis Crick and Jim Watson in Physiology or Medicine, shared with Maurice Wilkins (1916– 2004) at King's College, London; and John and Max in Chemistry. However, whereas Crick and Watson have since become household names, frequently mentioned in the popular press and the subject of a full-length film, *The Race for the Double Helix*, by comparison John and Max have remained in relative public obscurity.

Why all the attention for Watson and Crick and not for John and Max? Simply because DNA is the hereditary or genetic material in all living organisms that encodes all of the instructions for the development and functioning of organisms, and as such, has often been referred to as "the secret of life." The Watson-Crick model for its structure, published as little more than a one page article in the journal *Nature* in 1953, revolutionized biology and is considered by many as the most important discovery in the biological sciences in the 20th century. Some have concluded that it is one of the two greatest scientific discoveries of the 20th century, the other being the principle of relativity, and that Darwin (1809–1882), Mendel (1822–1884), Watson, and Crick would all be remembered in the same class by future generations.

Why then did John and Max decide to work on the structure of proteins and not on DNA? In the early 20th century it was thought by many scientists that proteins carried genetic information from one generation to the next. The composition of proteins, which consist of 20 different common amino acids, is far more complex than that of nucleic acids such as DNA, which consists of only 4 different nucleotides or bases. In 1938 William Astbury (1898–1961) and Florence Bell (1913–2000) detected the stacking of bases in DNA using X-ray diffraction, but they did not understand the biological significance of DNA and, consequently, did not continue to elucidate its structure. At the time it was thought that only proteins had the coding capacity required for functioning as genetic material and DNA was considered by many to be of only marginal biological importance. It was not until 1944 that Oswald Avery (1877–1955) and co-workers, Colin MacLeod (1909–1972) and Maclyn McCarty (1911–2005), at the Rockefeller University Hospital in New York provided evidence that inherited characteristics were somehow encoded in DNA. Consequently, a few scientists began to believe that chromosomes were made of DNA.

In 1936, Max asked his research supervisor, eminent physicist John Desmond Bernal (1901–1971), how to solve the secret of life. Bernal assured him that the secret of life could only be found in the structure of proteins and X-ray crystallography was the only way to solve protein structure. Similarly, in 1944 in the dense jungles of Ceylon/Sri Lanka during WW2, a young John Kendrew asked Bernal the same question and received the same unequivocal answer. Just like Max, John was convinced by Bernal that application of physical methods such as X-ray crystallography was the only way to obtain answers to fundamental questions about the function of proteins.

John confessed later on that he began his research in 1946 because at the time it seemed that the structure of proteins, not DNA, was the most important unresolved problem in what came to be called molecular biology. Consequently, Max spent nearly 30 years and John, his research student and then colleague, spent 15 years on the determination of the three-dimensional structure of proteins, Bernal's secret of life. It was not until the early 1950s that enough evidence was available to unequivocally exclude proteins and designate DNA as the genetic material. In this context, it is ironic that Crick who joined Max as a research student to work on protein structure and Watson who joined John as a postdoctoral fellow to do the same, instead joined forces at the Cavendish to solve the structure of DNA. This is testimony to the value of academic freedom to pursue ideas or projects that students and postdoctoral fellows feel are important.

John and Max

Since the lives of John Kendrew and Max Perutz are so closely linked it is impossible to write about one without writing about the other. John and Max were only three years apart in age, shared the 1962 Nobel Prize in Chemistry, and were colleagues for more than 25 years, first at the Cavendish and then at the LMB. They were founding members of what became the LMB, and for many years Max served as chairman and John as deputy chairman of the LMB Governing Board. Max and John also were founding members and officers of EMBO.

However, despite close professional ties John and Max came from very different backgrounds and had very different personalities, work styles, and lifestyles. Richard Dickerson (b. 1931), John's colleague in the late 1950s working on the high-resolution structure of myoglobin, commented that "one learned by talking with John, but by watching Max."[1] He characterized John as a mentor, guide, and organizer who was simultaneously involved in several other activities outside the laboratory. On the other hand, "Max was a hands-on bench biochemist whose center of gravity was always the laboratory itself."[2] In this connection, in 1974 John confessed to the *New Scientist* that after solving the structure of myoglobin "[I] never did want to continue churning out structures . . . I never felt myself to be a fanatic for lab bench research, and possibly because during the war I was involved in a quite different world, and continued to be interested in it, I moved more into the politics of science."[3]

Max was an immigrant from Austria who arrived in Cambridge as a 22 year old to be a research student with John Desmond Bernal at the Cavendish, whereas John was born in Oxford into a family that was part of the British professional middle class. He was schooled in the classics and taught the values of public service. Simply put, John was part of the British establishment and was a quintessential Englishman. Max was not. During WW2 Max was considered an enemy alien by the British and was arrested and held in an internment camp in Canada for several months. John, on the other hand, performed distinguished service during the war, both at home and abroad, and left the military as an Honorary Wing Commander and Scientific Advisor to the Allied Air Commander-in-Chief, South East Asia. For Max, his immediate family and hands-on research were central to his life, whereas for much of his life John was a bachelor with wide-ranging interests in the arts and compelling interests in the organization, administration, and promotion of science on an international scale.

In public the relationship between John and Max appeared to be friendly and supportive, but in private it could be cool and inhospitable. Apparently John felt that he played a subservient role to Max who headed the Molecular Biology Unit at the Cavendish and then the LMB on Hills Road. John believed that he could do a much better job than Max at running the unit. On the other hand, Max was annoyed that John, not he, was first to solve the three-dimensional structure of a protein and was envious of all the celebrity that came with it. In the late 1950s, he lamented that John was probably solving the structure of myoglobin, whereas he, having solved the phase problem, was getting nowhere on the structure of hemoglobin.

Although they were colleagues and collaborators in Cambridge for more than 25 years, John and Max came from very different worlds, were very different from one another, and to some extent competed with each other professionally. As a result, they had a rather complex and at times difficult personal relationship over the many years they were together. In retrospect, Max and John should not be thought of as close friends. At the very end of his life, while confined to the Evelyn Hospital in Cambridge with a terminal illness, John absolutely refused to see Max for what would have been a final goodbye.

Structure of Myoglobin

John achieved enormous success and celebrity in research while in his early 40s. His research, begun while a research student at the Cavendish, culminated in 1958 in publication of the low-resolution (6 Å), three-dimensional structure of the protein sperm whale myoglobin. It was the first protein structure ever solved by X-ray crystallographic analysis. Two years later in 1960, John and his colleagues published the high-resolution (2 Å) structure of myoglobin. With the passage of time, however, some have championed the misconception that Max Perutz, not John, determined the first protein three-dimensional structure. In 2002 there was a reminder to everyone from Greg Petsko that,

> Contrary to the implication in some obituaries, Max Perutz . . . did not determine the first 3-dimensional structure of a protein molecule. John Kendrew did that . . . Kendrew's low-resolution myoglobin structure predated the 5.5 Ångstrom-resolution hemoglobin structure by more than a year, and when Max published his low-resolution work . . . Kendrew was publishing his interpretation of myoglobin at 2 Ångstroms resolution, in full atomic detail.[4]

The information about proteins provided in John's two *Nature* publications represented an accomplishment that had been thought impossible only a few years earlier. One of John's collaborators sarcastically noted that John did not know it was impossible to solve a protein structure, and so he succeeded where others had failed. Aside from the myoglobin structure itself that revealed some of the basic principles of protein structure, John's research demonstrated the effectiveness of the heavy-atom isomorphous replacement method for proteins, as well as the importance of electronic computing and model building in solving protein structures. It was certainly a watershed moment in the history of structural biology. John continued to espouse the pure intellectual curiosity that motivated his research on myoglobin at the Cavendish and brought him a share of the Nobel Prize.

The early 1960s was the end of an era in which all protein crystallographers knew each other and everything going on in the field. The first international symposium on protein crystallography organized by Max Perutz was held at Hirschegg in the Kleines Walsertal, Austria, in March 1966 and had less than two dozen participants who made up nearly all of the protein crystallography community at that time. However, the number of solved protein structures has continued to increase dramatically over the years. In 1978, 18 years after the high-resolution structure of myoglobin was published by John, the structures of more than 200 proteins had been solved, and by 2014 the Protein Data Bank included close to 40,000 protein structures solved at 2 Å or higher resolution. Today protein structures are solved at a rate of several thousand each year, largely due to an enormous increase in the number of practicing protein crystallographers and to tremendous technological advances in the field.

Since the first Nobel Prize awards in 1901, 28 prizes have gone to 49 scientists whose achievements were directly related to, or involved the use of, crystallographic methods and techniques. There have been 17 Nobel Prizes in Chemistry, 10 in Physics, and one in Physiology or Medicine going as far back as 1901 when Wilhelm Konrad Röntgen (1845–1923) at the University of Würzburg in Bavaria was awarded the Nobel Prize in Physics in recognition of his discovery of X-rays.

Life Changing Decision

Surprisingly, in his early 40s and after less than 15 years of hands-on research, John made up his mind to seriously reduce his research commitments and pursue other avenues as an organizer, administrator, and spokesman for science. This decision was made at the very height of his success in research and only a short time before being awarded a share of the Nobel Prize in Chemistry. Perhaps John's responsibilities during WW2, and shortly thereafter his desire to join the scientific civil service, caused him to question the importance of further academic research for himself. Although John appeared to continue in research for another decade or so, in reality his attention was directed elsewhere, and he became almost invisible in the laboratory. By the summer of 1968, when his colleague Herman Watson moved from Cambridge to Bristol, John was no longer even indirectly involved in research. During the transition period from 1960 to 1968, John served as a very active fellow of Peterhouse, editor-in-chief of *JMB*, part-time deputy scientific advisor to the Ministry of Defense (MoD), chairman of the Defense Scientific Advisory Council, a key figure in debates on molecular biology in the UK, and was intimately involved in the founding of EMBO, EMBC, and EMBL. These and numerous other outside activities required John to be away from the LMB a great deal of the time.

In 1997 when John was asked by Ken Holmes (b. 1934) about the relatively early change in direction of his career, he explained:

Well, to tell you the truth and it's perhaps a surprising thing to say, I got a bit bored with research. See, I thought it was fascinating and it took me, what, ten years to solve the structure of myoglobin. I thought the first structure was fascinating but structures following that are not going to be so interesting. So I was sort of ready to do other things. And that's I think basically why I went to the Ministry of Defense for a while, why I sat on various Whitehall Committees after that, and then why I was ready to go to Germany to start the EMBL.[5]

John's assistant at *JMB* remembers asking him why he had given up research, to which he answered that "he couldn't be like Max and spend the rest of his life sitting on a stool looking at crystals."[6] However, perhaps another factor that contributed to John's early retreat from research had to do with Watson and Crick's discovery of the structure of DNA and the remarkable biological implications suggested by the structure. Their discovery overshadowed the structural work on proteins at the Cavendish since it contributed so significantly to the realization that DNA, not protein, was truly the secret of life. How deeply this affected John is unclear, but it is likely that it diminished his enthusiasm for determining the structure of yet another protein after myoglobin. He had demonstrated that it could be done and did not wish to do it again.

At the memorial for John in Cambridge in 1997, Jim Watson remarked that in the late 1950s "John saw that there were more important things to do than basic research; to convince governments to do the right things."[7] John realized that for post-WW2 Europe to be competitive with the United States in biological research, especially in molecular biology, it would require a much more unified effort from European countries. John hoped to catalyze this effort through EMBO, EMBL, and other international scientific agencies and societies. He took on many demanding administrative duties during this period such that little time was left for what possibly remained of his own research interests. John sincerely believed that science could make the world a better place.

The change in direction of John's professional life away from research to science policy was noted by many of his colleagues in Cambridge and elsewhere. Hands-on research was the hallmark of senior scientists at the LMB, so John was considered a traitor by some when he changed his career path. In some quarters, including among his peers at the LMB, John was roundly criticized. Some concluded that he should no longer be thought of as a passionate scientist, but instead as a bureaucrat, a civil servant.

Research scientists tend to have little sympathy for those who choose to leave the laboratory for an office and strictly administrative duties, especially early in their careers. John was a sensitive man and felt very uncomfortable when subjected to occasional sniping by other scientists about his abandonment of research. Since he continued to associate with practicing scientists throughout his life, many of whom were leaders in their field, for John the sniping was a recurrent theme and irritant. John was a loyal friend, expected loyalty from others, and did not take a lack of support or criticism lightly. On the other hand, he surely enjoyed the new challenges and life-style, including the extensive traveling and freely admitted that he should have changed the course of his life earlier.

In an interview with the *New Scientist* John insisted that "my decision in the early 1960s to take a back seat in scientific research was not influenced by my Nobel Prize. I had decided sometime before then to get into international affairs."[8] However, it must be said that without the benefit of a Nobel Prize John might not have been offered the many opportunities he enjoyed as an organizer and administrator of science and as an internationalist. John was willing and able to take on such jobs, and as a Nobel laureate he had cachet possessed by only a very few other scientists.

2

Life, Traits, and Predilections

[John Kendrew exhibited] an aloofness or elusiveness of temperament which sets certain limits to personal relations.

Jeannine Alton

A Full Life

John Cowdery Kendrew was born in Oxford on Saturday, March 24, 1917.
He was the only child of Wilfred George Kendrew, a climatologist and university reader in the School of Geography and dean of St. Catherine's Society,
Oxford, and Evelyn May Graham Sandberg-Vavalà, an Oxford-trained geographer who became a well-known Italian art historian and teacher of early
Veronese and Florentine paintings. John's parents separated permanently in
1921 when he was only four years old, and his mother emigrated to Florence,
Italy, to initiate her career as an art historian. John remained in Oxford with
his father and his father's family and was not to have any communication
with his mother for more than a decade.

John attended Dragon School, Oxford, Clifton College, Bristol, and
Trinity College, Cambridge; the last as a major scholar reading the natural
sciences. Shortly after graduation from Cambridge in 1939, as WW2 began,
John joined the Air Ministry after a brief stint as a research student with E.A.
Moelwyn-Hughes in the Department of Physical Chemistry at Cambridge.
From 1940 to 1946 he served with distinction in various capacities in Britain,
the Middle East, and South East Asia and left military service with the honorary rank of Wing Commander, Royal Air Force (RAF). As a young man,
John excelled at school, university, and during WW2, and was recognized by
his peers, teachers, and superiors as an extremely talented and cultivated individual certain to achieve great success in his chosen career.

At the conclusion of WW2, John was undecided whether to join the civil
service or return to academic research. However, during the war John was

greatly influenced by the eminent scientist J.D. Bernal who advised him to take up protein crystallography as a research student when he returned to civilian life. In 1946 John moved back to Cambridge as a research student at the Cavendish, with protein crystallographers Max Perutz and Lawrence Bragg, and as a fellow of Peterhouse. He retained the latter position for many years and at various times was director of studies for the natural sciences, teacher of chemistry and the history and philosophy of science, custodian of paintings and portraits, praelector, wine steward, steward, and librarian. John Meurig Thomas, a master of Peterhouse from 1992 to 2002, remembered that,

> As keeper of portraits, John Kendrew served his college with distinction. Not only did he catalogue all, and trace the provenance of many, of the pictorial possessions of the college, he also undertook to have several of them X-rayed and cleaned, and in this his love of Renaissance art helped him greatly. His exceptional knowledge of classical music, and in the best recordings, made him popular with the students and Fellows alike; some of them (now themselves retired) recall, as does Max Perutz, that it was in John Kendrew's room on C staircase, Old Court, Peterhouse, that they first heard "hi-fi."[1]

John received a PhD in 1949 and stayed in Cambridge for the next 25 years. In 1948 he married Mary Elizabeth Gorvin-Jarvie, widow of a close friend killed in WW2 and a medical student. John and Elizabeth did not have children, divorced in 1956, and John was to remain a bachelor for the rest of his life.

John published the first low-resolution and then the first high-resolution three-dimensional structure of a protein, sperm whale myoglobin, in 1958 and 1960, respectively. For this work he shared the Nobel Prize in Chemistry in 1962 with fellow protein crystallographer Max Perutz and subsequently was made a Commander of the British Empire (CBE) in 1963 and a Knight Bachelor in 1974, the latter honor bestowed in recognition of his service to the British government, particularly to the MoD and the Council for Scientific Policy. In 1964, using money awarded with his share of the Nobel Prize and inherited from his father, he purchased The Old Guildhall in Linton, an historic house and garden not far from Cambridge.

John's pioneering research on the three-dimensional structure of proteins contributed enormously to the launching of the discipline of molecular biology. It contributed as well to the establishment of the LMB in Cambridge as a foremost center for research in molecular biology, first at the Cavendish and then at the Addenbrooke's Hospital site on Hills Road. John remained deputy chairman and head of the Division of Structural Studies at the LMB until he moved in 1974 to EMBL in Heidelberg. He was an employee of the MRC for more than 35 years, even while living and working in Heidelberg.

John became founder and first editor-in-chief of *JMB* in 1958 and thereby provided an outstanding venue for the new discipline. At the time molecular biology was commonly thought of as biology done by non-biologists, since many of the early participants, such as William Astbury, J.D. Bernal, and Lawrence Bragg, were trained in either physics or chemistry. Their research emphasis was on structure, since they felt strongly that knowledge of structure would lead to an understanding of biological function. *JMB* was a major factor in bringing about growing acceptance of the term molecular biology among scientists. The journal attracted articles from the most outstanding investigators in the rapidly emerging discipline of molecular biology. In this context, Sydney Brenner commented that "Few present readers [of *JMB*] can appreciate how daring this venture seemed at the time; after all, the subject had only just begun, and the number of people willing to call themselves molecular biologists in public was still quite small. But as we now all know, the subject and, for that matter, the Journal, were doomed to success from the start."[2]

John was one of the founders of EMBO, EMBC, and EMBL in the 1960s, and stimulated internationalism and especially growth of European science. These efforts were made to counter what John recognized as a shift of the center of gravity of molecular biological research to the United States. He was concerned that universities in the United States grasped the promise of molecular biology, while European ones ignored it. John became secretary general of EMBO and EMBC and the first DG of EMBL.

Also in the 1960s, John became an advisor to the British government as a part-time member of the MoD. He was particularly effective in providing advice, devising British science policy, and fostering international cooperation on many aspects of science. John was a member and then deputy chairman of the Council for Scientific Policy, chairman of the International Scientific Relations Committee of the Council for Scientific Policy, and chairman of the Defense Scientific Advisory Council. He felt strongly that national boundaries were bad and science was so international in spirit that it was one of the most fruitful ways to break down boundaries. As an internationalist, John served as vice president, president, and honorary vice president of the International Union of Pure and Applied Biophysics (IUPAB); secretary-general, first vice president, and president of the International Council of Scientific Unions (ICSU); and president of the Confederation of Scientific and Technological Organizations for Development. He knew most senior figures in science policy throughout the world, and all evidence suggests that they were extremely impressed by John and his advice was valued highly in all quarters.

John was a very accomplished organizer and administrator of scientific matters. He was diplomatic, fair, and ethical, able to reduce complex topics under debate or discussion to a limited number of addressable points. John was admired for his breadth and depth of knowledge, his leadership and organizational skills, and his ability to communicate orally and in writing. Through his diplomacy and tireless efforts over more than a decade, John convinced many European governments to establish an international molecular biology laboratory, EMBL, in Heidelberg. It was pointed out by Conrad Waddington that, "John Kendrew did not give up his resolve to get the EMBO lab set up. I think he is the only man of my acquaintance lucky enough to have it reliably if privately reported, that he can maintain the rigidity of his attitude even beyond the point of satiety into a period of diminishing if not vanishing returns. In this instance his persistence was eventually rewarded."[3]

John was appointed the first DG of EMBL in 1974; he oversaw every detail of its design and construction, set its research goals, and hired its first employees. Many of his early hires recognized John's exceptional, farsighted views of scientific research, his firm but very personal leadership style, and his strong desire for them to achieve success. The highly successful EMBL exists today as a direct consequence of John's hard work and strong convictions. For some of its early staff members, EMBL will always be considered as John's monument. Since 2007, the John Kendrew Young Scientist Award has been given annually to a recent alumnus of EMBL in recognition of John's principal role in creating the laboratory.

In 1982 John was appointed 33rd president of St. John's College, Oxford, the first scientist and first person not educated at Oxford to occupy such a position. He served with distinction until 1987 when he retired at the age of 70.

The final decade of John's life was spent living in The Old Guildhall in Linton, the house and gardens that he treasured. But even in retirement John continued to participate frequently in various university activities, particularly those in Cambridge, Oxford, and London. Despite lingering illness he travelled abroad regularly and attended lunches, dinners, concerts, and musical recitals, often together with a female companion. John died in Cambridge on Saturday, August 23, 1997, and was cremated shortly thereafter.

Enigmatic Figure

John was an enigmatic figure to many people who knew him. For despite the great success of his professional life and his connection to so many people in high places, John was modest and reserved. Hugh Huxley, John's PhD student and then colleague for many years, felt that John "created a personality that suited his natural bent and that he refined it over the years."[4] As president of St. John's, some colleagues described John as reserved and shy, while other acquaintances considered John quite secretive; perhaps a quality he acquired during his military service in WW2 and carried into later life. In short, John was characterized by some as an intensely private man, while to others he often appeared distant and aloof.

A woman close to John in the late 1970s and early 1980s felt that "he was not happy in his own skin" and she found him to be "quiet, conflicted, and insecure."[5] Perhaps indicative of his modesty and insecurity, in photographs taken during his school days at Clifton and thereafter at various gatherings, John frequently is positioned at the very end of a row of participants, perhaps seeing himself as an outsider. Even as a 19 year old at Clifton College, John was characterized as a very good head of Watson house, but almost too aloof.

In describing John's extensive correspondence in the Bodleian Library's archives, Jeannine Alton concluded that "One touches here on the reticent element in his temperament; the correspondence is open and friendly yet rarely develops into long-term exchanges." She went on to say that the biographical section of John's archives "unsurprisingly, contains little of a personal nature" and Alton saw in John "an aloofness or elusiveness of temperament which sets certain limits to personal relations."[6] Consistent with Alton's assessment, John's secretarial assistant at EMBL observed that John was rarely emotional and always polite. She noted that John "didn't walk around EMBL very much during working hours because he didn't want to bump into people he didn't know."[7]

Kai Simons, a staff member at EMBL when John was DG, found that

> It was a great time gently managed by John Kendrew who sometimes was so distant that some of the staff did not realize he was there . . . John was a

shy man . . . I usually chose Saturday afternoons to talk with him . . . These Saturday afternoon discussions were fantastic, and I learned to respect him for his enormous knowledge and his insights into biology. The amazing fact is that as soon as a third person joined the discussion the tone changed completely and you could even get the impression that John Kendrew was boring because he did not say very much. His style of leadership therefore remained an enigma to most people who did not come into close contact with him.[8]

John's secretarial assistant for much of his term as president of St. John's in the 1980s recalled that during her first six months working for John her

first impressions of him were rather puzzling and contradictory . . . At my interview he was very pleasant and quietly engaging with a nice sense of humor, but when I first started working for him, my impression was of someone with a rather cold and indifferent manner . . . if my path ever crossed John's, he would studiously avoid eye contact, even to the point of turning away . . . This situation continued for some six months or so. . . . Over time, watching him in all the various situations required of the President and of his many other roles, I realized that he was essentially very shy, but was able to do a wonderful job of covering up this with the "the mantle" of each particular role.[9]

These comments are consistent with those of Ross McKibbin, John's colleague at St. John's in the 1980s, who found that "Anyone at St. John's during his Presidency, particularly anyone of his vice-presidents, was aware of his social unease; the reluctance to get involved."[10]

Overall, people who knew John liked and respected him, recognized his civility and breadth of cultivation, and enjoyed his company. But they also saw in him an elusiveness and social discomfort that seemed uncharacteristic for such an accomplished individual who took on so many significant leadership roles and received so many accolades. John was extremely well connected. Perhaps one of the more remarkable aspects of John's life was his close connection with so many highly successful and notable people in the sciences, arts, government, military, and business. These primarily male relationships, established during his time at Clifton and Cambridge and during WW2, were nurtured and retained throughout his life and enhanced both his professional pursuits and private life.

Despite his elusiveness and uneasiness John was quite attractive to women, even to women considerably younger than he. From his divorce in 1956 to his death in 1997 John remained a bachelor who had close relationships with many women, for some of whom he expressed love. John had some long-term relationships and contemplated marriage from time to time, but remained a bachelor to the end. He was linked to many women, but even in these relationships John could be quite manipulative, and he applied considerable organizational skills to ensure that one woman would not know about the other. This led to a stir at the Cambridge memorial for John in 1997 when several of his female partners attended the ceremonies. As John lay dying in the Evelyn Hospital, euphoric due to administration of morphine to ease his pain, women surrounded his bed. When an executor of John's final will asked him "What about the ladies?" John immediately replied "After I'm gone they'll be your problem!"[11]

Perhaps it is an exaggeration to say that John mistrusted women, but his relationships with women were often awkward and uncomfortable and this may be related to his mother's permanent separation from his father when he was a youngster. Not surprisingly much of John's paradoxical nature can be attributed to his parents and early upbringing. His parents married in 1914 and separated in 1921 when John was only four and never reconciled. Apparently John was told that his mother had died, not an uncommon practice in the case of marital separation in the early part of the 20th century— his mother was a bolter and had emigrated alone to Florence. In time, she became a very respected art teacher and historian specializing in Italian primitives from the 12th to early 14th centuries. Consequently, John was raised solely by his father and his father's family, which included a grandfather, grandmother, and several paternal aunts and uncles.

John's father had a rather cold demeanor, not a particularly favorable characteristic for a father raising a four-year-old son on his own, and John had rather unhappy memories of his early childhood at home prior to going off to boarding school in Bristol. His father was an authoritarian and the source of many of John's unhappy memories of his childhood. However, the fondness shown John by his paternal aunts at least compensated to some extent for his father's behavior. John participated in walking and cycling tours with his father and became quite interested in natural history, architecture, and later on in photography. At the age of seven, John became a day-student at the Dragon School and at 13 a boarding student at Clifton College; later on he was to become a governor of both schools. It was only as a teenager at Clifton that he became reacquainted with his mother, and his relationship with her blossomed over subsequent years.

The sharp disparity in character of his parents undoubtedly accounted for much of John's own behavior. From early childhood he had a strained and adversarial relationship with his father, although they apparently reconciled before his father's death in Cambridge in 1962. John regarded his father, a climatologist and reader in Geography at Oxford, more as a late-Victorian naturalist than a scientist and found his father's right-wing political beliefs, particularly during the pre-war years of the 1930s, very galling. On the other hand, in his middle teens and onward John had a much more loving and nurturing relationship with his mother and his great love of the arts can be attributed largely to her. However, she too could be quite demanding of John and nearly always expected him to do her bidding. To people who knew John well he seemed to resent his father and idolize his mother.

Stickler for Accuracy

John was extremely concerned with the organization of details in all aspects of his life, from the way he dressed, to the automobile he drove, to the state of his garden in Linton. This trait can be readily seen in his school and university notes, approximately 9,000 pages from his time at Clifton and Cambridge, and extensive research records deposited in his archives at the Weston Library. Such compulsiveness surely contributed to the success of his X-ray crystallographic analysis of myoglobin structure. John was a problem solver who stuck with a problem until he had a solution.

At the Cambridge memorial for John in November 1997, Aaron Klug recalled that

[John] kept in his pocket a large expandable notebook, a precursor of Filofax . . . In this book he kept an extraordinary amount of information, easily accessible: the names of all undergraduates in Natural Science, their supervisors, their exam results: all the current papers submitted to *J. Mol. Biol.*, and their status: the members of staff at LMB: scientific references and a host of other items. If he needed to know something, he took out the book, and there it was, in an instant.[12]

Also in his notebook, in his characteristic handwriting, John kept names, addresses, and phone numbers of hundreds of people, as well as lists of wines, cars, books, antique shops, restaurants, garden plants, hotels, and airports that he considered to be the best. John's secretarial assistant at St. John's noted that "He expected perfection, both of himself and others . . . His attention to detail was phenomenal" and, commenting on the organization of John's desk at St. John's, remarked that "he would know if even a pen had been moved a few inches by a careless cleaner during his absence."[13]

Always a stickler for accuracy in everything, as a second-year undergraduate at Cambridge John wrote to the editor of the *London Times* to correct an article by W.L.A. Derby, "The Tall Ships Pass":

> May I draw your attention to an error in a review of The Tall Ships Pass by W.L.A. Derby. It is stated that the barge Herzogin Cecile grounded "on the rocks of Cornwall." It is possible that popular imagination may conceive of Cornish rocks as being the most suitably rugged ones available for such a romantic, even if sad, purpose. But in actual fact the Herzogin met her end on the rocks of Devon, near Bolt Head and Bolt Tail, which headlands already have a number of scalps to their credit. I feel sure that all good Devon men will resent such implied slight on the quality of their cliffs and natives of Cornwall will suffer no serious loss by giving up this claim to fame, or notoriety; for they too cannot feel themselves hardly used where shipwrecks are concerned.[14]

The *Times* thanked John for his letter and apologized for the error.

A further example of John's compulsion for detail is revealed in a letter he wrote to a friend and classmate from Clifton College while he was stationed in Ceylon/Sri Lanka during WW2:

> I now come to the concrete reason for writing to you at this particular moment. The fact is that a few days ago, while attending some trials in the jungle, I had the enormous misfortune to lose my precious Parker fountain pen, which had been with me for six years. It is like losing a right hand, and in this theatre it is quite irreplaceable. I therefore now address to you a most urgent SOS. Can you spare me the time, to buy for me a new fountain pen, . . . and dispatch the whole lot East Express Air Mail to the above address?

As you know I like a moderately fine, moderately (but not too) hard, straight (i.e., not oblique) nib, and I enclose a specimen of my writing done with the old pen. I also like a pen with the largest possible ink capacity, discreet in coloring (preferably black). I should prefer to try one of these new-fangled devices . . . (Parker 51??), although [the] nib is a bit hard for my liking; I can also bear the grey and silver color of this pattern. I hear however that these may be difficult to come by, these days, and if so my second choice would be a Parker of the standard pattern, plunger filling, large capacity, name I believe Vacumatic, black. And if this is impossible, any other damned thing, best possible quality. Money no object—I will send you a check on my Cambridge bank on receipt. On second thoughts, I think you can be even more extravagant on my behalf, and buy me two pens of the same pattern (but slightly different in colors). Well, . . . if you can do something for me, and speedily, I shall be eternally in your debt. Incidentally I should also be grateful if you can tell me of any other gadgetry now available in Washington, D.C.—I know your taste in such matters is quite impeccable.[15]

Political Views

Rather conservative in background, appearance, and demeanor, John was to the left politically and strongly supported middle-class ideals. In 1981 he was a founder-signatory of the centrist Social Democratic Party in the UK, which attempted to create a more equal society that rejected prejudices based upon sex, race, color, or religion. The Social Democratic Party subsequently gave rise to the Liberal Democrats in 1989. Apparently John was elated when Tony Blair (b. 1953) was elected prime minister of the UK in May 1997.

As an undergraduate at Cambridge in the 1930s, John participated in the Cambridge University Socialist Club, a section of the University Labor Federation that included left-leaning Cambridge students and academicians as members. There were Sunday teas and public meetings on such topics as the "Theory of Socialism," "Communist Party Lectures," and "Who Are the Enemies of the People."

John said about himself that "I have always been slightly on the left politically, but certainly nothing like as far left as Bernal [a Communist] . . . He would have regarded me as a good Labour voter, hopelessly bourgeois right wing, and probably immovable."[16] John truly believed that science could help society and certainly improve international relations. This belief is recognized today at Oxford by the Kendrew Scholarship offered annually to a new graduate student at St. John's who comes from an economically depressed country. John stipulated in his will that a portion of his bequest to St. John's be used for the support of students and scholars from those countries of the world whose gross national product per capita falls outside the highest 25 percent.

Religious Views

Throughout his lifetime John was well read and curious about religion. He was particularly interested in the relationship between religion and science and many books on religion were included in his personal library. However, John did not believe in the existence of deities and could be considered an atheist. As president of St. John's he habitually avoided attending chapel, ordinarily an important function of the president, and although his house in Linton was just across a narrow lane from St. Mary's, the parish church, he never attended religious services there and rarely entered the church on other occasions.

At an Oxford Memorial for John in 1997, Ross McKibbin commented on John and religion:

> Kendrew's world-view . . . was remorselessly secular and rationalist. Furthermore, he retained his secularity to the last hour with a determination few of us possess. When life is over, life is over, and that is it. In fact, this was a position he reached not without effort. As a young man he was clearly interested in religion and human spirituality, even in something as recondite as Tibetan Buddhism, and in the relationship between science and religion. Nor, it seems likely, was he interested in them merely as intellectual problems. I suspect that his rejection of religion and his adherence to a rather positivist rationality was not instinctive but came after something of a mental struggle. And we might conjecture that "spirituality" was never banished from his personality: his intense preoccupation with art and music, especially music, was surely not entirely secular, and we know that he continued to be engaged with the relations between religion and science.[17]

John insisted in his final will that upon his death he was to be cremated in a private ceremony, there were to be no religious ceremonies of any kind and no funeral. Three friends from St. John's deposited his ashes in Bagley Wood, Kennington Village, Oxford, and then went for a pub lunch. Bagley Wood is a beautiful, 500-acre, ancient woodland owned by St. John's since 1557—but from 955 to 1538 it was owned by Abingdon Abbey, a Benedictine monastery. It was often visited and very much enjoyed by John, and he paid regular visits to Bagley Wood while living in Linton after his presidency at St. John's had ended.

Love of the Arts

Although always drawn to science and technology, John also was a devotee of classical music, literature, painting, photography, architecture, and archeology. John freely admitted that he might have preferred a career in archeology or as an orchestra conductor to one in science. A friend commented that although John's head was in science, his heart was in the arts.

Throughout his life John was an avid reader of an eclectic mix of literature and poetry and was fluent in several foreign languages, including French, German, and Italian. Fluency in languages undoubtedly was an inherited trait since John's paternal uncle, Alfred Kendrew, was known to be fluent in French, Italian, Spanish, German, Russian, Hungarian, and Hebrew and was learning Turkish at the time of his death. John was introduced to photography at an early age by one of his aunts and a camera, Rollei or Nikon equipped with several lenses, almost always accompanied him during his frequent travels.

As much as John was fond of photography, he was truly passionate about classical music and attributed this love to his mother. He remarked that Mozart's Great Mass in C minor, written as a celebration of Mozart's marriage, should be playing when he was about to die. John especially loved the works of Austrian composer Franz Joseph Haydn (1732–1809) and claimed to have multiple copies of every record and compact disc ever produced of Haydn's compositions. He always owned the latest and finest high-fidelity equipment for playing his recordings; when he played them they had to be in a certain order and could not be interrupted. He developed his own classification system for his extensive music collection. Throughout his adult life, John frequently attended live concerts of classical music usually accompanied by a female companion and was on friendly terms with numerous concert and other performance artists. Despite this obsession he could neither read music nor play a musical instrument and freely admitted that he could play only his hi-fi.

John's detailed knowledge and love of classical music is revealed in a letter he wrote to a colleague who was to be a visitor in Cambridge about what to bring:

> The other is for my own amusement; namely gramophone records . . . there is an album of 3—Bach's English Suites, played by Valenti, Westminster album WAL-305 (this should include a miniature score). Alternatively, singles if you are loaded down—Bach Well-Tempered Clavier Book II nos. 41–49, played by Landowska on Victor LM (number unknown—this record may even not yet be issued—I already have all the earlier part of this work, so do not bring any of it except this one); or Bach trio sonatas 3 and 6, played by Noehren, Allegro 71; Beethoven Quartet Op. 130, Budapest Quartet, Columbia, ML-584; Handel suites for harpsichord, played by Pelleg (there are 3 records—any one would do) . . . please do not take too much . . . And please do not load yourself down with these.[18]

When John died he bequeathed his vast compact disc and long-playing record collections, as well as reel-to-reel tapes, cassettes, and catalogues, to the Royal College of Music, London, established by royal charter in 1882. The reference librarian there considers the roughly 2,000 items, consisting primarily of 16th- to 18th-century music, an important addition to the audio-visual library in terms of quantity and quality.

In recognition of John's well-known passion for classical music, St. John's currently hosts the Kendrew Chamber Music Recital Series each year and appoints a Kendrew Music Assistant who is a graduate student that has substantial experience in music research and/or practical music-making. The appointee assists in the promotion and organization of a variety of musical and music related events in the college.

A producer of the BBC's television program featuring John, "The Thread of Life," got to know him quite well during production of the series and described him in 1964 as

> far removed from the conventional stereotype of the unworldly scientist out of touch with life beyond the laboratory walls. He enjoys driving fast cars,

has a lively appreciation of good food and wines, and is fluent in French, Italian, and German. He counts among his principal pleasures sitting in the sun, looking at paintings, and listening to early music . . . In summer he gets into his French car and you don't see him for perhaps six weeks. He never says where he is going, but he maintains contact with everything while he is away . . . His interests range far beyond his own particular field . . . He is very much the fully-informed man.[19]

3

Family, Parents, and Separation

Wilfrid was an elusive man, very difficult to know.

Victor Brook

Evelyn will be remembered as a personality, original sometimes to the point of eccentricity, as well as a scholar.

Hugh Honour

Great-Grandparents

Kendrew is a Scottish name derived from the Gaelic, MacAndrew or son of Andrew, and means manly or brave. The name appears first in Inverness, Scotland, and there has always been a relatively large population of Kendrews in Yorkshire, England. So it is not too surprising that John's great-grandfather, John Cowdery Kendrew, was born in Cowthorpe in 1798/99 in the Harrogate district of North Yorkshire; he died in 1864. John's great-grandmother, Mary Taylor-Kendrew, was born in 1803/04 in Newton-on-Ouse, in North Yorkshire, about seven miles from York; she died in 1877. John and Mary farmed 95 acres of land and had five children: two sons, William (b. 1835) and Thomas (b. 1847); and three daughters, Ann (b. 1840), Charlotte (b. 1843), and Thomasin (b. 1844). Their second son, Thomas, was John's paternal grandfather.

Grandparents

In 1871, 24-year-old Thomas Kendrew (1847–1930) married 22-year-old Janet Halcro (1849–1936), the daughter of George and Janet Halcro of Stromness in the Orkney Islands, Scotland. Janet, John's grandmother, was born there in 1849, where her father sold textile fabrics and her mother was a housewife. When she married Thomas, he was a supervisor with Inland Revenue, that is an excise officer, in the Burgh of Stromness. The Board of Inland Revenue, a department of the British Government, was created under the Inland Revenue Act of 1849 following the amalgamation of the Boards of Excise and of Stamps and Taxes and was responsible for the collection of income, inheritance, capital gains, corporation, excise, and other taxes.

Thomas and Janet had seven children: four sons, Augustus (1874–1937), Thomas (1875–1929), Wilfrid (1884–1962), and Alfred (1893–1972); and three daughters, Leonora (1877–1954), Annabella (1879–1968), and Janet (1895–1983). The five oldest children were born in Scotland and the two youngest were born in England. Wilfrid George Kendrew was John's father.

Thomas Kendrew's job as a civil servant took him and his family from the Orkney Islands to Banffshire (Fife Keith, Station Road) in northeast Scotland, to Wiltshire (Devizes, Landsdowne Road) in southwest England, to Dublin (Carlisle Terraces, Eblana Road), Ireland, and finally in 1902 to Oxford (Iffley Road), England. At each location the professional middle-class family was wealthy enough to employ a young girl as a general servant to assist with household chores: in Fife Keith, Ann Pirrie, age 17; in Devizes, Alice Kent, age 19; in Dublin, Agnes Swanborough, age 19; and in Oxford, Ellen Shillon, age 21.

Located in Oxford from 1902 onward, much of the time Thomas Kendrew's Inland Revenue office was at 15 St. Giles', very close to St. John's College where his grandson, John, would serve as president from 1982 to 1987. The building Thomas worked in dates from about 1800 when it belonged to Balliol College, but for more than 30 years, 1904 to 1936, it housed an Inland Revenue Office. Today the building serves as an accommodation for St. John's staff and students.

The Kendrew family moved from Dublin to Iffley Road in east Oxford with its many Victorian and Edwardian style multi-family, terraced, and semi-detached houses on narrow plots of land. They moved first to a house at 195 Iffley Road, close to the Iffley Road Track where in May 1954 Roger Bannister (1929–2018) set a world record by running a mile in just under four minutes. Four years later the family moved to an attached house at 18 Stanley Road, just off Iffley Road, and three years after that to an unattached house, Dor-a-Gor Cottage, immediately next door at 14 Stanley Road, which would be owned by the Kendrew family for nearly 30 years.

Thomas died on September 16, 1930, at the age of 83, when John was 13 years old, and left his family about £19,000 (about $92,000)—the average annual salary in Britain at the time was about £200 (about $970). In 1937 the house was sold following the death of Thomas's wife Janet, John's grandmother, on May 4, 1936, at the age of 88. Thomas and Janet, John's paternal grandparents, are buried in the same grave at Rose Hill, one of four large Victorian cemeteries in Oxford. The former Kendrew home, Dor-a-Gor Cottage, is currently owned by Kate Wilson and her husband, Richard. It was purchased in 1947 by Kate's parents, her mother Anne Bradby-Ridler (1912–2001), a well-known British poet and secretary to T.S. Eliot (1888–1965) at the publishing house Faber and Faber between 1935 and 1940, and her father Vivian Ridler (1913–2009), a printer for Oxford University Press. The Kendrew family tree is presented in figure 3.1.

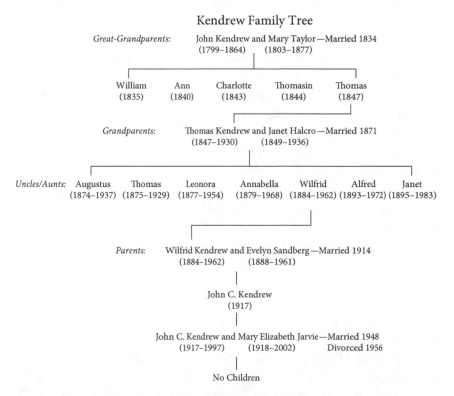

Kendrew Family Tree

Great-Grandparents: John Kendrew and Mary Taylor—Married 1834
(1799–1864)   (1803–1877)

William   Ann   Charlotte   Thomasin   Thomas
(1835)   (1840)   (1843)   (1844)   (1847)

Grandparents: Thomas Kendrew and Janet Halcro—Married 1871
(1847–1930)   (1849–1936)

Uncles/Aunts: Augustus   Thomas   Leonora   Annabella   Wilfrid   Alfred   Janet
(1874–1937) (1875–1929) (1877–1954) (1879–1968) (1884–1962) (1893–1972) (1895–1983)

Parents: Wilfrid Kendrew and Evelyn Sandberg—Married 1914
(1884–1962)   (1888–1961)

John C. Kendrew
(1917)

John C. Kendrew and Mary Elizabeth Jarvie—Married 1948
(1917–1997)   (1918–2002)   Divorced 1956

No Children

Fig. 3.1 The Kendrew family tree including John Kendrew's great grandparents, uncles, and aunts, his grandparents, uncles and aunts, his father and mother, and his wife

Paternal Uncles

Thomas and Janet's oldest son, Augustus Kendrew, was a graduate of Oxford University and practiced as a dental surgeon in Oxford from 1918 to 1936, first at 65 St. Giles' and then at 42 Broad Street. The latter address, across the street from the Bodleian Library, was the center of 13 houses dating from the first half of the 17th century, but in 1937 all of the houses were demolished to make room for the New Bodleian Library. The building, now called the Weston Library, is where John's extensive archives are housed today. In 1921 Augustus married Florence (Flo) Winbury, daughter of Walter and Sarah Winbury from Walsall, six miles east of Wolverhampton, Staffordshire, about 90 miles north of Oxford. They had no children. Augustus died in 1937 at the age of 63 and Flo died 48 years later in 1985 at the age of 92. They are buried in Wolverhampton. John corresponded with Flo until her death, and a few of her letters are in John's archives at the Weston Library.

Thomas Thackeray Kendrew, the second son, emigrated to America in 1899 and trained as a veterinary surgeon at McKillip Veterinary College in Chicago, Illinois, graduating in March 1902. The college opened in 1892 and closed in 1920 having graduated nearly 1,400 veterinarians. About one month after receiving his MDV degree, Thomas married Emily Stahl Johnson on April 28, 1902, in St. Joseph, Michigan, about 90 miles from Chicago. They lived at 3907 North Ashland Avenue, and he practiced at a veterinary hospital located at 3039 Sheffield Avenue in Chicago. Thomas became a naturalized US citizen in 1908 and registered in 1918 for military service in WW1. Thomas and Emily had no children. Thomas died of pneumonia on December 4, 1929, at the age of 54, one year before his father died, and is buried in Ridgewood Cemetery, Des Plaines, Illinois.

Alfred Kendrew was a graduate of Oxford University, Christ Church College, and was a medical student at University College, London, and Bonn University, Germany. He served in WW1 and spent considerable time in Russia and Hungary on special medical missions. Alfred was a general medical practitioner from 1927 to 1954 and then, at the age of 61, became a Medical Officer at a private clinic, Ticehurst House, in East Sussex. Alfred married Nancy Hollis in 1946. They had no children. Alfred died in 1972 at the age of 79. John kept a newspaper clipping of Alfred's obituary in his archives.

Paternal Aunts

Leonora Kendrew worked as a nurse for a time in London at the St. Marylebone Infirmary, which became the St. Charles Infirmary. She married Benjamin Robinson of Dublin, a clerk for the railway, in 1906. In June 1919 Leonora boarded the ship Iyo Maru in London bound for Singapore intending to be a nurse there. Three years later she emigrated to San Francisco, California, as the wife of William O. Gropius, an Englishman. The couple moved to Ross in Marin County just north of San Francisco in 1924 and lived at 5 Woodside Way. Leonora became a naturalized US citizen in 1947. She died unexpectedly of a heart attack in February 1954 at the age of 77. She and William had no children. John stayed with Leonora and William in Ross when he was making his way back to England via the United States at the end of WW2.

Annabella Kendrew remained unmarried and at home throughout her life. When her mother died in 1936 and Dor-a-Gor cottage was sold, Annabella purchased a house at 45 Henley Avenue, Oxford, not too far from Stanley Road, where she lived until a few years before her death in 1968 at the age of 89. When Annabella died she was living at Barton End, a private house built in 1880 in Headington that was taken over by the Oxford City Council Services Committee in the late 1940s as accommodation for elderly people. Annabella was cremated and her ashes spread around Rose Bed number A1 at the Oxford crematorium.

Janet Maud Kendrew was a graduate of Cambridge University, Girton College, where she received a degree in mathematics in 1918. In South Africa, sometime between 1927 and 1931, she married John Henry Parkyn Lamont, an author and teacher who published under the nom de plume Wilfred Saint-Mandé and had emigrated to South Africa in 1927. Following their marriage, Janet taught at the Roedean School in Johannesburg and Henry was a senior lecturer teaching French at the University of Pretoria. Following Henry's dismissal from the university in 1932, the couple returned to England. However, a notice in the *Kenya Gazette* in September 1947 indicates that Henry was appointed an education officer in Kenya in August 1947. He died in June 1967 at the age of 71 while he was a retired schoolteacher living at 34 Upland Park Road in Summertown, Oxford. Janet lived at Long Cottage, King Henry's Road, Lewes, Sussex, for many years and died in 1983 at the age of 88 in Gloucestershire. Janet and Henry had a daughter,

Janet Lamont (1937–2001), who became a teacher. Janet had a son, Timothy, in 1973 and married Ernest Ineson in 1981. She died in 2001 in Cornwall at the age of 64.

Kendrew Offspring

Notably, John's paternal grandparents had seven children, but only two grandchildren, John and Janet, came from their children's six marriages. John and Elizabeth had no children and Janet and Ernest Ineson had a son. A summary of Kendrew marriages and offspring is presented in table 3.1.

John's Father

Wilfrid George Kendrew (fig. 3.2), who preferred to be called WG, was born on September 14, 1884, in Banffshire, Scotland. At the time he was the youngest of five children. Upon his family's move to Dublin in about 1900 he attended the Mountjoy School, a boarding school located on Mountjoy Square in the city center. Mountjoy School amalgamated first with the Hiberian Marine School in 1968 and then with the Bertrand and Rutland School in 1972 to become the Mount Temple Comprehensive School. During the family's stay in Dublin there were four children living at home, Annabella,

Table 3.1 Kendrew family marriages and offspring

Uncles/Aunts	Wife/Husband	Children
Augustus	Florence Winbury	None
Thomas*	Emily Stahl	None
Leonora*	Benjamin Robinson/William Gropius	None
Annabella	Unmarried	None
Wilfrid	Evelyn Sandberg	John
Alfred	Nancy Hollis	None
Janet	Henry Lamont	Janet
Offspring		
John Kendrew	Mary Elizabeth Jarvie	None
Janet Lamont	Ernest Ineson	Timothy

*Emigrated to the United States and became a naturalized US citizen.

Fig. 3.2 Photo of Wilfrid George Kendrew, father of John Kendrew.
Reproduced from a photo in Joan M. Kenworthy's "Meteorologist's profile—
Wilfrid George Kendrew (1884–1962)" that appeared in *Weather*, vol. 62, no. 2,
February 2007

Wilfrid, Alfred, and Janet/Maud. A general servant, Agnes Swanborough,
also lived with the family at Carlisle Terraces on Eblana Road.

Wilfrid was 18 years old in 1902 when his family moved to 195 Iffley Road,
Oxford. In October of that year he matriculated as a non-collegiate student
at Oxford University, undertaking a Pass degree in Classics in three years. He
did not enroll for the four-year honors degree in Classics, apparently something he would have greatly preferred. Non-collegiate or unattached students
were first admitted to Oxford University in 1868, were not members of any
college or hall, but had full rights of membership in the university without
the costs. Such students had to reside within about two miles of the center
of Oxford and were overseen by a delegacy consisting of senior members of
the university. In 1962 the Delegacy for Non-Collegiate Students became St.
Catherine's College for male students only, but in 1974 the college began to
admit female students.

Wilfrid passed examinations of the Pass School in Greek, Latin, mechanics, physics, and chemistry in 1904 and in French language and literature
and elements of political economy a year later. He was awarded a Bachelor of

Arts (BA) degree in 1905 and a Master of Arts (MA) in 1909. Shortly after receiving his degrees, Wilfrid was employed by the university to lecture to non-collegiate students. He was appointed junior tutor two years later and joined the teaching staff of the St. Catherine's Society as a Don in the Classics. While teaching the classics, mathematics, and French, he attended lectures in the School of Geography at Oxford; established in 1899 under the direction of A.J. Herbertson (1865–1915), in time this school became acknowledged as the best university course in geography in Britain. Wilfrid obtained a certificate in Regional Geography and a diploma in Geography in 1911, both with distinction. Subsequently, he began to lecture on aspects of meteorology, climate, oceanography, mapping, and survey in the School of Geography where he was considered an excellent teacher. In 1915, following the death of Herbertson at the age of 49, H.O. Beckit (1875–1931) was appointed acting director of the school.

Wilfrid was elected a fellow of the Royal Meteorological Society in 1913 and a year later his first publication appeared, a chapter titled "Climate," on the climate of the British Isles, in *The Oxford Survey of the British Empire*, edited by A.J. Herbertson and O.J.R. Howarth, and commissioned by the former. Wilfrid published a book of his own, *The Climates of the Continents*, in 1922, which went through four more editions between 1927 and 1961, all published by Clarendon Press. Robert Beckinsale (1908–1998) of Oxford praised the book in *Nature*, saying that "The book aimed at filling a gap caused by the lack in English of any adequate description of the actual climates of the countries of the Earth, considered regionally. It was hailed as a pioneer work of its class and of great value as bringing together in one volume a vast amount of information." By the early 1960s more than 30,000 copies of the book had been sold, providing more than £40,000 in royalties for the author.

In 1930 another of Wilfrid's books, *Climate: A Treatise on the Principles of Weather and Climate*, was published by Clarendon Press and new versions appeared in the 1950s—more than 10,000 copies were sold. C. Gordon Smith (1921–1999), one of Wilfrid's students at Oxford, concluded in *Geographers: Biobibliographic Studies* that "The value of Kendrew's writing is that he was able to combine clear and vivid description with a basic knowledge of the fundamental principles of atmospheric physics." Wilfrid's books brought him international recognition within the geography and climatology communities.

Wilfrid joined the Oxford University Officer Training Corps (OTC) in 1914, for which he undertook practical courses and field excursions. This was also the year that he married Evelyn May Graham Sandberg. Until 1917 he continued to lecture at Oxford on climatology, climatic and vegetation regions, oceanography, and meteorology, and conducted field classes and gave practical instruction in surveying. That summer, only a few months after John's birth, Wilfrid was commissioned into the Royal Irish Fusiliers as a Second Lieutenant and began his active service as an intelligence officer on the staff of Field Marshal Sir Douglas Haig, the Chief of Staff of the British Expeditionary Force. Wilfrid served in France in 1917 and 1918 in the Censor's Department in Boulogne, but there is no evidence that the army utilized any of his knowledge of climatology during his service in WW1. In 1922 he applied for and received medals for his military service. Upon his return to Oxford, Wilfrid was appointed tutor and elected junior proctor for the delegacy for one year. He became dean of non-collegiate students in 1927 and dean of St. Catherine's Society in 1932. As such, he served as a member of the delegacy for non-collegiate students and played a major role in establishing an Honors School of Geography at Oxford and designing its courses. He became a lecturer in the School of Geography in 1933 and university lecturer in climatology in 1935. The latter appointment came when he was over 50 and already an eminent climatologist.

Always an ardent traveler, Wilfrid went to Brazil (1932), New Zealand (1933), Burma (1934), Canada (1938), and to Cameroon, South Africa, and Jamaica. Beckinsale remarked that Wilfrid "travelled widely, and for him there was always, besides meters and gauges, the human recorder, if possible in the form of Wilfrid George Kendrew." In his mid-50s, Wilfrid served as a lieutenant in the Royal Navy Volunteer Reserve during WW2, his duties taking him to Africa and the Indian Ocean and was primarily responsible for the collection and analysis of meteorological data from Africa. In 1942 he wrote *Weather—An Introductory Meteorology for Airmen*, published by Oxford University Press as one of the Oxford Air Training Manuals for British student pilots.

From 1940 to 1950 Wilfrid was university reader in climatology at the School of Geography at Oxford and became director of the Radcliffe Meteorological Station, remaining so until he retired at the age of 65. However, Wilfrid remained very active after retirement, taking up appointments at the University of British Columbia in 1950 and with the Defense Research Board and the Meteorological Branch of the Canadian Government Department of Transport in 1951. During the latter period he worked with Canadian colleagues on the climates of Central Canada, British Columbia, and Yukon

Territory. Wilfrid spent a semester at the University of Colombo, Ceylon/Sri Lanka, before returning to Canada for a couple of years.

Suffering from hypertension and various other ailments in the late 1950s, Wilfrid moved to a flat in Cambridge at 26 Station Road to be closer to John. While residing with John on Tennis Court Road he died of cerebral thrombosis at the Chesterton Hospital in Cambridge on April 4, 1962 at the age of 77. He was cremated five days later.

In view of Wilfrid's international reputation as a climatologist, following his death obituaries appeared written by several eminent geographers, most notably by Robert Beckinsale at the School of Geography, Oxford. Beckinsale's obituary for Wilfrid and the announcement of the opening of the LMB on Hills Road in Cambridge, with John as deputy chairman and head of the Division of Structural Studies, appeared on back-to-back pages in the June 2, 1962 issue of *Nature*. Wilfrid died just a few months before it was announced that John had been awarded the Nobel Prize in Chemistry. Beckinsale's obituary for Wilfrid began with "The death on April 4 of W.G. Kendrew will be deeply regretted by climatologists and meteorologists throughout the world. For more than fifty of his seventy-seven years he took a great interest in climatology and his name soon became a household word in climatological circles." An obituary for Wilfrid also appeared in the *London Times* and recalled that "he always stood aloof from the working of the school and faculty, preferring to direct his comments, and often criticisms, to individuals." In addition to these obituaries, Gordon Smith in the School of Geography, Oxford, and Joan Kenworthy in the Department of Geography, Durham University, published biographies of Wilfrid in 1997 and 2007, respectively.

Another obituary appeared in the *St. Catherine's Association Chronicle*, written by the Reverend Victor Brook (1887–1974), at the time a fellow at Lincoln College and censor of the non-residential St. Catherine's Society, Oxford. Reverend Brook knew Wilfrid well, and the obituary provides a very insightful look into Wilfrid's temperament and character:

> Wilfrid was an elusive man, very difficult to know. He was friendly, and yet always seemed to thwart, perhaps unconsciously, any attempt to get near him at all intimately. Whether this was due to sheer modesty (he was always reluctant to talk about himself; it is typical that in Who's Who neither his parents nor his date of birth is mentioned), or to a dislike of what he might regard as impertinent curiosity, or to simple shyness it is impossible

to even guess. But certainly there always seemed to be a barrier of some sort; and yet, with it all, he was a friendly man. He did not like large or noisy parties, but very much enjoyed a day in the country with a friend or a small theatre party. On such occasions he was full of happiness and talk.

Of course, he was always critical, sometimes over-critical to the extent of being awkward. But it was not just "bloody-mindedness." He was determined that no sort of personal attachment should interfere with the complete disinterestedness of his judgment. He seemed at times to lean over backwards to avoid concessions, and never hesitated to disagree. But it was all genuine, and in his view had nothing to do with personal relationships. And nothing pleased him more than to feel that his help was needed and that he could give it. That satisfied him completely.

He was a loyal and devoted son to St. Catherine's. His teaching for Pass Mods [Honor Moderations] he enjoyed, and over it he took endless trouble—probably because he felt that it provided a chance of helping with dull work which no one else really wanted to do. Of course, as an individualist, he was not enthusiastic (to say the least) about any changes in the Society towards a more collegiate or corporate spirit. He liked it as it had been when he was an undergraduate—but he spared no effort to make it, as such, as good and useful as it could be.

That was all the stranger because he always had a grievance that he had been forced to take a Pass degree (he used to say that it was because that was more financially profitable to the Delegacy!). Naturally, such a course did not satisfy someone of his ability, and he at once set himself, with typical individuality, to follow a line of his own. In the study of climatology he made a distinguished career for himself by his ability and determination. Once, when he needed information about the remote North Sea, he sailed on a fishing trawler to get it—and as he could not legally be taken as a passenger, he signed on and served as part of the crew. Still, even when he was a well-known authority (one of his books is in its fifth edition) he was always modest about it, but rather cross if he were referred to as a geographer! To him that seemed to be an insult to climatology.

When the Second World War broke out, he was determined to do his part. Though over-age for the RNVR [Royal Naval Volunteer Reserve], he managed to wangle himself a place (and "wangle" is the correct word) because of his special qualifications. He thoroughly enjoyed himself, and the Navy was one of the few established institutions (if not the only one) of which he would speak with unbridled enthusiasm. He thought it really competent—and it satisfied his deep passion for travel and adventure. After retirement, he found an outlet for the same longings by his work in Canada on climatology.

He was an odd mixture—angular and difficult at times because of his critical mind, but full of enthusiasm for what interested him, utterly sincere and honest, plucky and hard-working, as well as able—and at the end of it all, despite his eagerness to help, a rather lonely figure.

When Wilfrid died John inherited his considerable estate that included a relatively large sum of money, more than £60,000 (about $165,000), and various investments valued at more than £20,000 (about $55,000). In a letter from aunt Janet (Maud) to John following Wilfrid's death, she remarked that it was unbelievable that Wilfrid should have amassed so much money; she found it tragic that a rather rich man should have chosen to live like a poor man. Wilfrid's frugal nature remained unabated to the very end of his life. Just like his father, John maintained a closeness with money throughout his life.

John's Mother

Evelyn May Graham Sandberg-Vavalà (fig. 3.3) was born in 1888 in Compton in the Wantage district of Berkshire in southeast England, at the foot of the Berkshire Downs. Her father, Reverend George Alfred Sandberg, was born in Benares, India, in 1848 and was Clerk in Holy Orders. The family lived for a time at the parsonage of St. Mary and St. Nicholas Church on Aldworth Road in Westhide parish. George died in Herefordshire in November 1910 at the age of 63. Evelyn's mother, Annie, was born in 1858 and died in Herefordshire in 1894 when she was 36 and Evelyn was only six. As a child, Evelyn lived at 7 Barton Road, Hereford, and attended the Castle Street School from 1897 to 1904. She moved with her father to Cavendish House, Christchurch Road, Bournemouth, and attended Bournemouth High School from 1908 to 1910.

In October 1910, at the age of 22 and just prior to the death of her father, Evelyn enrolled at Oxford University as a member of the Society of Oxford Home Students that became St. Anne's College in 1952. Home students were women who wanted to study at Oxford but did not choose to join one of the four existing residential women's colleges opened between 1879 and 1893

Fig. 3.3 Photo of Evelyn Sandberg-Vavalà, mother of John Kendrew.
Reproduced from a photo in Wikipedia from the Fondazione Federico Zeri,
University of Bologna, Italy

(Somerville, 1879; Lady Margaret Hall, 1879; St. Hugh's, 1886; and St. Hilda's, 1893). In 1878 women were admitted to Oxford University and sat exams, but it wasn't until 1920 that they were admitted as full members of the university and granted degrees. Women were admitted for the first time into a few all male colleges in 1974. Home Students lived either in family homes, lodgings, or hostels, and studied the same range of courses as other women enrolled at Oxford colleges. The courses were organized by the Association for Promoting the Higher Education of Women established in 1878. While attending Oxford Evelyn lived first at 203 Iffley Road, then at 18 Staverton Road, and then at 41 Wellington Square.

Evelyn studied geography at Oxford, receiving the Certificate in Geography (Physical Geography) and the Diploma in Geography (Geomorphology) in 1911 and 1912, respectively. Her tutors were A.J. Herbertson and a Miss Byrne, and she probably took courses taught by Wilfrid Kendrew, although that is unclear. She left Oxford in summer 1912 and in September was geography mistress at the Girls' Grammar School in Bradford, West Yorkshire.

Then in 1914 she married Wilfrid Kendrew and from 1915 to 1916 lectured on geography at University College, Reading, as a replacement for a professor who was on temporary government service during WW1.

On March 24, 1917, Evelyn gave birth to John. Four years later Evelyn left Oxford and moved to Florence where she worked for more than 35 years under the nom de plume Evelyn Sandberg-Vavalà; the last name was that of a military officer with whom she was romantically linked for a time. Initially Evelyn trained in the history of Italian art at the Uffizi Gallery with some guidance from Bernard Berenson (1865–1959), an eminent American art historian specializing in the Renaissance. By 1926 Evelyn had acquired enough training and expertise to be able to publish her first book, written in Italian, on Veronese primitive painting of the 14th and 15th centuries. Two more books in Italian followed in 1929 and 1934; the former a survey of Italian painted crosses and iconography of the passion and the latter on the iconography of the virgin and child in the 13th century.

Evelyn was a quite prolific writer on early Italian art. She published two books written in English, *Uffizi Studies: The Development of the Florentine School of Painting* in 1948 and *Studies in Florentine Churches* in 1959. She published many scholarly articles between 1927 and 1960 in the *Burlington Magazine,* a prestigious monthly publication founded in 1903 as the *Burlington Magazine for Connoisseurs* and devoted to the fine and decorative arts. Evelyn also published several articles in the *Art Bulletin* between 1929 and 1938, as well as articles in a few other magazines.

Evelyn left Florence and returned to England for a time during WW2 where she obtained employment organizing the Oxford University Gramophone Society. What began as an undergraduate club with the aim of acquiring records and scores of music of permanent value eventually came under the supervision of the university, however, general management of its affairs remained with a society committee. When Evelyn worked for the society it had acquired a large number of limited editions and foreign records, a substantial lending library of classical records, and a circulation of more than 100,000 records annually to more than 300 members.

Eugene Thaw (1927–1990), well known in the international art world for his activities as an art dealer, collector, and author, reminisced about Evelyn when interviewed by James McElhinney in 2007:

I met Richard Offner [1889–1965], who taught at the Institute of Fine Arts and who was the rival to Berenson in authenticity matters on early Italian

paintings and lived in a house just outside of Florence. And I had an introduction to him from Millard Meiss [1904–1975], who was his colleague and
had once been his student. So I met people. I attached myself to a woman
who rode her bicycle around Florence, and she was a kind of marvelous
English woman who had married briefly a Yugoslavian military officer.
And she was called Evelyn Sandberg-Vavalà, with a hyphen in between. She
was poor but had lived in Florence through the war. She had written a guide
to the Uffizi, a small thing you could buy for a couple of dollars. And she
was afraid of dogs, so I would run alongside her bicycle to chase away the
vicious dogs, and we became great friends.[1]

Thaw remembers that Evelyn was a significant figure around Florence who
knew everyone, was warm, delightful, and eccentric, and likened to the well-
known English actress Margaret Rutherford (1892–1972).

In addition to Italian art, Evelyn loved classical music and Agatha
Christie mysteries, and she kept a large collection of both in her apartment.
She had very little money and usually struggled financially, but did receive
help from her friends and John, and had many students who paid her for
their training in the history of Italian art. Nicholas Clifford recalled that
"My father [Henry Clifford, 1904–1974] was at the Philadelphia Museum,
where he would later become curator of painting, and in those first years
immersed himself in a study of Italian painting [in Florence] under
the tutelage of the redoubtable English art historian, Evelyn Sandberg-
Vavalà."[2] Another student, Marvin Eisenberg (1922–2016), a renowned
scholar of Italian Renaissance painting, presented a choir book page from
the late-14th to early-15th century Italy to the Michigan Museum of Art
"in memory of Evelyn Sandberg-Vavalà, beloved teacher and mentor
in Florence."[3] There is a Sandberg-Vavalà bequest, the Contini-Volterra
Photographic Archive, at the Fondazione Giorgio Cini on the island of San
Giorgio Maggiore in Venice. The bequest consists of photographs, cuttings,
prints, and manuscript notes that were organized by art connoisseur Count
Alessandro Contini-Bonascossi with Evelyn's assistance.

Evelyn guided people around the Uffizi Gallery, gave lessons on the history
of Italian art to students in her apartment in the Borgo San Jacopo between
Ponte Vecchio and Ponte Santa Trinita, and frequently accompanied students
to view and discuss Florentine 14th-century paintings. Many of her students
took advantage of the extensive photographic archive of early Italian art that
she assembled and kept in her apartment.

During the final years of her life, Evelyn kept busy revising a catalogue of all known Italian gothic and renaissance paintings for the Fondazione Giorgio Cini. John helped to support her financially and flew to Italy frequently to be with her, but was unable to fulfill Evelyn's request to bring her back to England due to his father's insistence that she not return. Evelyn was ill much of the time, and having converted to Catholicism and been received into the church by Fr. John Bligh, SJ, she tried to recuperate in a convent where she was looked after by the nuns. Evelyn died from lung disease on September 8, 1961, at the age of 73 and was buried a day later by the Camaldolese Oblates (Benedictine Oblates) in the cemetery of Moggiona, Comune di Poppi, in Tuscany, about 50 kilometers southeast of Florence.

John Pope-Hennessy (1913–1994), British art historian, scholar of Italian Renaissance art, and director of the Victoria and Albert Museum (1967–1973) and British Museum (1974–1976), remembered Evelyn affectionately in an obituary written for the *Burlington Magazine*:

The death of Evelyn Vavalà in Florence on 8th September will give rise, in a wide circle of friends, to a sense of personal impoverishment. In the 1930s many students, now middle-aged, went to her apartment in the Borgo San Jacopo for their first lessons in the history of Italian art, visited the great monuments of Florentine Trecento painting in her company, were initiated by her in the Academia into the mysteries of the Orcagna school, and consulted the unique photographic archive she had built up for their use. Not only was she a remarkable teacher—fragments of conversation from thirty years ago still hang about one's ears as one goes into the Baroncelli Chapel now—but a woman of exceptional courage, integrity, and warmth, and there was not one of her pupils who did not, imperceptibly, become her friend. She was not an easy friend, but if she exacted sympathy she also returned it, with great self-effacement and single-mindedness, and it had the added value that she was incapable of insincerity. As a result no one had warmer friends than she, as the stream of visitors first to the little house in the Erta Canina in which she lived after she returned to Florence in 1946, and then to the flat in Via Maggio to which she moved when she was discouraged from bicycling daily to the German Institute and returning by bicycle at night after dining near Or San Michele, was there to testify.

Her first book, La pittura veronese del trecento e del primo quattrocento, appeared in 1926. A pioneer work, its conclusions have not been superseded, and when, years ago, an exhibition of Veronese primitive painting was held at Museo di Castelvecchio at Verona, her book alone stood up to retrospective scrutiny. Thereafter, on the suggestion of Bernard Berenson, she began work on a survey of Italian painted crosses. This, her most important book, appeared in 1929 as La croce dipinta italiana, and is, in the most literal sense, definitive. In the 1930s her interests veered towards dugento painting, and in 1934 she published an excellent short study of the iconography of the Virgin and Child in the thirteenth century. In the normal course this might have been followed by the general introduction to dugento painting she was ideally well equipped to write, but in 1940 the war forced her back to England, where she spent four fallow years organizing the Oxford University Gramophone Society. Much as she valued the new contacts which this brought her, it was, through no fault of her own, time misapplied, and when, after the war ended, she moved once more to Florence, financial pressure, which she faced with her customary fortitude, prevented her from following a continuous line of original research. None the less, these years witnessed the publication of a number of short, semi-popular books, 2 of which, Uffizi Studies (1948) and Studies in the Florentine Churches (1959) are of notably high quality, and if published in a worthier, less unimaginative fashion, would have secured the success they deserved. A visitor to Evelyn Vavalà was often deluged with trivial complaints, but it is a measure of her character that never once did she allude to the predicament of which she must have been unceasingly aware, that her real talent for research, her heroic tenacity in organizing bodies of intractable material, and her sane, unpretentious attitude to attribution, were frustrated by lack of the means which enabled less sensible, less gifted students to pursue their chosen study undisturbed. Yet it was some compensation to feel, as she was amply justified in doing, that very many scholars of a younger generation than her own owed part of their achievement to the early help and to the critical encouragement she had so generously given them.

Hugh Honour (1927–2016), an eminent British art historian, also prepared a perceptive obituary for the London Times upon Evelyn's death:

The death in Florence of Evelyn Sandberg-Vavalà (Mrs. Kendrew) on September 8 marks a period in the history of Anglo-Italian cultural relations. For she was perhaps the last of a long line of English people who settled in Italy and, inspired solely by a deep love for the country, devoted herself to work of enduring scholarly value on the history of its art.

As an historian Evelyn Vavalà will for long continue to hold the respect and earn the indebtedness of students of Italian painting. But in Florence she will be remembered as a personality, original sometimes to the point of eccentricity, as well as a scholar. An ardent Italophile, she won a place in the hearts of numerous Florentines who shared her interest in art and music. Indeed, so completely did she become absorbed by her adopted city, and so completely did she estrange herself from the international society of the larger villas, that she can hardly be described as an Anglo-Florentine. To all who knew her it was obvious that her love for Italian art derived from her love for the Italians themselves, and she sometimes found it difficult to disguise her contempt for those English residents who failed to share her predilection. The fact that her scholarly reputation stood far higher in Italy than in England seemed to be a source of pleasure rather than disappointment to her—both her major books were written in Italian and never translated into English.

Many Florentines, besides many English and American lovers of Italy, will long remember her with affection—ensconced behind a huge pile of books in the library of the German Institute, imparting to young visitors her enthusiasm for the art of the Duecento and Trecento, or at home in her flat, playing some new recording of music by Monteverdi or Vivaldi. Appropriately enough she lived in Via Maggio, a few steps from Casa Guidi, the home of Robert Browning whose poetry first fired her with the desire to visit Italy. He also, she liked to remember, "haunted the dim Santo Spirito" and "loved the season of Art's spring-birth so dim and dewey."

Marital Separation

Evelyn Sandberg and Wilfrid Kendrew married in 1914 in the Warminster District of Wiltshire when she was 26 and he was 30 years old. Evelyn, an only child, had lost both her parents at about the time she moved to Oxford and possibly viewed Wilfrid and his extended Oxford family as a secure harbor. As a student in the School of Geography at Oxford and initially a lodger on Iffley Road, Evelyn probably knew, was taught by, and admired Wilfrid. However, they were not at all well suited to each other and had an empty shell of a marriage almost from the beginning.

Prior to their permanent separation in 1921, Evelyn attempted to escape from England to the Continent with John, but Wilfrid caught up with them and demanded that either Evelyn return with John or that she continue on alone. On that occasion the three of them returned to Oxford. But soon after Evelyn left for Italy alone, leaving John in the sole care of his father. For the rest of her life, Evelyn remained very bitter about her experiences in England and what she saw as the elitism and pomposity of the academic community there. Evelyn did not become reacquainted with John for more than a decade, when he was 15 years old and a boarding student at Clifton College.

No matter the problems of her marriage it is hard to understand how Evelyn could desert a four-year-old son who had to grow up in the absence of a mother's love and attention. However, Evelyn had very few alternatives once she made up her mind that she could no longer remain with Wilfrid. Divorce in Britain remained the preserve of the rich in the early 1900s, such that among the peerage as many as one-third of marriages ended in divorce. On the other hand, in 1921 divorce was extremely rare among the general population, with less than one-tenth of a percent of the population (16,000 out of 38 million) in England and Wales divorced. Although the Matrimonial Causes Act of 1857 had modernized divorce law in Britain, broadening the availability of divorce beyond the privileged few, it remained a very unusual and infrequent action since there were no acceptable grounds other than adultery. Divorces only began to proliferate after the Matrimonial Causes Act of 1923, which established an equalization of divorce criteria between men and women and made divorce more accessible to the middle and lower classes.

Evelyn's sole choice was separation from Wilfrid, although implicit in this action was the understanding that John would remain with his father. Generally, children were considered legally the husband's and in the case of separation the husband had first claim to any children unless he had been convicted for assault. Financial support for children was the major reason behind this and, since most married women were housewives without any income of their own, children of separated couples remained with the wage earner, the husband.

With a small inheritance from her parents, Evelyn emigrated to Florence where, except for a few years during WW2, she resided and worked as an Italian art historian for the remainder of her life. John stayed in Oxford in his father's care and had no contact or communication with his mother for more than a decade. When John reconnected with his mother as a teenager he made many visits to Italy, referred to Florence as his second home, became fluent in Italian, and shared his mother's passion for the arts and classical music.

4

Formal Education (1924–1939)

[A]t 15 I quite clearly made up my mind that I wanted to do science.

John Kendrew

Birth

John was born on Saturday, March 24, 1917, at 223 Iffley Road in Oxford, just two weeks before the United States entered WW1, a war that saw Britain mobilized under Prime Ministers Herbert Henry Asquith (1852–1928) and David Lloyd George (1863–1945). Oxford was undergoing significant changes at this time due largely to William R. Morris (1877–1962) who began building cars at the Military Training College at Cowley in 1913. Beginning with a labor force of about 300 men, the Morris car manufacturer grew to nearly 10,000 employees by 1938 and produced its one millionth car in 1939. Morris accumulated a fortune that ultimately was dispersed largely to the academic and medical communities.

After John's birth in 1917, Wilfrid and Evelyn moved from 223 Iffley Road, very close to Wilfrid's parents on Stanley Road, to 18 Staverton Road in north Oxford. A nanny, Elizabeth Pimm, was hired to care for infant John and remained with the family for three years until she left his parents employ, married, and moved to Ramsden, about 15 miles northwest of Oxford. When John was awarded a Nobel Prize in 1962, Pimm wrote to him and noted that John resembled his mother, but his color and profile were very like his father as she remembered him. In 1921, about one year after Elizabeth Pimm left, his parents separated permanently leaving John in the care of his father and his father's family.

Dragon School, Oxford

In January 1924, two months prior to his seventh birthday, John became a new boy at the Dragon School, Oxford, located at the northern edge of the university near the River Cherwell. Its motto is *Arduus ad Solem*, "Reach for the Sun." The Dragon is considered one of the best preparatory schools in England, its students well prepared for entrance to the best English public schools.

When he entered Dragon as a day student, John and his father were living in Bladon near Woodstock, about seven miles north of Oxford, where Winston Churchill, his wife Clementine, and his father and mother are buried in St. Martin's churchyard. However, within a few months their address became 74 High Street, Oxford, at the corner of Merton Street where John's father was a tutor to non-collegiate students in the School of Geography and had rooms on the second floor of the building. The large stone building was erected in 1888 for the Delegacy of Unattached Students on a site previously occupied by three shops: a confectioner, a butcher and greengrocer, and a bootmaker. The delegacy, which started in 1867 in order to open the university to a larger and less well-off class, became St. Catherine's Society in 1931 and remained in the High Street building for five years. The building served in part as a hospital during WW1; in 1975 the Ruskin School of Drawing and Fine Art moved there from the Ashmolean Museum and is located there today.

The Dragon School had opened in Balliol Hall on Blackhall Road as the Oxford Preparatory School in 1877. It was founded by a committee of Oxford University academicians, including four heads of colleges and seven professors, with the aim of providing a strong academic grounding for children of university professors. Reverend Arthur Clarke, an experienced teacher and half-fellow from Magdalen College, was hired as headmaster and a Visiting Council was formed and led by Henry Liddell (1811–1898), the dean of Christ Church. Lessons began for 14 students in two rooms at Balliol Hall. Two years later the school expanded and moved to 7 Crick Road. Five-year-old J.B.S. Haldane (1892–1964), destined to become the foremost British authority on genetics and biochemistry, became a new boy at the Oxford Preparatory School in 1897. Today the Dragon is one school located at two sites in Oxford and consists of a Preparatory School for children 8–13 and a Pre-Preparatory School for children 4–7 years old.

The name Dragon originated with the entering class of boys at the Oxford Preparatory School. They were seeking an identity suitable for a badge to wear on their caps and hatbands at sporting events. Among the school's establishment was a very active and influential member of Visiting Council, the Reverend H.B. George, a fellow of New College. Soon the students were calling themselves dragons after Reverend George and the heroic legend of Saint George and the dragon. Hence the official change in the name of the school, but not until 1921.

Clarke served as the first headmaster of the school from 1877 to 1886. In 1887 Charles Cotterill "Skipper" Lynam (1858–1938), a teacher at Oxford Preparatory School, bought the school from Arthur Clarke's widow for £2,000 (about $10,000), and in 1895 the school moved to its current location on Bardwell Road. Several new buildings were completed over the next 15 years at a cost of about £4,000, which was contributed by parents of students at the school. Lynam was a skilled and enthusiastic sailor and sailed his yacht around the Hebrides every summer. He did not attach much importance to overt politeness, confessing that he hated to be called sir and preferred to be called Skipper.

Skipper Lynam served as headmaster for 34 years and from 1886 to 1965, nearly 80 years, the headmaster of Dragon was a member of the Lynam family: in succession, C.C. "Skipper" Lynam (headmaster 1886–1920), then Skipper's younger brother Alfred E. "Hum" Lynam (1873–1956; headmaster 1920–1942), and then Hum's only son Jocelyn H.R. "Joc" Lynam (1902–1978; headmaster 1942–1965). Consequently, the school was often referred to as Lynam's Preparatory School. During John's time at Dragon the headmaster was Hum Lynam, so called because he hummed under his breath all the time so as to warn the boys that he was approaching. Hum also wandered around Dragon in the evening playing his violin.

In view of his father's relatively low income, fees at Dragon were adjusted down so that John would be able to attend. Such action was in line with the desire of the previous headmaster, Skipper Lynam, to open Dragon to those unable to pay full tuition but who would benefit from the exceptional education offered. Skipper also promoted less supervision for the boys and introduced more freedom for them: unlike boys attending other schools, Dragon boys played cricket and football on Sunday afternoons. Skipper felt that a boy would never learn to use his time properly unless allowed to waste it and would never learn self-control unless he had the opportunity of getting into mischief. Dragon was quite progressive and was co-educational from the 1890s onward; in 1923, when John entered the junior school, there were five girls in his starting group of 39 pupils, all with brothers at Dragon. It was common for pupils to refer to female staff as "Ma" and male staff as "Pa," a tradition that continues even today.

Throughout his time at Dragon, John was both younger and smaller in stature than most of his classmates. In 1930 when he left Dragon for Clifton College, John was 13 years old, 4 feet 9 inches in height, and weighed less than 80 pounds. Headmaster Hum Lynam found that for John drill was good, but he did not care for team games. From a young age John suffered from very bad eyesight that eventually was recognized by his aunts and corrected by thick eyeglasses that he wore for the rest of his life. John's slight build and very poor eyesight contributed to his aversion for team sports such as football and hockey since he simply couldn't follow the movement of the ball without eyeglasses. However, John did participate in skating on the frozen River Cherwell and Port Meadow in winter and was an accomplished skier and ice skater. His skill as an adult skater was noted by onlookers in Cambridge when the River Cam would occasionally freeze in winter. At Dragon John also participated in swimming races across the River Cherwell in summer term, and it was noted by a teacher that John swam most keenly in the form relay races in 1930. In his second year at Dragon, John played a small role as the messenger in the Dragon production of *Macbeth* (fig. 4.1).

In most school terms John took English, French, mathematics, and science, with students organized into groups in each subject based on ability not age. Since he was an accomplished student in each subject, John was frequently one of the youngest in his group. He was evaluated as quite good to very good in all subjects in the Junior School and, after promotion to the Big School in 1927, was evaluated by the headmaster as maintaining his early promise. In his classes as a 10 year old, John placed first in mathematics and French, third in English, and eighth in classics. With the exception of Greek, a subject he failed in his last year, John performed well; he continually excelled in mathematics and science and was considered a champion statistician by his teachers.

In his last year at Dragon, John produced the best paper on the science examination and his teacher, William Stradling (1898–1936), found that his performance on exams was excellent and that he observed accurately and recorded correctly. In evaluating John's final examination in science, Stradling went on to say that John's was the best paper and that other students were not so accurate in details as he. That year John received the Stradling Prize for Science. It is clear that John's interest in science was awakened at Dragon, and his success in the subject as a young boy was a portent of things to come.

Fig. 4.1 Photo of John Kendrew (ca. 8 yoa) as a member of the cast of *Macbeth* at the Dragon School. Contributed by Gay Sturt, Archivist at the Dragon School, Oxford

In his final year at Dragon, John wrote an essay, titled "Nothing Doing," that was published in the *Draconian*, a magazine for alumni. It is likely that the essay was inspired by his father's occupation as a climatologist and suggests that even as a 13 year old John already leaned toward science as his career:

Nothing Doing

One day last holidays it poured with rain, and by afternoon it showed no signs of stopping. I was half thinking of putting down "Nothing doing" in my diary, but then, as I gazed out of the window at the teeming rain, an idea struck me. I would trace the life history of a raindrop. This was what I thought:

One of the raindrops which are pouring down today first soaks into the dry earth through all the porous limestone. Then suddenly, it can sink no more. It has come to clay, which is a barrier as effectual to water as a stone wall is to us. So our drop runs along the clay until it bubbles out in a clear spring on the hillside. Here it is carried off in a bucket by a village boy. The boy, being careless, spills some water, and our drop is among those spilt. But as soon as our drop is exposed to the warm sun it turns to vapor and sweeps up into the skies, and so ends its terrestrial existence.

Once it gets up into the air, where it finds many of its companions, it has a fine time. It is swept eastwards by the westerly wind and carried over Siberia. There it gets caught in a storm, and joins with its brothers to make a drop so big as to form a cloud. Then it gets bigger still, and falls through the air, but gets so big that it splits up. And each little bit feels a thrill when it splits, caused by being electrified. This process goes on again and again, until our drop is so highly charged that it cannot stand it any longer. Suddenly there is a blinding flash and a roar, and our raindrop swoops down to earth and soaks in. Then it starts again, and does something like what it did before. For what a great distance a particle of moisture must travel! At one time it steams in the marshes of Central Africa; at another, it is part of an iceberg.

I went out for a tramp in the rain and had a very good time, although I was soaked.[1]

In later life John maintained strong ties with Dragon and in 1968 opened the new science block, the Cartwright Building, named for Fred Cartwright who was then chairman of South Wales Steel Board (fig. 4.2). At this occasion, among other things John revealed was that he had enjoyed doing illegal experiments on his own during his time at Dragon.

Clifton College, Bristol

The early 1930s was a time of depression and mass unemployment in Britain, reaching a high of 23 percent unemployment. In September 1930, at the age of 13, John moved from the Dragon to Clifton College, Bristol, as a boarding student (fig. 4.3). John was awarded a scholarship of £100 (about $485), a relatively large sum since the average annual wage in England at the time was less than £200 (about $970). The scholarship enabled him to attend Clifton and reside in Watson House with about 50 other boys; today Watson House is a residence for girls.

Clifton had been founded by a group of Bristol's leading citizens in 1862 as a boarding school for boys and it took more than 100 years for it to become

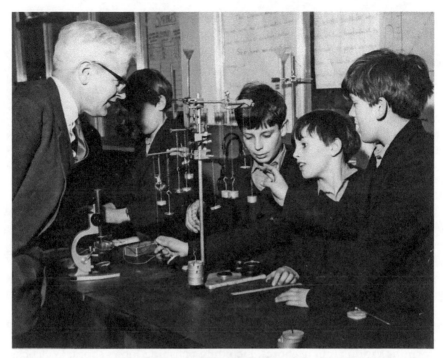

Fig. 4.2 Photo of John Kendrew (ca. 51 yoa) and students at the opening of the new Science Block, the Cartwright Building, at the Dragon School in 1968. Contributed by Gay Sturt, Archivist at the Dragon School, Oxford

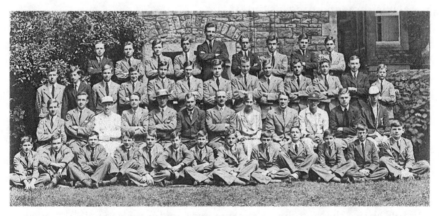

Fig. 4.3 Photo of members of Watson House at Clifton College in 1933. John Kendrew (ca. 16 yoa) is standing at the right-end of the third row. Contributed by Charles Knighton, Principal Assistant Keeper of Archives at Clifton College, Bristol

co-educational. Modeled after Rugby School founded in 1567, Clifton was created by Bristol citizens for Bristol citizens. The first class consisted of 76 pupils, 50 of whom came from Bristol. Today the co-educational upper school has more than 700 students between 13 and 18 years old.

The first headmaster of Clifton was 27-year-old Charles Percival (1835–1917) who had a rather unique attitude among British headmasters, actively promoting close ties between masters and boys, as between friends. This was a very liberal attitude during a period of strict conservatism in Britain. When Percival left Clifton in 1879, he became president of Trinity College, Oxford.

Percival's years at Clifton, 1862 to 1879, were marked by the appearance of many new school buildings including seven boarding houses in the upper school where all of its members ate, thereby creating a feeling of belonging to a closely knit group. Polack's House only took Jewish boys and provided both kosher dining facilities and a synagogue, but it closed in 2005 due to the small number of members. Watson House where John served as head was founded in 1879 and had a long tradition of academic excellence, although it routinely ranked last in athletics. John sometimes referred to a lower classman in Watson House, David Willcocks (1919–2015), who had served as his fag performing various minor chores, including polishing John's shoes. Willcocks was a music scholar at Clifton and went on to King's College at Cambridge, and subsequently became an internationally recognized composer and one of the most influential choirmasters of his generation.

A student at Clifton between 1935 and 1941 summarized the state of corporal punishment during his time there:

It was taken for granted that there were rules, and if you transgressed those rules, you were punished—not necessarily with the cane, though it was always there as a possibility, standing as it were like a slender threat in the corner of the house-master's study. And it did not happen very often; the severity varied with the crime, and was usually limited to three stinging strokes, six strokes being reserved for particularly loutish boys whose record of misdeeds was too awful to contemplate; six of the best, as it was known, was in fact very infrequent. It was obviously painful to be beaten, and one had to screw one's courage up beforehand, but afterwards you could always show off the wheals on your bum for the general enjoyment, sympathy, and admiration of your friends. Above all, the sin was forgotten and forgiven, there were no hard feelings, and you probably remembered not to break that rule again.[2]

There is nothing to suggest that John was ever subjected to such punishment.

In 1864 the Public School Commission had reported that natural science was frequently excluded from the education of individuals that came from the professional class in England. But the study of science at Clifton was established at the very outset, and as far back as the 1860s teaching of science was a hallmark of a Clifton education. A laboratory for physics and chemistry was opened in 1867, biology laboratories were added two years later, and a scientific society was established soon after.

An article in *Nature* in 1871 praised the teaching of science at Clifton in glowing terms:

> Foremost, if not positively the first among the schools in which the sciences are thus taught stands Clifton College, under the able direction of the Rev. J. Percival, in which scientific study is introduced to the utmost, and keenly pursued by the boys, with the encouragement of all their masters, the latter a most important consideration, and which, we are sorry to say, we cannot assert in reference to other schools of equal pretensions . . . Science is much indebted to Mr. Percival for the magnificent example he has set in science education . . . Natural Science at this College is not a voluntary subject, but forms a regular part of ordinary school work . . . In the Junior School Botany is taught, in the Upper School Chemistry and Physics . . . Special classes are formed for those who wish to go deeper into these subjects, or to take up others. Thus, there are special classes studying Chemistry, Physics, Zoology, Physiology, Botany, Physical Geography, and Civil Engineering . . . Facilities are also afforded for learning science practically . . . In the Chemical Laboratory about twenty boys study analysis. A Physical Laboratory has been built and will open in September . . . A large workshop fitted with carpenters' benches, vices, and lathes has been opened this term, and is exceedingly popular. There is also a physiological laboratory, in which a few of the elder boys receive instruction in Practical Zoology and Physiology . . . The taste for natural history is developed by means of the Natural History Society, the School Museum, and the Botanic Garden . . . The society is subdivided into sections, which hold special meetings and make excursions for the study of different branches of science . . . They publish transactions. A conversazione was given last month to celebrate the opening of the Museum and Botanic Garden . . . .[3]

In 1919 a new phase in Clifton science began with the appointment of 28-year-old Eric Holmyard (1891–1959) as head of science and chemistry. The excellence of Clifton science during the 1920s and '30s is associated with Holmyard and his team of teachers, which included William Badcock (1880–1951) as head of physics. Holmyard had read history and natural sciences at Cambridge, taught at Bristol Grammar School and Marlborough, and wrote several textbooks, including *Elementary Chemistry*, that were used in schools throughout the English-speaking world. Holmyard and Badcock were rather formal as teachers, but loved science and are believed to have taught very well. Badcock was an inspiring teacher despite having sarcastic sense of humor. John recalled that "I daresay there are quite a lot of people here [Clifton] who never knew them [Holmyard and Badcock] personally, but I have always felt that it was to the eternal credit of Clifton that it was able to secure those two people to run Chemistry and Physics here."[4]

During the Holmyard-Badcock era a new science school was built and opened by the Prince of Wales in June 1927. It was an ambitious undertaking conceived by headmaster Norman Whatley (1884–1965) and achieved by an appeal organized by John Whitley (1866–1935), speaker of the House of Commons, that raised £50,000 (about $230,000) from Clifton alumni. The new science school had excellent, well-equipped laboratories for teaching biology, chemistry, and physics and a new Stone Science Library that was considered as significant as the laboratories.

John's interest in science was fully sustained in Clifton's science school in the 1930s. At the time of the science school's jubilee in 1977, he reminisced about his experiences at Clifton:

> I must say that, to be in the Sixth Form [studying for A-level examinations] at that time, was an experience—stimulating and challenging but, at the same time, rather tough and in some ways a little alarming. With Badcock control was very close. For example, in these days of pocket calculators, one has to remember, that such things did not exist. Badcock wouldn't even allow you to use a slide rule. He would challenge you to a race in which he would use log tables and you used a slide rule, and he always won the race, and of course got one more place in decimals. There was no mercy if you made a mistake in using logs or adding figures. You were simply not permitted to make any error in arithmetic . . . Holmyard left you entirely to your own devices. We were given a textbook in organic chemistry written by Karrer. It was in German with no English translation

and we had not learned any German. This book [which consists of about 1,000 pages and weighs 10 pounds!] would normally be used in third-year university studies.[5]

John coped with this by using a German bible in chapel and teaching him-self German: "After about six months, we could read scientific German as well as we could read English . . . But the more important thing about the teaching of Holmyard and Badcock, was what one might call the inculcation of good habits—the habits of what I might call hard thought. Hard thought which characterized good physics and good chemistry."[6] In recalling biology at Clifton, John commented that "Well, I would like to pay a tribute . . . to another splendid teacher here, who was A.G. Graham, running the biology teaching at that time. He was appropriately known to us all as "Bunny" Graham—appropriate because, under his guidance we used to dissect rabbits, having graduated from earthworms and frogs."[6]

By the age of 15 John, at Clifton, had made up his mind to do science. It should be noted that until he was 15 years old and required to specialize, John read classics and only a little science. In 1974 John commented on what led him to a career in science: "at 15 I quite clearly made up my mind that I wanted to do science . . . Science has always interested me in a number of ways. First of all, I rather liked working with my hands; and secondly I liked it intellectually and philosophically—I was interested in the kind of world view one might have, and science seemed to be a way of getting to grips with that."[7]

John Pinkerton (1919–1997), a well-known computer designer, attended Clifton and was a member of Watson House (1932–1937) when John was head. In the late 1950s Pinkerton's sister, Mary, was to assist John's group in their work on myoglobin structure at the Cavendish. In 1937 Pinkerton followed John to Cambridge and in 1997 reminisced about their student days together:

Kendrew exhibited a certain absentmindedness, was occasionally some-what reserved, and from time to time switched off his attention from whatever it was that you wanted to talk about; then you just had to wait for another occasion to raise the topic. I was rather more conscious of this trait at Cambridge both before and after the war when, as an undergraduate and research student, I saw more of him socially. Kendrew was regarded at school [Clifton] as a sort of science swot [one who studies hard and is con-sidered boringly studious; a bookworm/grind/nerd] and was known for awhile as "Binks" [a studious boy absorbed in his own lonely excellence] on account of his studious appearance behind his thick spectacles. He told me he had a private "reading place" overlooking Avon Gorge to which he would go on fine days to study undisturbed . . . Holmyard gave his best students much individual attention, running a special practical class in organic chemistry during what would have otherwise been one's free time; Kendrew was one of his star pupils . . . His talent for administration was already evident when he became head of house and had to assign boys travelling in the school buses at specified times across the Clifton suspension bridge to the games fields to play practice games with other houses at rugby or cricket . . . He was not notably good at games but must have performed well enough to have escaped my own reputation as nearly useless . . . Until the end of his life, his handwriting never changed. It seemed to me later that X-ray crystallography demanded precisely that combination of imagination with attention to and recall of innumerable minute details which he brought to the subject.[8]

John hated organized games at Clifton primarily because of his very bad eyesight, so he did not excel at games there. However, because of his academic prowess as a science swot he was not ridiculed by the other boys and garnered a considerable degree of admiration and respect from classmates usually reserved only for star athletes. In summer term 1934, John became honorary secretary of the Clifton Scientific Society and was responsible for producing the society's minutes until he left school in 1936. The minutes reflect both John's intellect and his organizational skills and are a detailed account of all aspects of the society, which flourished with over 300 members. John's notes include "Those who went to the Gas Works were presented with copies of "The Romance of a Lump of Coal." "An interesting, if odorous, afternoon was spent by everyone," and "Those who visited George's Brewery managed to obtain half a glass of beer, chiefly owing to the absence of any masters." There were additional excursions to facilities and works near Bristol, such as Fry's Chocolate Factory, the General Post Office, and Bristol Aeroplane Co., Ltd.

The many essays John wrote at Clifton addressed such topics as "the ethics of punishment," "ugliness," "the arts receive their perfection from an ideal beauty, superior to what is to be found in nature," and "we should do our utmost to encourage the beautiful, for the useful encourages itself." These and other essays reveal John's interest in the arts, but an interest colored by his predisposition toward science.

Significant insight into John's character as a 19 year old (fig. 4.4), as he prepared to leave Clifton and attend Cambridge, can be gathered from excerpts of a long essay he wrote titled "The British Public Will Endure Anything": "The British public is notorious (abroad at any rate) for a complete lack of understanding of all the most important problems of life. It has no taste, it has no principles . . . It cannot appreciate good music, good architecture, or good art . . . ."[9] It is a scathing essay, in which John is hypercritical of everything British, from politicians to the BBC, from football to the pronunciation of words. He labels Britain's association with the League of Nations as "political stupidity and myopia," condemns the "rule of bureaucracy known as the Civil Service," bemoans the removal of India "from our jurisdiction," and criticizes Britain's lack of preparation for another war. John's opinions of such things was sparked by his travels, particularly several trips to Germany between 1934 and 1939, and reunion with his mother in Italy. The reunion

Fig. 4.4 Photo of John Kendrew (ca. 19 yoa) standing on a pile of rubble at the Roman Site of Caerwant in Monmouthshire, Wales. He is lecturing to members of the Archaeological and Architectural Section of the Scientific Society of Clifton College in June 1936 shortly before matriculating at Cambridge University. Contributed by Charles Knighton, Principal Assistant Keeper of Archives at Clifton College, Bristol

with his mother further instilled in John a great love of the arts and dismissal of a variety of British traditions. He accuses the British public of despising "the highbrow" and refers to a writer who defined the highbrow "as anyone who ever read a reputable book." In this essay John anticipates the coming of another great war and suggests that, as in 1914, it may be with Germany. By the age of 19, John was already leaning to preciosity, broad interests in and strong opinions about subjects other than science, and a burning desire to bring about change in Britain.

Throughout the remainder of his life John maintained strong ties with Clifton. He was the first Clifton graduate to be awarded a Nobel Prize, although subsequently two other graduates, John Hicks (1904–1989) and Nevill Mott (1905–1996), were awarded Nobel Prizes in Economics (1972) and Physics (1977), respectively. In 1964 John was made a Governor of Clifton and at the Golden Jubilee of the opening of the new Science School in November 1977 he gave a lecture, "50 Years of Biology: The Impact of Physics and Chemistry." Following John's lecture there were presentations by Nevill Mott (Clifton 1918–1923; Poole's House and School House) and Brian Pippard (Clifton 1930–1938; School House). John was about to become president of the Old Cliftonian Society at the time of his death. John Pinkerton, a lifelong friend from their days together at Clifton and Cambridge in the 1930s and '40s, died at the age of 78 in December 1997, only four months after John.

Cambridge University

Cambridge University is located about 55 miles north of London in East Anglia on the banks of the River Cam. It is the second oldest university in the English-speaking world and one of the world's oldest surviving universities together with the universities of Al Karaouine (founded 859, Morocco), Bologna (founded 1088, Italy), Oxford (founded 1096, England), Salamanca (founded 1134, Spain), and Paris (founded 1160, France). Cambridge University's origins date to the early 13th century when it was founded by a group of disgruntled scholars who fled from Oxford during riots that took place following the murder of a local woman by students. Today, Cambridge consists of 31 colleges with more than 10,000 staff and 18,000 students; it was ranked fourth out of 980 universities worldwide in the Times Higher Education World University Rankings in 2016.

In December 1934 and 1935 while a student at Clifton, John sat exams for entrance scholarships to Clare, Corpus Christi, King's, Trinity, Trinity Hall, and Magdalene Colleges at Cambridge. His first attempt at a scholarship failed, but a year later he won a major scholarship to Trinity College in the natural sciences, reading chemistry, physics, biochemistry, and advanced mathematics. John recalled his experience taking a scholarship examination at Cambridge, where he took part in a practical physics exam in the Cavendish, which he thought compared most unfavorably with the facilities at Clifton. He recalled that "Clifton had excellent, very modern, science labs . . . the old Cavendish and the old Chemistry Department were awful. The Chemistry one was lit by gas, the Cavendish . . . benches were ancient."[10]

Trinity College, Cambridge, was founded in 1546 when Henry VIII amalgamated two medieval foundations, Michaelhouse and King's Hall, into the college. Trinity has enjoyed a distinguished record in the sciences and mathematics, from Francis Bacon (1561–1626) in the 16th century and Isaac Newton (1642–1726/7) in the 17th century until today. Thirteen of its members had been awarded a Nobel Prize in Physics, Chemistry, or Physiology or Medicine when John entered Trinity in 1936, and by the end of the 20th century the number of the college's Nobel laureates in science had doubled to 26.

John entered Cambridge in 1936 at the age of 19. From 1936 to 1938 he lived at G5 Whewell's Court and from 1938 to 1939 at K8 Great Court. In 1997 John reflected back on why he chose to go to Cambridge:

Well, in those days, if you wanted to be a scientist it was pretty obvious it had to be Cambridge because Oxford was really known to be the place for humanities and Cambridge for science, so there was really no real choice. It would be much more difficult now that Oxford science has become, I think, at about the same level as Cambridge . . . So in spite of being born in Oxford and going to school in Oxford it was clear it had to be Cambridge.[11]

When asked about the move from Clifton to Cambridge, John confessed that "Well you know, having come from an English public school where your every movement was controlled I thought it was absolutely marvelous to have my own rooms in a College and you could lock the door to keep other people out, you could do exactly what you liked."[12] This admission clearly reflects John's desire for privacy.

John's tutors at Trinity in 1937 and 1938 were Frederick George Mann (1897–1982), a reader in organic chemistry, and Norman Feather (1904–1978), a university lecturer in physics. Each academic year at Cambridge consisted of three terms, called Michaelmas (October–December), Lent (January–March), and Easter (April–June). John had an intense schedule, taking physics (four terms; 1936, 1937), organic chemistry (three terms; 1936, 1937), inorganic chemistry (two terms; 1938), theoretical chemistry (two terms; 1939), physical chemistry (four terms; 1937, 1938, 1939), biochemistry (three terms; 1937, 1938), and mathematics (two terms; 1937, 1938). Through these courses John came in contact with many stellar Cambridge scientists, including F.G. Hopkins (Nobel/Physiology or Medicine, 1929), J.J. Thomson (Nobel/Physics, 1906), R.G.W. Norrish (Nobel/Chemistry, 1967), E. Rutherford (Nobel/Chemistry, 1908), W.L. Bragg (Nobel/Physics, 1915), J.D. Bernal, J. Needham, E.H.F. Baldwin, F.J.W. Roughton (John's biochemistry supervisor), E.K. Rideal, A. Neuberger, J. Lennard-Jones, J.A. Ratcliffe, W.J. Pope, and M. Dixon, among others.

In an interview in 1974, John spoke about his days at Trinity:

> The Cambridge system was very good for me because I started by choosing my school subjects for Part 1—chemistry, physics and maths. But these only counted as 2.5 subjects and you had to do 3; so I decided to do biochemistry too, and this, I suppose was a turning point, my first interest in biology. It was a marvelous time to do that course—Gowland Hopkins was still active, people like Szent-Györgyi and Carl Cori [Nobel/Physiology or Medicine, 1947] were passing through, and they all gave a few lectures each. It was rather bitty but very stimulating."[13]

Begun at Clifton and continued at Cambridge, John constructed a code book for his course notebooks in physics, chemistry, biology, and mathematics. For chemistry the code was (A) systematic inorganic chemistry, (B) theoretical and general chemistry, (C) physical chemistry, and (D) organic chemistry. For physics (a) electricity, (b) light, (c) heat, (d) mechanics, (e) sound, (f) atomic physics, (g) crystal physics, and (h) philosophy and scientific method. The Greek alphabet and Roman numerals were used to code for biology and mathematics, respectively. These notebooks, consisting of nearly 9,000 pages with about 1,500 pages from Clifton and 7,500 pages from Cambridge, are in John's characteristic script that hardly changed over ensuing years.

Jeannine Alton who organized John's archives at the Bodleian commented that "The total sequence [of notebooks] thus provides an exceptionally full conspectus of the intellectual training then available [at Cambridge], though it may be doubted how many took such diligent advantage of it." She continued:

A modern historian of intellectual or educational development will be grateful for the scrupulous indexing of topics, the very full notes of lecture courses, and the careful identification of lecturers. The latter included most of the leading figures in Cambridge science immediately before and after the Second World War and some visiting lecturers. The total sequence therefore provides an exceptional comprehensive picture of the education available at that time at a well-run school (Bristol) science department and a major "science" university (Cambridge).[14]

During his undergraduate days at Cambridge, John was a member of the Trinity College Natural Sciences Society under J.J. Thomson and became treasurer of the society in 1938, replacing A.F. Huxley (Nobel/Physiology or Medicine, 1963), and then secretary in 1939. John was also a member of the Cambridge University Chemical Society under Professor E.K. Rideal and attended lectures by J.D. Bernal in January 1937 ("X-rays in organic chemistry") and Lawrence Bragg in February 1938 ("The chemistry of intermetallic compounds"). At the same time John was a member of the Trinity College Mathematical Society and Natural Sciences Society, Cambridge University Old Cliftonian Society, and Cambridge University OTC. As a member of the latter, John received certificates of proficiency in the Signal Corps in March and November 1937. He also attended lectures sponsored by the Cambridge University Socialist Club and by the Communist Party; the latter being open to members of the Section of University Labor Federation.

John was appointed a senior scholar in the Natural Sciences in June 1938 having taken Part I in physics, chemistry, biochemistry, and mathematics, and received a first in Part II chemistry in 1939. He graduated in June 1939 with first class honors and immediately joined the laboratory of E.A. Moelwyn-Hughes (1905–1978), a physical chemist at Cambridge, to initiate research on reaction kinetics. However, shortly thereafter John's plans were altered by the outbreak of WW2.

5

World War 2 Service (1939–1946)

[T]he end of the war in Europe has left a lot of people wondering what they do next, and I am no exception.

John Kendrew

Role of Science

In the fall of 1940 Prime Minister Winston Churchill (1874–1965; Nobel/ Literature, 1953) wrote to his cabinet that "our supreme effort must be to gain overwhelming mastery of the air . . . we must regard the whole sphere of RDF [range and direction finding] . . . as ranking with the Air Force of which it is an essential part. The multiplication of the high class scientific personnel . . . should be the very spear point of our thought and effort."[1] In fact, the scientific personnel to whom Churchill referred played a major role in the Allied victory in WW2. Henry Dale (1975–1968; Nobel/Physiology or Medicine, 1936), an English pharmacologist and physiologist who served on the Scientific Advisory Panel to the Cabinet during WW2, concluded that "science played a much larger part [in WW2] than in any earlier war, and the fact that scientists of the Allied Nations were able to overtake those working for the enemy, and then to keep ahead of them, played a very important part in our ultimate victory."[2]

Chronology of WW2 Service

1939–1940—John was appointed a junior scientific officer at the Air Ministry Research Establishment in December 1939. He was a temporary civil servant with an honorary commissioned rank throughout WW2. He served in Operations Research, RAF Coastal Command, with special reference to anti-submarine warfare, bombing accuracy, and radio aids.

1940–1941—John was promoted to operation research officer and then to scientific officer. In the early months of 1940 he served as an officer of the Telecommunication Research Establishment and was then seconded to Coastal Command to study and advise on the installation and operation of Air to Surface Vessel (ASV) equipment.

1941–1943—John was granted an honorary commission as squadron leader to the Middle East Command in September 1941 and served as deputy officer-in-charge. In 1942 he was promoted to senior scientific officer (acting) and in 1943 to principal scientific officer (acting). In October and December he visited Malta to instruct aircrew and controllers in the use of ASV equipment. He was stationed in Cairo at Headquarters RAF and promoted to senior scientific officer. In the Middle East, he was concerned with sea reconnaissance, use of a radar-based system, and army cooperation matters.

1943–1945—In December 1943, after a period of home leave, John's rank was raised to wing commander (honorary) on a move to South East Asia Command. He served in New Delhi, India, and Kandy, Ceylon/Sri Lanka, as an officer-in-charge of operation research and as scientific advisor to the Allied Air commander-in-chief, South East Asia. He was concerned with signals traffic, supply drops, air transportation, and explosives trials.

1945–1946—John's overseas service ended in February 1945. Beginning in March he returned to England via Australia, New Guinea, the Philippines, and the United States (March–May 1945). In June 1945 he rejoined the Air Ministry in London. On January 10, 1946, he resigned from the Air Ministry and became a research student at the Cavendish Laboratory, Cambridge.

Joining the Air Ministry

In March 1938 German troops occupied Austria; a year later Germany annexed all of Czechoslovakia, Italy took Albania, and German troops invaded Poland on September 1, 1939. In response, on September 3, Britain, France, India, Australia, and New Zealand declared war on Germany and mobilized their forces. Canada and South Africa declared war on Germany only a few days later.

In June 1939 John was a research student working on reaction kinetics with physical chemist E.A. Moelwyn-Hughes in the Department of Physical Chemistry at Cambridge. However, in the months that followed the declaration of war many of John's friends and acquaintances were recruited into the war effort. Conscription in Britain was imposed on all males between 18 and 41 years of age, but John's father, Wilfrid, enlisted in the Royal Navy at the age of 55. John's commitment to his research gradually diminished due to strong feelings that he too should enlist. His initial attempt with the recruiting board to enlist failed, and he was advised to stay in Cambridge until some important war work was found for him.

In December 1939 John received a letter from the Air Ministry offering him a temporary appointment as a junior scientific officer at the Research Department, Dundee, Scotland. However, he would be liable for service at any Air Ministry establishment or station at home or abroad. His salary would be £275 (about $1,300) annually, at a time when the average annual salary in England was £215 (about $1,000). This letter was followed shortly thereafter by another telling him that he must resign from the Officer Cadet Reserve and report to the superintendent, Research Department, Training College, Park Place, Dundee, on February 1, 1940. John was to be a temporary civil servant holding an honorary commissioned rank throughout WW2.

In February 1940 John was recruited to work on development of airborne radar as a junior scientific officer in the Air Ministry Research Establishment, that was soon renamed the Telecommunication Research Establishment (TRE). His recruitment may have been the consequence of a letter John wrote to Wilfrid B. Lewis (1908–1987), a physicist at the Cavendish who had worked with Ernest Rutherford. In the same year John's first research paper, with Moelwyn-Hughes, "The Kinetics of Mutarotation in Solution," was published in the *Proceedings of The Royal Society*. John referred to the paper later on as being rather boring. In December, Moelwyn-Hughes was recruited to the Ministry of Supply.

During his initial period of wartime service, John was assigned to work on oscillators, aerials, and valves for early airborne radar applications. Development of radar in Britain had been initiated in late 1934 by the Committee for the Scientific Survey of Air Defense. Asked to develop a microwave oscillator with enough power for airborne radar John made little progress with the materials available to him. Many years later John recalled this early period of his involvement in the Air Ministry:

Being a contrary character, at my public school [Clifton] where almost everyone joined the OTC [Officers Training Corps] I refused to do so; at Cambridge, where almost no one joined, I proceeded to do it [in November and March 1937 he obtained Certificates of Proficiency at OTC examinations]. I joined the Signals Unit, of which W.B. Lewis was—I think—Commanding Officer. Having joined, a friend (Jim Wilkins) and I discovered that Lewis had a contract with the War Office to develop ultra-shortwave two-way radio, and that he needed spare-time help with this. The work was non-secret and had no connection with radar; the wavelengths involved were 3 m and 66 cm [100 and 450 MHz]—very short by the standards of those days; but it meant that we could evade some uniformed parades and various tedious chores by enrolling ourselves with him in what I think was known as the Research Section—and as far as I remember we were the only members of it. As a consequence I had my picture in the newspaper relaying results from the track to the scoreboard at an athletic meeting at the White City [the headline read "Meet the Walking Wireless Man"]. When the War came I was interviewed by the Cambridge Recruiting Board and told to continue my academic research, which I did not find very satisfying with all my friends going off to the armed services. I had no idea about radar but I heard a rumor that Lewis was involved in something interesting and I wrote to him with the result that I was recruited into the TRE.[3]

The TRE to which John was assigned originated in 1934 when the Air Ministry set out to evaluate the application of science to problems of air defenses. The appointment of Adolf Hitler as Chancellor of Germany and the rise of the Nazi Party was recognized as a direct threat to Britain. The Committee for the Scientific Survey of Air Defense was established to consider new technologies that the RAF might use to defend against attack by bombers. John's experience with two-way radio communication while a member of the OTC at Cambridge was clearly a major factor in his recruitment to the service.

The newspaper picture that John referred to was taken in July 1938 when he served as a radio-man at an athletic meet in White City, a district of London (see fig. 5.1). *Wireless World* reported that "Results of recent AAA Championship races at the White City were for the first time relayed to the scoreboard by means of this portable 3-meter transmitter."[4] Four days later the *Daily Sketch* followed up with a more detailed description of John's role:

Meet the walking wireless-man. Thousands saw him for the first time at the international athletic meeting at the White City—and wondered what

Fig. 5.1 Photo of John Kendrew (ca. 21 yoa) as the "Wireless Man" relaying results to the scoreboard using a portable transmitter at an Amateur Athletic Association of England track and field meet between Oxford and Cambridge at White City, London, 1938. The photo on the left appeared in the *Daily Sketch*, August 8, 1938 and the photo on the right appeared in *Wireless World*, July 28, 1938. Reproduced from items in the Kendrew Archives, Weston Library, Oxford

was it all about. They marked his aerial (from the back of his neck); his earphones, mouth-piece, transmitting gear, receiving apparatus and all the gadgets that go to make a human radio. His hot job yesterday was to wireless the results, times, etcetera, to the scoreboard people, and very efficient he proved, too. The Americans were particularly interested.[5]

Joining Operational Research

In May 1940 John moved from the RAF St. Athan, South Wales, to Worth Matravers, Dorset, where a new center for radar development had been established. He was with a team working on Airborne Interception and ASV radar. John spent most of his time at RAF Christchurch carrying out flight trials on TRE equipment, but in September was reassigned to Operational Research

(OR) duties concerned with the study of weapons, tactics, and strategy for air and naval warfare based on statistical analysis of intelligence data. As many as 1,000 men and women were engaged in OR duties during WW2.

Radio devices for detecting and locating aircraft were developed by the mid-1930s and the technology soon became known as radar (radio detection and ranging). Before WW2 electronic systems were used to find the position of targets by transmitting radio pulses and estimating the time delay in receiving an echo back from the target. During WW2 radar was developed rapidly by the British and Germans, and it became a vital tool in combat. Radar stations were established along the east and south coasts of England in time for the outbreak of WW2. This represented the first radar to be organized into a complete air defense system for use in wartime, and in 1940 it provided advanced information to the RAF Fighter Command during the Battle of Britain.

Attached to the staff of Robert Watson Watt (1892–1973) as a member of the RAF Coastal Command, John participated in some of the initial flight trials over Northern Ireland testing air interception radar equipment. He was advising on the installation and use of ASV radar equipment that permitted assistance in directing ships, convoys, and submarines at long distances in the dark or poor visibility. John recalled that during one test flight instead of following an RAF target indicated by the radar they mistakenly followed a German plane across the coast of France.

From November 1940 to March 1941, John dealt with a wide range of topics on radar and coastal air defenses, prepared diagrams of circuits and installations, and assessed air stations and their equipment. He also visited coastal squadrons, primarily in Scotland and Northern Ireland, to assess personnel and equipment. In particular, John provided advice on the installation and operation of ASV equipment.

Assignment to the Middle East

In September 1941 John was reassigned from the RAF Coastal Command to build up an OR Section at the RAF Middle East Command Headquarters in Cairo, Egypt. In connection with John's new assignment, a letter was sent by the vice chair of Air Staff to the air officer commander-in-chief, Middle East Headquarters, emphasizing John's great expertise in OR matters and the vital importance of his new assignment to the war effort. The letter concluded that the success of ASV Wellington bombers in the Middle East would depend to a large extent on John's efforts and expert knowledge. Consequently, he was to be looked after and given more latitude than the usual RAF officer of his rank. It was an unqualified endorsement of 24-year-old John by a senior officer.

John paid brief visits to Malta in October and December where he instructed aircrews in the use of radar aboard ASV Wellingtons fitted out with anti-submarine weapons. John arrived in Cairo in December, with the rank of squadron leader, cadet class, the deputy officer-in-charge stationed at the RAF Headquarters, Middle East. He had orders to assess and improve the success rate of air attacks on shipping in the Mediterranean and to deal with army cooperation matters. John's rank entitled him to exercise executive powers of command under the Air Force Act and soon he was authorized to take, develop, and have in his possession films and photographs of RAF aircraft, equipment, and installations throughout the Middle East. However, the appointment carried with it no entitlement to RAF pay or allowance. John's conditions of employment and benefits remained those of a civilian, but he was allotted £45 for the purchase of a uniform.

In a letter to a friend in the UK John described his living conditions in Cairo:

> I suppose . . . you find the austerity of the U.K. rather a trial. I think if I ever return I should feel much the same; for certainly in Cairo one lives in fair comfort. [We] inhabit a flat presided over by two Arab servants who for once are remarkably efficient; food and drink are plentiful and good even by peacetime standards. It is true that almost everything costs between two and three times its English price; but one has to put up with little things like that in wartime.[6]

While John was stationed in Cairo he produced the *Handbook for Aircrews and Controllers* that described the use of ASV to locate objects at sea—the manual that became known as "the bible." In this connection, he wrote to OR Headquarters, Coastal Command, in May 1942 that

> I am just about to produce a sort of "Bible" or data-book for pilots, operators, and controllers, dealing with most aspects of A.S.V. It is going to be a rather lengthy document, and as you will imagine I am seriously hampered by lack

of good experimental observations. I will arrange for copies to go to you as soon as it is available. I hope to issue it in such a form that it can be brought up to date by amendments from time to time. I must confess that some of the data it contains comes from your own excellent Reports; I hope you will forgive a certain amount of plagiarism necessitated by my present isolated situation . . . .[7]

John was particularly concerned with the use of long-range Wellington bombers for dropping torpedoes directed against enemy ships, including submarines. He introduced the use of a portable radio-beacon system developed by TRE, called REBECCA, an airborne receiver that served as a navigational guide for ground-based transponders, called EUREKA.

As officer-in-charge of an OR section John decided that he had three main tasks to carry out: to maintain good relations with Air Marshals and other senior officers; to organize and administer the scientific direction of his staff; and to maintain the quality of their work. He felt that he was pretty good at most of his duties, but in some areas his own short research experience was somewhat of a handicap. On the other hand John's detailed reports from Cairo were greatly appreciated by OR higher command for their comprehensiveness and clarity and were thought to be of material value for future operations.

After months of overwork with little assistance in the extreme heat of Cairo, John was in ill health and complained to the Air Ministry. He was briefly reassigned to the Far East, but the order was countermanded by the air chief marshal who considered John his key man on ASV and REBECCA and felt strongly that he must not be reassigned at this time.

Romance in Cairo

In Cairo John met and fell desperately in love with a very attractive 20-year-old university student, Shoshana Ambache (1921–2011), known as Suzy, who was born into the European community of Ismalia, a small Egyptian town on the west bank of the Suez Canal. Suzy was Jewish and a member of a staunchly Zionist family. Her father's family had emigrated to Palestine from Russia and during WW1 were among the thousands of Ashkenazim who left Palestine. Suzy's father was born in Palestine, trained as an engineer, and became a consultant to the Suez Canal Company in Ismalia. When Suzy was 15 the Ambaches moved to Cairo and eventually she began studies at the Fuad-el-Awal University of Cairo and American University.

John had known Suzy's older brother Nachman Ambache in England at both Clifton and Trinity. Nachman had taken a double first in natural sciences at Cambridge and was doing research in physiology and a period of clinical work at Kasr-el-Aini Hospital in Cairo. He brought friends from Cambridge to Suzy's house, among them John and Abba Eban. In her biography, *A Sense of Purpose: Recollections*, written when she was 87, Suzy recalled that "He [John] joined our family circle with ease, and we were pleased to have the opportunity to reciprocate for the summer long ago when his father had invited Nachman to stay with him and John for a holiday on the Devonshire coast. For his part, coming from a broken home, John had never been so close to family life before so this was a brand new experience for him."[8]

John and Suzy would go on sightseeing expeditions through Cairo and end up at Suzy's home

for drinks and dinner, listening to the BBC or playing classical records. Intellectual and emotional attachments were forming between us, and after the family dinner we would sometimes walk along the nearby western bank of the Nile, a walk we specially liked on luminous summer moonlight nights . . . I was attracted to the "otherness" in him—his scientific outlook, and his English characteristic of being calm and tactful, disciplined and proper.[9]

John wrote to his mother in Florence alerting her to his romance with Suzy and, unexpectedly, Evelyn was very supportive of the relationship since "she felt that neither she nor her son would ever find completion in England" and that with Suzy he would "dip into another layer of emotion." John's mother saw "the hesitation in giving one's children Jewish blood and Jewish disabilities, yet what a fine heritage they have too."[10] Regarding Suzy's Jewish background, Evelyn felt that "to those who have none and want one—an adopted background can also be important." She saw that "a mixed marriage was one of the most enlightening, widening processes that existed, and that once a person becomes a part of two worlds, he can never again be the denizen on only one. For her part, she would be glad if he was to definitely cut loose from English middle-class street-ism and from a donnish exclusiveness."[11]

However, a letter from John to Suzy strongly suggests that the two of them viewed the Middle Eastern world quite differently:

> I walked through town and all I could see was the filth and disease. Every face almost a revolting travesty of everything that is human: avaricious, cunning, greedy. Even the Europeans are corrupted by it too. I saw it all as a sort of foetid manure-heap, crawling with vermin: the sun pouring down and making that vermin fester and multiply—and just as you can only grow marrow on a dung heap—marrow with its glaring yellow flowers and its bloated over-developed fruits—so the only culture that can come out of this town is represented by the corruption and the wealth which flashes past in its large motor-cars—all the materialism which you know so well and which even I have caught glimpses of.
>
> This life did seem a horror. I wanted you out of it, clearly away. I think more strongly than ever before. When I compare it with a civilized country like England, the contrast was amazing. Just look at the book I gave you—On Foot—and you will see hundreds of examples of what I mean. The miracle of it is that you—you have kept your purity, your virgin freshness, your quality of spring—in spite of it all. It is a miracle. It is.
>
> I would give anything to show you now, just one glimpse of a Scottish moor under the summer sun, springtime by the Thames with the long lush grass and the flowers—sunset from the cliffs of Devon—a dripping, foggy, November day when every object more that 10 meters away is a veiled mystery![12]

In summer 1942 Rommel's forces took Tobruk in Libya near the border with Egypt and the Germans approached Alexandria. Under the circumstances, Suzy, her mother, and sisters were evacuated to Johannesburg, South Africa, where they remained for six months and only returned to Egypt after the Allies were victorious at the Battle of El Alamein in November. With the Germans pushed back to Tunisia, Suzy's family was reunited in Cairo and John and Suzy could resume their relationship. However, Suzy remembered that "On my return, our amorous friendship continued, but we gradually slid into conflict, first over each other's autonomy and eventually over the question of our respective religion."[13]

To allay any reservations that Suzy and her family had about a mixed marriage, John decided to convert to Judaism, started to learn Hebrew, and wrote to his mother about his decision. John was eager to have a wife and children. In response to John's letter, his mother "decided that John had to be rescued from losing his soul and should not be sidetracked from his future career by rushing to marry and taking on the problem of a new religion."[14] Suzy found the letters on the subject of religion and career quite offensive and the divide between her and John slowly widened. She found that "Months passed and we lost the happy spontaneity of our relationship."[15]

During this final phase of their relationship, John wrote to Suzy about the view of most scientists that "the place of the scientist, like that of the monk, is in exclusion from the outer world: the watertight compartment of the laboratory . . . The pure scientist, narrow in his outlook, is not good . . . ." John felt strongly that it was time for scientists to change their view, to increase their breadth, and he "must work for change." John related this objective to his feelings for Suzy by saying: " . . . For you have things I have not got: you are a complement to me. Together we can do much. That is why we have so much work ahead for us to do together. I must teach you about scientific thought. You must give me insight into your spiritual world."[16]

With their romantic relationship effectively ended, in 1943 Suzy was to fall in love with Abba Eban (1915–2002), a captain in the British Army, whom she married two years later. Eban would serve Israel as its first representative to the United Nations, in several ministerial posts, including minister of education and culture and deputy prime minister, and as president of the Weizmann Institute of Science. In the years that followed the end of their relationship John would occasionally refer to Suzy as "the love of his life."

John was to maintain very close ties with Israel for the remainder of his life. Among his many Israeli friends were Michael Weizmann (1916–1942), son of Chaim Weizmann (1874–1952) first president of Israel (1949–1952); Ephraim Katzir-Katchalsky (1916–2009), fourth president of Israel (1973–1978); and Aharon Katzir-Katchalsky (1914–1972), a pioneer in the study of the electrochemistry of biopolymers. At the invitation of Lord Rothschild, John became a governor of the Weizmann Institute in 1964 and remained so for more than 30 years. In 1966 he joined the executive of the Weizmann Institute Foundation, served as chairman of its Scientific and Academic Advisory Committee for eight years, and, in 1969, became an honorary fellow of the institute. John served as a member of the Executive Council, as vice chair of the Appointments and Promotions Committee, and on the Nominating Committee for Members of the Board of the Institute. In

May 1986 John chaired a symposium at the Weizmann in honor of Katzir-Katchalsky's 70th birthday. It is very likely that these long, very close ties with the Weizmann Institute and with Israel can be attributed, at least in part, to his close relationship with Suzy Ambache during WW2. Suzy wrote her memoirs, *A Sense of Purpose: Recollections*, in 2008 and died in September 2011 at the age of 90.

In 1985 John gave a speech at the 50th anniversary celebration of the Weizmann Institute:

> Congratulations to the Weizmann Institute on its fiftieth anniversary. I personally first made the acquaintance of the Institute in 1963 when I attended a scientific meeting there; it was already then a flourishing institute of world renown, so I was delighted when shortly thereafter Suzy Eban and Victor Rothschild suggested that I should become a member of its Board of Governors, because this gave me the opportunity to visit it frequently for scientific discussions and to see many old friends, some dating from 1963 and some from long before. At the beginning I was a member of an exclusive club—of those Governors who were scientists, who were not Jewish, and who were not rich—there were only three of us, the others being Robert Robinson and Chris Anfinsen.
>
> Since that time the Institute has gone from strength to strength, with scientific enterprises in the world class in many fields, with many new buildings and excellent equipment, and with a growing responsibility for training students from Israel and abroad. In the last years financial stringencies have made enormous difficulties for the Weizmann as for the other institutions of higher learning in Israel. But the effects of these stringencies have been minimized in a remarkable way, and the scientific output is still of the topmost quality and the morale of its members as high as ever.
>
> It is a pleasure and an honor for all of us to be associated with so distinguished and so friendly an institute. So we all wish it well for the next fifty years.[17]

Transfer to South East Asia

In late 1942 John was informed that he might be transferred to India to set up an OR station (ORS) in response to Japan having entered the war in

December 1941. He considered the transfer premature and ill advised, and wrote that

> We have recently had a signal requesting my release to India to start Operations Research there. The situation is still somewhat obscure because the Command do not wish to release me until a replacement is forthcoming. Further, I am reluctant to go to India unless I have positive information of staff arriving, because of some personal experience I am pessimistic about the value of "one-man" ORS's. I also feel I should like the opportunity of a quick trip back to England to bring myself up to date, for I already feel sadly behind the times, and I think that particularly for a new ORS recent knowledge of new developments is essential.

He felt that "The best work of a scientific nature cannot be expected of scientific personnel if their movements are to be regulated without regard to their health or personal inclinations."[18] A response from his command agreed that if John moved to India all work on anti-shipping operations, anti-submarine operations, smoke tactics, and so on, be dropped. They considered other available personnel fully occupied and unqualified to take on John's duties.

Sometime later, in a moment of great self-confidence, John changed his mind and decided that he was prepared to go to South East Asia if he was made a wing commander, officer-in-charge, and scientific advisor to the Allied air commander-in-chief. As a result of his bravado, in December 1943, after a short period of leave in England, John moved to the South East Asia Command in New Delhi as a wing commander in charge of OR and scientific advisor to Louis Mountbatten (1900–1979), Allied air commander-in-chief. Although he had confidently requested the promotion while stationed in Cairo, John was somewhat embarrassed by his senior rank and felt slightly ridiculous every time he used the title, since he confessed that his experience and age were not commensurate with it.

In Delhi, John was to be concerned for the most part with bombing accuracy and effectiveness as well as radio aids. Before leaving London for India he summarized his time in the ORS of the Middle East Command in a report, *The Air Sea War in the Mediterranean Jan 42–May 43*. The report emphasized that sinking of the enemy's fuel supplies had the desired effect on the enemy's actions in battle because it left the enemy with the choice of either battling on or retreating, not both.

Initially the headquarters for the South East Asia Command was in New Delhi, but in April 1944 the command moved to Kandy, Ceylon/Sri Lanka. In September John moved to Kandy where trials were being carried out to evaluate the effectiveness of high explosives in clearing the dense vegetation of jungle terrain. John appreciated the move to Kandy and wrote to a friend that "I have never worked in Headquarters more pleasantly situated than our new site in Kandy, and the change from the Delhi climate has been a great relief. Some day you should come out to visit us and see how we make the best of living in the middle of jungle."[19] In Kandy John would participate in the jungle trials together with J.D. Bernal whom he had met on several occasions in Egypt and India. After their first meeting in Cairo in 1943, Bernal became more and more impressed with John's intellect and ability.

John Desmond Bernal

Scientists contributed enormously to the Allied victory in WW2, including many outstanding British scientists who were recruited from both academia and industry, among them John Desmond Bernal who John greatly admired. Bernal more than anyone else would influence John to pursue protein crystallography as a research student in Cambridge at the end of WW2. However, Bernal's influence certainly was not limited to John. Maurice Goldsmith concluded that Bernal's

> strength lay in causing other minds to light up. Most of the leading names in Britain working on molecular biology and the analysis of protein crystals were either Bernal's associates or students. Wherever he went he left behind intellectual "fall out," providing more than sufficient for a lifetime of scientific work. If one traced back almost any fruitful line of crystallographic work it would be found that Bernal assisted at its conception, but, significantly, left the child to be brought up by foster-parents.[20]

In civilian life Bernal was a highly successful X-ray crystallographer who had a very wide range of research interests. Bernal was one of the founders of OR during WW2 and after joining the Ministry of Home Security in 1939 served as scientific advisor to Mountbatten, chief of Combined Operations from 1942 onward. Bernal also was a political activist who joined the Communist Party in 1923 and who felt that the Soviet Union would be a catalyst for creation of a socialist scientific utopia in the 1950s. Home Secretary Herbert Morrison (1888–1965), was severely criticized for recruiting a Communist like Bernal to work as an explosive expert during WW2, but Morrison claimed that he would employ Bernal even if he was "as red as the flames of hell."

Bombing Trials in Kandy

Bernal travelled together with Wing Commander Solly Zuckerman (1904–1993), of the Administrative and Special Duties Branch of the RAF, to evaluate the effects of bombs on buildings and their occupants and the relationship between the number of bombers used and the population of target areas. Such analyses were carried out in order to determine the number of casualties and level of destruction achieved under a particular set of conditions. This enabled the British to evaluate how much bombing and what type of bombing was required to achieve a specific objective. Following WW2, Zuckerman would become chief scientific advisor to the MoD and as

Fig. 5.2 Photo of John Kendrew (ca. 27 yoa) in 1944 standing in a crater produced by a 500 pound medium capacity bomb in the jungle in Kandy, Ceylon/Sri Lanka. Reproduced from an item in the Kendrew Archives, Weston Library, Oxford

a result of their interactions during WW2 he would recruit John to the MoD on a part-time basis in the early 1960s.

Stationed in Kandy, John was searching for the best type of bomb to use to clear the jungle (fig. 5.2). The RAF had run out of fragmentation bombs and was using depth charges instead. Since the latter made a much louder noise on explosion than fragmentation bombs the army thought them much more effective at clearing the jungle. Mountbatten was being pressured by the army to use depth charges even when fragmentation bombs were made available. Consequently, he asked John and Bernal to arrange bombing trials in the jungle.

Bernal arrived back in Kandy in November, and he and John carried out test trials of bombs and depth charges to obtain evidence of the suitability of these weapons to increase visibility in the jungle, enable personnel to penetrate through the jungle, and assess their lethal effect on personnel. Elephants were used to transport the heavy bombs to the jungle test site and rats gathered by the official rat catcher of Kandy and housed in wire cages were used to test the lethality of the explosives. In the neighborhood of Naula, about 12 miles south of Dambulla, they tested 500-pound, medium-capacity bombs filled with Amatol, a mixture of TNT and ammonium nitrate, as well as depth charges filled with either Amatol or Torpex. Explosion of the bombs or depth charges in the jungle produced a crater about 6 feet in radius and 3 feet deep. Rats 15 to 30 feet away from the blast site were killed, or escaped, and those 45 feet away or more were unhurt. John and Bernal concluded that 500-pound bombs or depth charges would clear jungle up to a radius of about 60 feet, improving visibility within this area, and that human beings within 20 feet of the blast would be killed. The bombs, owing to fragmentation, brought down trees and branches making movement of personnel through the area more difficult. They were convinced that their results should do away with exaggerated and baseless tales that credited enormous lethal distances for such weapons.

Years later John recalled an incident with rats and depth charges in which Bernal pulled out his slide rule and calculated that he and John were at a safe distance from the site of the explosion. "There followed the most enormous explosion and we rapidly dived into the nearest ditch with rocks and earth flying overhead. Bernal emerged brushing earth out of his hair and looking very puzzled, saying that he must have got the decimal point in the wrong place."[21]

In the late 1960s John reminisced about some of the complications of jungle warfare to the *Listener*, saying:

I think what one's up against in the jungle all the time is nature getting in the way. I mean, it's rather like the difference between playing tennis and playing golf. Tennis is a kind of geometry—it's flat and it's well marked out. Whereas in golf you've got a lot of natural obstacles, so it becomes a different kind of operation. Perhaps the naval war and conventional bombing are a bit more like tennis and the jungle war is a bit more like golf. And so what you're up against is these awkward bits of nature, like tropical vegetation, and the problem is how do you see through them, and how do you get round them, and how do you see what the other man is doing?[22]

Post-War London

John's overseas service during WW2 ended in February 1945, and in March he began a three-month return trip to England via Australia, New Guinea, the Philippines, and the United States. In California John visited relatives and met with Linus Pauling (1901–1994; Nobel/Chemistry, 1954; Nobel/Peace, 1963) at the California Institute of Technology to discuss protein structure and possible directions for John's postwar career. John found Pauling very enthusiastic and encouraging about structural biology in general. However, when John told Pauling that he might take up protein crystallography as a research student after the war, Pauling discouraged him by saying that it was nonsense since John had never solved the structure of anything. Recalling this reprimand at a symposium more than 40 years later, Pauling conceded that John had appropriately ignored his advice and indeed had succeeded in solving the first protein structure.

While in New York City John attended *The Hasty Heart*, a new play at the Hudson Theater, after dinner with the play's author John Patrick (1905–1995), who would win a Pulitzer Prize in 1954. The following evening John attended a New York Philharmonic concert at Carnegie Hall accompanied by Laura Boulton (1899–1980), at the time a well-known ornithologist, ethnomusicologist, and film maker. Throughout his life John often befriended performers in the arts, such as writers, painters, and musicians, who had achieved some celebrity status.

By June 1945, shortly after VE day had been declared, John had returned to the Air Ministry in London and come down with a recurring bout of malaria that lasted a couple of months and kept him from work. In a letter to a friend he commented that malaria is "Not a very serious disease, but an annoying one; the methods of treatment have advanced not one whit in the last 3,000 years or so, and there is no guarantee I shall not go down with it periodically until the end of my life. Evidently in my case Atebrine [quinacrine] merely suppressed the disease until I stopped taking it."[23]

In London John made several proposals that he felt would help to organize the scientific manpower resources of the country and provide a central organization of science. However, he lamented that science was very badly organized in the UK before the war and nothing significant had been done by those in authority to improve the situation. John insisted that there was a need to increase the number of UK scientific personnel to a proper level and to identify areas of science that should be strengthened.

Near the end of John's service in South East Asia he had been offered the opportunity to return to the UK as officer-in-charge of the ORS Fighter Command. However, he felt that the commitments of the command had been so reduced that the posting was not particularly attractive to him, and he turned it down. So with WW2 at an end, John remained very ambivalent about his future, but seriously considered remaining in government as part of the scientific civil service. His war responsibilities and experiences left him with a serious sense of purpose; he felt that science would play an important role in post-war Britain and, in view of the experience he had gained during the war, he had a moral obligation to continue in government service.

However, career choices in the civil service remained fairly ambiguous and John confessed that

the end of the war in Europe has left a lot of people wondering what they do next, and I am no exception . . . My own future is still completely vague . . . On the other hand Bernal has mentioned me to Bragg, who seems disposed to take me on as a senior DSIR student [Department of Scientific and Industrial Research] (though I have not yet been to Cambridge to see him); so the question is, whether the gamble of two more academic years, at the end of which I might or might not achieve a satisfactory academic career, is worthwhile. Alternatively, there is the Government; for since the election result (to celebrate which I became mildly drunk with one or two kindred spirits) things are definitely looking up in that line . . . So at the moment I am pursuing a policy of masterly inactivity, stalling in all directions hard . . . I have no idea at all—nor has anyone else.[24]

John's frustration that summer about which way to turn in his career is made painfully clear in a letter he wrote to a friend at the Air Ministry:

You said it might be helpful if I told you some of the problems facing me at this time when . . . I have to reach some decision about my future career. I am writing you on account of these problems in the belief that my own situation may be typical of that of many others, and that my remarks may for this reason be of value to those whose duty it is to organize the future of Science in Government Service . . . you will see that I, and others like me, are in a considerable dilemma. We can see something of the great need for scientific work and scientific thought in the Government of the country, and the possibilities for the future are so vast as to form a source of considerable inspiration; moreover we feel it is up to us who have gained experience of Government work during the War, to play our part. But all this is frustrated by the complete lack of central organization in the Government at large, and even of clear prospects in our own Ministry which has probably made more use of science than any other during the war years. Thus while the needs are clear, lack of mechanism makes the future so obscure, not to say gloomy, that we are falling back on an alternative—namely the return to academic research—where the prospects are for many of us very doubtful. Academic work and teaching are of course of fundamental importance too, and a fair balance must be struck in allocations of manpower between them and Government service. The trouble is that there is no balance at all; both ends of the see-saw carry large queries and the fulcrum (a national scientific organization with responsibilities outlined above) is missing. I can only ask, what should I, and others like me, do? And even to ask is a waste of time, for there is no one to hear our question . . . Let us hope that the Government and our senior scientists will get together and provide the necessary organization. And it must be soon, for time is running short and before many weeks are out many of us will have been forced to make decisions which are irrevocable.[25]

Bernal had informed John that Lawrence Bragg (1890–1971), Cavendish Professor in Cambridge, was inclined to take him on as a research student and would provide financial support for him with a grant from the Department of Scientific and Industrial Research (DSIR). So in September 1945 John sent a letter to Bragg to inquire further about the possibility of his return to academic life:

> Professor Bernal suggested I should write to you: and I hope you will forgive me for taking up your time. The subject on which I wished to ask your advice, is my possible return to academic life: for after graduating in June 1939 and doing six months research in the Department of Physical Chemistry at Cambridge, I joined the Ministry of Aircraft Production and have been engaged ever since on war work—first of all at TRE and later in Operational Research.
>
> Although I was trained in the main as a pure chemist, with some emphasis on the organic side, I had always wished, and wish still, not to remain such: my object has been to use my chemical training in research on borderlines either of physics or of biology, and I should be best of all pleased if I could find work at the "triple point" between all these sciences. With this aim in view I spent as much time as possible, while at Cambridge, in studying Physics, Mathematics and Biochemistry. As you are aware the standard demanded in the Part I of the Tripos is little higher than in Entrance Scholarships: during the two years of Part I work I was therefore able to spend a certain amount of time in reading subjects outside my immediate special subject of chemistry.
>
> During that short period I was able to do research before being called to war work, I was working on reaction kinetics with Dr. Moelwyn-Hughes: but even at that time it was not my intention to remain on purely physico-chemical problems: I had not, however, decided what the next step should be. My width of interest, although in many ways a great advantage, did also have the disadvantage of making it difficult to see where one could most usefully work, and on which particular problem.
>
> After 3.5 years overseas, doing work only remotely related to academic research, I find myself very much out of touch with recent developments and very ill-qualified to make any decision about the most useful sphere of activity for a person of my particular training and interests. Professor Bernal suggested that I should write to you because he knew you were particularly interested in borderline problems of the kind I mentioned above: and because he felt you would be in touch with work of this kind either in progress or in prospect in Cambridge.

I now feel anxious to return to Cambridge if possible, but I have very little knowledge of what opening there may be. My only present status at Cambridge is that of a fraction of my Senior Scholarship at Trinity remains unused. I should perhaps mention that I do not anticipate enormous difficulties in obtaining release from the MAP [Ministry of Aircraft Production]: operational research is so intimately dependent upon active military operations that a considerable reduction in staff has begun since the end of the European War. I should perhaps mention that the MAP is anxious to retain my services permanently in its . . . establishment: but possibly this prospect does not greatly appeal to me from any but the financial point of view, for I feel the government work in the immediate postwar years in a Ministry whose activities were directed toward preparing for another war, would be particularly unsatisfying . . .

I should be extremely grateful for your advice on these matters: should you wish to see me I could come up to Cambridge at any time after about the end of the month (until which time I am on side leave, recuperating from malaria contracted during a recent trip through the Pacific). With apologies for taking up your time.[26]

In October John was informed by M.E. Bonewell at the Ministry of Labor and National Service that Bragg's application for a Senior DSIR grant for him had been approved. John responded that he had decided to accept the DSIR grant and give up his post with the Air Ministry. The grant would permit John to return to Cambridge, provided that he was released from his present employment with the Air Ministry. But John continued to remain undecided about whether or not to accept a government job since he had been advised by some that he should remain in government service rather than return to Cambridge.

At about the same time John also wrote to Civil Service Commissioner C.P Snow (1905–1980), also a noted physicist and novelist, about his future plans and to ask for Snow's advice:

In general I feel enthusiastic about the prospects in the Government Scientific Service, particularly for operational research in its civil applications . . . Under existing circumstances I feel that I should accept this grant [DSIR] as soon as my release comes through, and return to Cambridge . . . When government scientific plans have become more crystallized, and there are openings for which you feel I might be considered,

I should be most grateful if my name could be put forward: and in that case, and providing I was found to be suitable for the appointment, I suppose I might reckon on returning to Government service in perhaps 6 months' time . . . These decisions have not been arrived at without many qualms of conscience towards DSIR and particularly toward Professor Bragg. Should the plans you have in mind mature within, say, 6 months it is clear that, while I should have received the mental stimulus of working once more in an academic institution, my research would not have made much progress. While this is very unfair on the laboratory in which I work, I see no real alternative, just so long as there are no definite openings in Government service of the type I should like and under conditions under which I should be ready to work. Failing these I must take steps to effect a new entry into what would have been my normal field had I been fortunate enough to be born 10 years earlier . . . I should be most grateful for any comments you may care to make on this plan, for if you can suggest any reasonable alternative I am only too happy to consider it.[27]

Snow sent a very short and rather unhelpful response to John saying only that he found his letter most interesting and that he would remember him.

Final Decision

Although offered a place at Birkbeck College with Bernal, John was strongly advised by Bernal to go to Cambridge because he was a communist and, consequently, found it very difficult to raise money. So in late autumn, wearing his wing commander's uniform, John paid a visit to Bragg and was introduced to Max Perutz at the Cavendish Laboratory. As a result of the visit he made up his mind to return to Cambridge as a research student, initially with two years of support from the remaining portion of a scholarship he held at Trinity College prior to WW2 and a DSIR grant obtained by Bragg.

However, John remained reluctant to resign his appointment in the Air Ministry and in January 1946, after moving to Cambridge, was offered a government job at the Board of Trade. The board intended to set up a small organization on an experimental basis for the application of the principles of OR to its functions with regard to industry and commerce. John was invited to join the staff of the organization as a senior assistant with the title of statistician at a salary of about £900 (about $3,690); the average annual salary in Britain at the time was £255 (about $1,025). This represented somewhat of an advance over what he earned from the Air Ministry and significantly more than he would earn as a research student. But John declined the board's offer and finally resigned from the Air Ministry effective January 10, 1946.

Postscript

John was a significant figure among those individuals who initiated OR in various commands at home and abroad. Among these individuals were P.M.S. Blackett, B.G. Dickens, H. Lardner, G.A. Roberts, A.F. Wilkins, E.C. Williams, and E.J. Williams. By March 1945 there were about 200 officers engaged in OR in the home and overseas commands.

In September 1961 John was invited to speak on "The Impact of Science on Defense Administration in the Next Ten Years" at an international conference on defense administration at Queen's College, Oxford. Similarly, in March 1962 he was invited to speak at the Royal United Services Institution (RUSI), London, on "Science and the Services in the Next Ten Years."

John's lecture in Oxford included sections on "the deterrent," "nuclear weapons in field warfare," "complexity of weapons systems," "the increasing power of the offensive relative to the defensive," "the role of the scientist," "scientists in their technical capacity in the laboratory," and "scientists outside the laboratory." He ended the lecture by saying that "Scientists have no monopoly of rational thought, but by virtue of their special knowledge and of their professional training they have a unique and essential role to play in finding the solutions to the grave problems with which we are all concerned, and which must be solved if our society is to survive."[28]

John's lecture in London was quite similar to the earlier one in Oxford, but spawned some press coverage in the *London Times* about his comments on the possibility of accidents with nuclear weapons. John responded to the coverage by saying "I never said that aircraft carrying H-bombs were very liable to blow up; on the contrary, I said that the probability was very small—but that if this sort of activity goes on long enough, eventually an accident is bound to happen, possibly with serious consequences."[29] However, the London lecture did not escape the notice of the MoD and less than one week later John received a very strong reprimand from Chief Scientific Advisor Solly Zuckerman.

6

Cavendish, Peterhouse, and Myoglobin
(1946–1962)

I believe him [John Kendrew] to be of the type that will make a real
leader in the future, and I think he is worth watching.

Lawrence Bragg

Cavendish Laboratory

The Cavendish Laboratory is the oldest and most prestigious laboratory for
physics research in Britain, perhaps in the world. Inscribed over its doors
is the biblical quote, *Magna opera Domini exquisita in omnes voluntates
ejus*, translated as "Great are the works of the Lord, sought out by all who
delight in them" (Psalm 111:2). Since the founding of the Nobel Prize in
1895, 30 members of the Cavendish have been awarded the illustrious prize
in Physics, Chemistry, and Physiology or Medicine. Today the Cavendish is
the Department of Physics at Cambridge University and has been located in
West Cambridge since 1974.

The Cavendish was located originally at the New Museums Site on Free
School Lane in Cambridge, on land used for a Botanic Garden. It was founded
in 1871 with approximately £6,900 (about $30,000); £5,000 for the building,
£1,300 for instruments, and £600 for personnel. The funds were donated by
William Cavendish (1808–1891), 7th Duke of Devonshire and Chancellor of
the University. William Cavendish was distantly related to the 17th-century
chemist Robert Boyle (1627–1691) and to the 18th-century chemist Henry
Cavendish (1731–1810) for whom the laboratory was named.

At the same time a chair in experimental physics was established and
the electors first considered William Thomson, Lord Kelvin (1824–1907),
in Glasgow, Scotland, and then Hermann von Helmholtz (1821–1894) in
Berlin, Germany, for the position. However, since both Thomson and von

Helmholtz refused a move to Cambridge the position was finally filled by James Clerk Maxwell (1831–1879) the great Scottish physicist and mathematician who is considered the father of modern physics. One of Maxwell's biographers concluded that Maxwell's importance in the history of scientific thought was comparable to Einstein's. Einstein himself felt that Maxwell's contributions to physics were the most profound and fruitful since the days of Isaac Newton.

Over the next 60 years Maxwell was followed as Cavendish Professor by John Strutt in 1879 (1842–1919; Nobel/Physics, 1904), John Joseph (J.J.) Thomson in 1884 (1856–1940; Nobel/ Physics, 1906), Ernest Rutherford, who had been Thomson's research student at the Cavendish, in 1919 (1871–1937; Nobel/Chemistry, 1908), and Lawrence Bragg in 1938 (1890–1971; Nobel/Physics, 1915). Since Bragg's time, the Cavendish Professorship has been held successively by Nevill Mott (1905–1996; Nobel/Physics, 1977) from 1954 to 1971, Brian Pippard (1920–2008) from 1971 to 1984, Sam Edwards (1928–2015) from 1984 to 1995, and Richard Friend (b. 1953) from 1995 to the present time.

On June 16, 1874, the chancellor formally presented the Cavendish to the university in a ceremony in the Senate House. Two days later the Cavendish opened and, although intended to be used for experimental physics research by Cambridge graduates, it soon became a place for practical instruction in physics and for graduates of universities other than Cambridge; although the latter graduates were often referred to as "aliens" there. McKenzie noted that "The Cavendish laboratory had then [prior to 1900] such meager financial resources that the purchase of an instrument costing £5 was preceded by long and careful deliberation and it acquired the nickname of the 'string and sealing-wax' laboratory."[1] When Max Perutz arrived at the Cavendish he was surprised to find the famous laboratory in poor shape compared to what he was used to at university in Vienna. He was "disappointed to find the famous laboratory poorly equipped and some of its members making a virtue of necessity by boasting of the great discoveries that had been made with no more apparatus than string and sealing wax."[2]

Upon the opening of the Cavendish, the journal Nature predicted that "The genius for research possessed by Professor Clerk Maxwell and the fact that it is open to all students of the University of Cambridge for researches, will, if we mistake not, make this before long a building very noteworthy in English science."[3] Indeed, this proved to be a very accurate forecast. Cavendish member John J. Thomson discovered the electron in 1897;

Francis W. Aston (1877–1945; Nobel/Chemistry, 1922) identified isotopes using mass spectrometry; Charles T.R. Wilson (1869–1959; Nobel/Physics, 1927) invented the cloud chamber to visualize electrically charged particles in 1911; Ernest Rutherford (1871–1937; Nobel/Physics) foretold the existence of the neutron and James Chadwick (1891–1974; Nobel/Physics, 1935) discovered it in 1932; Edward Appleton (1892–1965; Nobel/Physics, 1947) provided evidence for the ionosphere that led to the development of radar; Patrick M.S. Blackett (1897–1974; Nobel/Physics, 1948) identified the positron; John D. Cockcroft (1897–1967; Nobel/Physics, 1951) and Ernest Walton (1903–1995; Nobel/Physics, 1951) split the lithium atom using a particle accelerator in 1932; John Kendrew and Max Perutz used X-ray diffraction to solve the three-dimensional structure of proteins; Jim Watson and Francis Crick solved the structure of DNA and recognized its profound genetic implications; radio astronomy began at the Cavendish under Martin Ryle (1918–1984; Nobel/Physics, 1974); and Didier Queloz (b. 1966; Nobel/Physics, 2019), together with Michel Mayor, discovered the first exoplanet around a main star.

Arrival of Bragg

In October 1938 William Lawrence Bragg assumed the Cavendish professorship of experimental physics as a replacement for Ernest Rutherford who had died a year earlier. Rutherford, a New Zealander, had been J.J. Thomson's successor to the Cavendish professorship in 1920, "but only after long negotiations because the University authorities considered the honor of a Cambridge professorship to be worth a substantial drop in salary."[4] Bragg had succeeded Rutherford previously in 1919 as Langworthy professor and chair of physics at the University of Manchester. As Cavendish professor, Bragg's annual salary was £1,400 (about $7,000), about seven times the average British worker's annual salary at the time.

Lawrence Bragg was born in Australia in 1890 and emigrated to England in 1909. Together with his father, William Henry Bragg (1862–1942; Nobel/Physics, 1915), Lawrence Bragg was the co-founder of X-ray analysis and its use in the study of complex structures. This followed upon the discovery in 1912 by German physicist Max von Laue (1879–1960; Nobel/Physics, 1914) and his colleagues, Walter Friedrich and Paul Knipping, that X-rays are diffracted by crystals and the diffraction pattern reveals the symmetrical arrangements of atoms in the crystal. When X-rays impinge on atoms in a crystal their paths are bent into a regular pattern that can be recorded on photographic film or some other detector. However, Bragg was the first to understand and mathematically define how diffraction of X-rays from crystals occurred and how such information could be used to determine the exact arrangement of atoms in a crystal. That is to say, Bragg discovered how X-ray diffraction could be used for the determination of molecular structure. As Max Perutz put it, "Sir Lawrence Bragg . . . had the unique distinction of having himself created the science to which he devoted his life's work, and lived long enough to experience its revolutionary impact."[5] Just 25 years old in 1915, Lawrence Bragg was the youngest person ever to be awarded a Nobel Prize in physics.

Bragg's appointment at the Cavendish was made exactly one year after the death of Rutherford in October 1937. Rutherford had accidentally fallen from a ladder in his garden and died unexpectedly due to complications from a strangulated umbilical hernia and delayed surgery. In only a decade or so, under Rutherford's direction the Cavendish had become a leading laboratory in nuclear physics and Rutherford became known as the father of nuclear physics. He received the Nobel Prize in 1908, was knighted in 1911, and made a peer in 1931. A biographer of Rutherford wrote that "he is to the atom what Darwin is to evolution, Newton to mechanics, Faraday to electricity, and Einstein to relativity."[6]

Rutherford died in the Evelyn Nursing Home because of a delay in the arrival in Cambridge of Thomas Dunhill (1876–1952), a knighted surgeon from Harley Street, London. At the time peerage protocol required that a Lord, Rutherford's title, be operated on only by a titled physician. Unfortunately, the in-house physicians at the Evelyn were not permitted to operate on Rutherford in a timely manner and he died. His ashes were interred next to the graves of Isaac Newton and William Thomson, Lord Kelvin, in Westminster Abbey on October 25, 1937. Samuel Devons (1914–2006), a Cambridge physicist and science historian, recalled that "To every member of the Cavendish Laboratory, as we stood in Westminster Abbey amongst all those who came to express their respect, admiration, and affection and sorrow, it was a family bereavement. For the Cavendish, Rutherford, like J.J. [Thomson] before him, had been in person and in spirit a true father."[7]

One can only imagine how the course of events at the Cavendish would have differed had Rutherford not died tragically. He was just 66 years old, in good health, and as the father of nuclear physics undoubtedly would have remained as Cavendish professor for many more years. Had Rutherford lived, Bragg would not have replaced him, protein crystallography would not have taken root at the Cavendish, and Max Perutz would not have become the Cavendish professor's research assistant and been supported when Bernal left for Birkbeck College. Without Lawrence Bragg, Max, and protein crystallography at the Cavendish it is very unlikely that John would have chosen to return there at the end of WW2. It is anyone's guess what effect all of this would have had on the future of molecular biology in Cambridge and the rest of the world.

Bragg was a Nobel laureate, a pioneer in the application of X-ray diffraction to crystal structure, and was offered the Cavendish professorship unanimously and enthusiastically by a group of nine Cambridge electors. However, it was an unexpected appointment and not everyone was pleased with it. The latter is clear from an assessment of the appointment by Brian Pippard 50 years later: "W.L. Bragg's election to the Cavendish chair of experimental physics in Cambridge was taken by many as a threat to the great tradition of fundamental physics research established by J.J. Thomson and, especially, Rutherford . . . The choice of a crystallographer, however distinguished, was a blow to many hopes."[8]

Certainly Bragg was not considered a suitable replacement for Rutherford by everyone at the Cavendish, especially by the nuclear physicists. In 1981 William Cochran (1922–2003) and Samuel Devons (1914–2006) commented that "Bragg was a crystallographer, and that subject's second lease of life and crop of Nobel prizes lay well out of sight some 20 years in the future."[9] During an interview with Horace Judson in 1971 John also elaborated on feelings about Bragg's appointment at the time: "on the whole everybody thought it was absolutely terrible, the great days of the Cavendish had ended, that they had appointed this man who knew nothing about the main subject the Cavendish did [nuclear physics], the worst appointment in the whole history of the place."[10] Many members of the Cavendish felt that the MRC unit was not doing "real physics" and they considered it just a personal enthusiasm of Lawrence Bragg.

What then was the rationale for making the appointment? Under Rutherford resources for nuclear physics had dwindled substantially and this delayed investment in the next generation of accelerators. This led to the departure of many of Rutherford's most experienced Cambridge colleagues to other universities. J.D. Bernal moved in 1937 to Birkbeck College, University of London, Patrick Blackett to the University of Manchester in 1937, James Chadwick to the University of Liverpool in 1935, and Mark Oliphant (1901–2000) to the University of Birmingham in 1937. In part Bragg and crystallography were chosen by the electors because it was thought that Cambridge could not afford the sophisticated equipment that would permit nuclear physicists at the Cavendish to remain at the forefront of their field. In this context, John wrote in 1990:

> Bragg was under criticism from the nuclear physicists for not supporting their own subject more strongly in a laboratory world-famous for its reputation in nuclear physics under J.J. Thomson and Rutherford—and indeed for not being a nuclear physicist himself. Of course, the criticisms should have been directed at the electors to the chair, rather than to the incumbent, but in any case, it would have been financially impossible for the Cavendish to continue after the war as a major center of nuclear physics research, and Bragg's great contribution was to foster two quite new subjects—radio astronomy and molecular biology—both of which became very important during the years he was in Cambridge.[11]

When Bragg arrived at the Cavendish it consisted of three primary research areas: nuclear, low-temperature, and atmospheric physics. In addition, there was a rather small contingent of crystallographers. The crystallography group included the inimitable Bernal who had been one of William Bragg's research students in the Davy Faraday Laboratory at the Royal Institution, London. Bernal had been transferred from the Mineralogy Department in Cambridge to the Cavendish in the early 1930s. The transfer of Bernal to the Cavendish as assistant director of research in crystallography did not take place with Rutherford's blessing. Max recalled that

> the conservative and puritanical Rutherford detested the undisciplined Bernal who was a Communist and a woman chaser and let his scientific imagination run wild. He had wanted to throw Bernal out of the Cavendish but was restrained from doing so by Bragg. If Bragg had not intervened, Bernal's pioneering work in molecular biology would not have started, John Kendrew and I would not have solved the structure of proteins, and Watson and Crick would not have met.[12]

In 1934 Bernal and his research student, Dorothy Hodgkin (1910–1994; née Crowfoot; Nobel/Chemistry, 1964), working with hydrated crystals of pepsin, a digestive enzyme, demonstrated for the first time that good X-ray diffraction patterns could be obtained from protein crystals. Hugh Huxley noted that "this was the first defining moment in protein crystallography associated with the Cavendish."[13] Concomitantly, the work of English physicist William Astbury (1898–1961) at the University of Leeds, using X-ray diffraction to study pepsin crystals, and his later work with fibrous proteins such as keratins of wool, led to concepts about protein structure that were to be taken up by Linus Pauling and Robert Corey in the 1950s. Astbury's crystallography laboratory in Leeds was held in such high esteem that Max Perutz referred to it as "the X-ray Vatican."

Bragg brought his X-ray crystallographic laboratory with him from the National Physical Laboratory, London, where he had been director for only about one year, thereby establishing crystallography as a primary research area at the Cavendish. At about the same time, Bernal moved from the Cavendish to Birkbeck College, London, as chair of physics. Hugh Huxley reminisced that Bernal "had insufficient and unsuitable space in Cambridge in which to house enough people to explore the range of projects that sprang from his fertile mind. He had also been starved of resources by Rutherford, who disapproved of his untidy and permissive lifestyle, and his tendency to rely on other people to follow up and confirm his ideas, rather than doing the experiments himself."[14] Fortunately, when Bernal departed the Cavendish he left behind his research student, Max Perutz, who was destined to play a major role in the future of protein crystallography specifically and molecular biology generally. In addition, he was to play a major role in John's life and career from 1946 onward.

Nature of Proteins

The word "protein" was coined by the Swedish chemist Jacob Berzelius (1779–1848) in 1838 and comes from the Greek word *prota* that means "of primary importance." Enzymes, hormones, antibodies, receptors, and structural proteins are only a few of the many different kinds of proteins present in all living organisms. Proteins make up 75 percent or more of the dry weight of the human body. Friedrich Hünefeld (1799–1882) provided the first report of protein (hemoglobin) crystals in 1840; prior to about 1930 it was debated whether or not proteins were true molecules. Harvard biochemist John Edsall (1902–2002) reminded everyone that "there was a school of thought, among the colloid chemists, that proteins were not true molecules, but simply heterogeneous aggregates of various small molecules, presumably peptides of moderate size. Even the organic chemists found it almost incredible that true macromolecules [like proteins] could exist."[15] On the other hand, in the 1930s two Rockefeller Institute chemists, Max Bergmann (1886–1944) and Carl Niemann (1908–1964), proposed that proteins were constructed from periodic sequences of amino acids, the so-called "Bergmann-Niemann hypothesis."

Today we know that all proteins are composed of amino acids linked together by peptide bonds into polypeptide chains. There are 20 common amino acids in living organisms. Assuming a polypeptide length of x amino acids, the number of possible sequences is 20 to the power of x; so for a sequence of only three amino acids there are 20 to the power of 3 or 8,000 possibilities. Most proteins range in length from hundreds to tens of thousands of amino acids, consequently, there is an enormous number of possible amino acid sequences for them. In 1951 Linderstrøm-Lang (1896–1959), a Danish biochemist, proposed that protein structures were organized at four different levels; primary, secondary, tertiary, and quaternary. Primary structure referred to the amino acid sequence of a protein, secondary structure to folding of regions of a protein, tertiary structure to overall folding of a protein, and quaternary structure to the number and arrangement of polypeptides of a protein made of more than one polypeptide.

In 1926 James Sumner (1887–1955; Nobel/Chemistry, 1946) demonstrated that enzymes were proteins and that proteins could be isolated and crystallized. However, it was not until 1955 that it was demonstrated by Fred Sanger (1918–2013; Nobel/Chemistry, 1958, 1980) that each protein has a unique sequence of amino acids or a unique primary structure. Christian Anfinsen (1916–1995; Nobel/Chemistry, 1972) demonstrated that it is the amino acid sequence of a protein that determines how its polypeptide chain will fold up into a three-dimensional structure. From the mid-1930s to the early '50s several predictions had been made as to the overall shape or conformation of polypeptides. Among these was the so-called "cyclol-type" conformation for polypeptides championed by Dorothy Wrinch (1894–1976). However, the most significant prediction was that of Linus Pauling who proposed the alpha-helix and beta-sheet as the fundamental conformations of polypeptides, or protein secondary structure. This was a proposal that in time was proven to be correct.

The basis of Pauling's proposal for the secondary structures of proteins came about in 1948 while he was a visiting professor in Oxford lecturing on the chemical bond. What occurred while he was convalescing in bed with a cold was described by Jim Watson in his autobiographical account, *The Double Helix*: "The discovery of the alpha-helix, a fundamental element in the structure of proteins, was the first great triumph of model building. Pauling told of how he did it: While on a sabbatical visit to Oxford, he was confined to bed with a heavy cold. Bored with his detective novel, he amused himself by cutting out the form of a protein (polypeptide) chain from a sheet of paper and he then twisted it to make a fit between the units along the chain."[16] From such model building, using only paper, pencil, and ruler, Pauling was the first to predict the presence of the alpha-helix in proteins. He proposed that there were 3.6 amino acids per turn of the helix and that the helix was stabilized by extensive hydrogen-bonding between amino acids (that is, a chemical bond formed between an electropositive hydrogen atom and an electronegative atom such as oxygen). This model was consistent with certain reflections in X-ray diffraction photographs of proteins. In 1951 Pauling, Corey, and Branson published a paper on the alpha-helix, "The Structure of Proteins: Two Hydrogen-Bonded Helical Configurations of the Polypeptide Chain," in the *Proceedings of the National Academy of Sciences, USA*.

At first Cavendish crystallographers were skeptical of and annoyed by Pauling's proposal since, if correct, they had been scooped. John, Max, and Bragg had been attempting for some time to determine the folding of polypeptides using Astbury's X-ray data on keratins and model building. However, Astbury's X-ray data and the failure to recognize the planar configuration of the peptide bond, the linkage between amino acids in a polypeptide chain, led them astray. With the appearance of Pauling's publication in 1951, in a very short time Max confirmed the presence of the alpha-helix in proteins by X-ray diffraction. His results were consistent only with the alpha-helix and so excluded all alternative models put forward. But it was not until the late 1950s when John and co-workers solved the high-resolution, three-dimensional structure of myoglobin that alpha-helices were actually observed. Nearly 80 percent of the myoglobin polypeptide consists of right-handed alpha-helices.

Arrival of Max

Max Ferdinand Perutz was born in Vienna, Austria, in 1914 and in 1936, at the age of 22, he emigrated from Vienna to Cambridge. With financial assistance from his father in Austria, Max joined J.D. Bernal's group as a research student at the Cavendish. Turned down as a research student at King's, Trinity, St. John's, and Gonville and Caius Colleges, Max was finally admitted to Peterhouse. One of Bernal's associates, crystallographer William A. Wooster (1903–1984), had suggested Peterhouse to Max because, in his opinion, it was the college that had the very best food. Max was made an honorary fellow of Peterhouse in 1962 shortly after he had been awarded a Nobel Prize in chemistry.

Max described his initial impressions of the crystallographic laboratory at the Cavendish in an article in the *New Scientist*, "A Sagacious Scientist":

> In September 1936 I moved from the stately chemistry laboratories of the University of Vienna to the Crystallography Laboratory at Cambridge. I was disappointed to find it housed in a few dark and dirty rooms of a dilapidated grey-brick building or sort of outhouse, in more senses than one, of Lord Rutherford's famous Cavendish Laboratory. These dingy quarters were turned into a fairy castle by the brilliance of their director, John Desmond Bernal.[17]

Max wanted to use X-ray diffraction to determine the three-dimensional structure of biologically important molecules such as proteins. He attributed to Bernal his overwhelming desire to solve the structure of proteins:

> When I was a student, I wanted to solve a great problem in biochemistry. One day I set out from Vienna, my home town, to find the Great Sage at Cambridge. He taught me that the riddle of life was hidden in the structure of proteins, and that X-ray crystallography was the only method capable of solving it. The sage was John Desmond Bernal, who had just discovered the rich X-ray diffraction patterns given by crystalline proteins. We really did call him Sage, because he knew everything, and I became his disciple.[18]

Nearly a decade later Bernal was able once again to impart the desire to solve the three-dimensional structures of proteins, but this time to John Kendrew in the jungles of Ceylon/Sri Lanka during WW2.

Having trained as a chemist in Vienna and knowing practically nothing about crystallography, Max initiated his research by studying mineral crystals. As he would put it later on, he studied "a nasty crystalline flake of a silicate mineral picked off a slag heap." But within a year, Max was adept at crystallography and had begun to use X-ray diffraction to analyze crystals of two different proteins, hemoglobin, an abundant oxygen carrier in blood, and chymotrypsin, a digestive enzyme in the duodenum. Max's interest in hemoglobin had been aroused first in 1925 by biochemist Felix Haurowitz (1896–1987) in Prague who was married to Max's cousin, Gina Perutz. Max had been provided with crystals of hemoglobin by Gilbert Adair (1895–1979) in Cambridge and crystals of chymotrypsin by John Northrop (1891–1987; Nobel/Chemistry, 1946) in New York. His research resulted in publication of a short report in the journal *Nature* in 1938, "An X-ray study of chymo-trypsin and hemoglobin," authored by Max, Bernal, and Isidor Frankuchen (1904–1964). Because of some problems with the chymotrypsin crystals Max decided to focus all of his future attention on hemoglobin.

In 1938 Bernal moved to Birkbeck College in London, but Max remained behind since he liked Cambridge and the Cavendish so much. Bragg took over as his supervisor and in 1939 Max was appointed research assistant to Bragg supported by a grant from the Rockefeller Foundation in the United States. This external support, £275 (about $1,100) annually, slightly higher than the average British worker's annual salary at the time, was essential since Max's status in Britain had been changed from guest to refugee, and as such he was not permitted to earn money in England. At the time Max was providing some financial support for his parents who had fled Austria in 1938 and settled in Cambridge. It was indeed fortunate for Max that Bragg, who appreciated the power and inherent difficulties of protein crystallography, had been appointed Cavendish professor and served as his mentor.

In 1970 Max reminisced about his early relationship with Bragg:

I waited from day to day, hoping for Bragg to come round the Crystallo-graphic Laboratory to find out what was going on there. After about six weeks of this I plucked up courage and called on him in Rutherford's Victorian office in Free School Lane. When I showed him my X-ray pictures of hemoglobin his face lit up. He realized at once the challenge of extending X-ray analysis to the giant molecules of the living cell. Within less than three months he obtained a grant from the Rockefeller Foundation and appointed me his research assistant. Bragg's action saved my scientific career and enabled me to bring my parents to Britain.[19]

In March 1938 Germany occupied and annexed Austria and, as a result, Britain reclassified Austrians living in England from refugees to enemy aliens. Two years later, in the midst of WW2 and a few months after receiving his PhD, Max was arrested together with many other so-called "enemy aliens." He was sent first to the Isle of Man and eventually transported by ship to Canada where he was held for several months in an internment camp with more than 1,000 others. Max eloquently described his extraordinary experiences during this period of internment in the *New Yorker* magazine in 1985. He returned to Cambridge in January 1941 and became involved in some research related to the war effort, the so-called Habakkuk Project. This was a project that involved development of ice floes as ocean landing strips for planes. However, by 1944 Max had returned to research on the structure of hemoglobin supported by an Imperial Chemical Industries Fellowship.

Max was awarded a PhD in 1940 and sometime later commented that he would have remained a research student for 23 years if his thesis examiners had insisted on a completed three-dimensional protein structure. Max continued to carry out research on hemoglobin and other proteins for more than six decades, right up to his death in February 2002 at the age of 87. His final research publication, "Amyloid fibers are water-filled nanotubes," was published in the *Proceedings of the National Academy of Sciences, USA*, two months after his death. It took Max 30 years to determine the three-dimensional structure of hemoglobin, prompting him to say in a self-deprecating manner that he could have done it in half the time if he had been a bit brighter.

Today Max is an iconic figure, revered by many in and out of science because of his unwavering dedication to hands-on, bench research and his grand achievements in research, leadership, writing, and humanitarian causes.

Arrival of John

In January 1946 28-year-old John Kendrew arrived at the top floor of the Austin Wing of the Cavendish where protein crystallography was housed. Funding for the Wing, its equipment and furniture, had been provided by the automobile manufacturer Herbert Austin (1866–1941), after whom it was named. Austin was an automobile designer and builder who founded the Austin Motor Company and for a time served as a Member of Parliament. In 1936 Stanley Baldwin (1867–1947), chancellor of Cambridge University, accepted on behalf of the university Austin's generous offer of £250,000 (about $1.2 million) for the new building, some renovation of the Cavendish, and various expensive equipment items. In summer 1940 construction of the brick building was completed on the site of the old Zoological Laboratory. The building consisted of more than 30,000 square feet of space distributed over three floors and a basement; a fourth floor was added later to house theoretical physics.

Many members of the Cavendish staff, as well as research students, were housed in the Austin Wing which had a library, seminar room, and a tea room set up by Bragg. By the time of the building's completion England was involved in WW2, so the site was taken over by the government for wartime investigations by an army ballistics unit and a navy signals unit. In 1941 Bragg joined the war effort full-time as head of the British scientific liaison office in Canada, but returned to Cambridge and the Cavendish later that same year. It was only in 1945 that the Austin Wing was made available for what it was intended, basic research in physics.

John became Max's research student, working under Bragg. However, since Max did not hold a permanent university appointment, William H. Taylor (1905–1979), a mineral crystallographer and head of the Crystallography Division, was appointed John's official supervisor. Soon after John's arrival Max wrote a letter to his German parents-in-law, Herbert and Nelly Peiser, mentioning that John had joined him to work on hemoglobin:

> Kendrew, who is only just joining me, will work on an ambitious project which Sir Joseph Barcroft and I want to embark on. Barcroft, who is a physiologist, discovered a difference between the oxygen affinities of adult and fetal sheep hemoglobin [fetal hemoglobin had a higher affinity for oxygen than adult hemoglobin]. We now want to try whether that difference can be detected and interpreted in the X-ray diffraction pattern of the two compounds.[20]

In an interview in 1997 John reflected on what it was like getting back into research in the mid-1940s: "The thing was, the field we were in was a totally new one and so there was no established technique so there was no problem. It was something quite new for me and Bernal was a bit surprised that I worked for Max because he said to me 'Look you never did a course in crystallography when you were a student, I warn you it's a very boring subject to learn.' So I had to learn it as I went along, but I never did a course in it."[21] However, John did attend lectures on crystal physics given by crystallographers William H. Taylor, William Cochran, Max Perutz, and others.

For several years, before and after WW2, Max had been trying to use X-ray diffraction to determine the structure of hemoglobin, a protein used to transport oxygen in the blood stream, but with very limited success. There are more than 300 million copies of hemoglobin in each red blood cell, which give blood its red color. A hemoglobin molecule consists of four polypeptide chains, each consisting of nearly 150 amino acids and having a heme group containing iron that combines with oxygen. Each hemoglobin molecule can bind four oxygen molecules. Max's early X-ray results suggested that hemoglobin was a spheroid with a well-defined atomic structure and that it consisted of two identical halves, but it was unclear how to make further progress on the structure.

At the time X-ray diffraction had been used to solve the structure of molecules made up of only a few dozen atoms, not thousands, so it was unclear how to proceed with proteins. In fact, in the 1940s and '50s there were only about half a dozen protein crystallography laboratories worldwide. In a lecture entitled "What Mad Pursuit," a title suggested by John, Francis Crick predicted that X-ray crystallographic studies of proteins were doomed to fail and many of Bragg's colleagues at the Cavendish felt that they were attempting the impossible. John's own assessment of the situation years later was that "my own total ignorance [of X-ray crystallography] was fortunate, in that it concealed from me the extent to which contemporary X-ray crystallographic techniques fell short of what was needed to solve the structure of molecules containing thousands of atoms; it was indeed a case of ignorance being bliss."[22]

Some years later Michael Rossmann (1930–2019), one of Max's colleagues working on hemoglobin, delineated some of the problems associated with protein crystallography in the early 1950s: "The likelihood of ever solving the 3-dimensional structure of a protein seemed exceedingly small even in the 1950s. There were too many data to be collected, no reasonable way of solving the phase problem, and no way of computing results to make an interpretation of the collected data. But maybe even more daunting was the nagging question of whether each protein molecule had a unique structure."[23] The determination of protein structure by X-ray diffraction had been described as similar to bouncing tennis balls off an intricate building and trying to decide what the building looks like from where the tennis balls finally end up.

In view of the difficulties Max and other protein crystallographers had encountered, for John to embark on such a research project for his thesis required a great deal of self-confidence. One of Max's colleagues recalled that when John first appeared in the laboratory his demeanor conveyed to everyone that he knew what he was doing and that he would get what he wanted. Max confessed that "I had no research students, because responsible dons advised graduates against joining such a forlorn undertaking, but Kendrew's spirit of adventure won."[24] Aaron Klug concluded that "John had joined Max Perutz on a voyage of discovery (building the ship as they went along) where the land sought was clear—the 3-dimensional structure of proteins—but with no route through the unchartered waters: the conventional wisdom was that the goal was unreachable."[25] In truth, John had been convinced by Bernal and Pauling that determination of protein structure by X-ray analysis was of fundamental importance and he intended to carry it off.

Only a few months after arriving in Cambridge John had achieved a reputation as a real up and comer. A letter to John's father Wilfred from an officer at the Clarendon Press in early spring 1946 revealed that John was already held in very high regard at the time. So much so that Clarendon Press hoped that when John had a book in mind he would follow in his father's footsteps and publish it with them.

In the summer, shortly after John started research at the Cavendish, Max wrote once again to his parents-in-law praising John's work and stating how fortunate he was to have John as a colleague:

> I am so glad you enjoyed John Kendrew's essay. This and his fellowship thesis which it was meant to accompany, are both very well written accounts of a first-rate piece of research which K [Kendrew] carried out last winter. He knew no X-ray crystallography when he came last January, learnt the elements of the subject in two weeks, set to work on his problem which required great experimental skill, and solved it in a few months' concentrated work. Besides he is a charming fellow, and it is a pleasure to talk to him; so I consider myself very lucky to have him with me.[26]

Joining Peterhouse

In 1947 John obtained a research fellowship at Peterhouse (St. Peter's), the oldest college in Cambridge, founded in 1284 by Hugh de Balsham, bishop of Ely, and modeled after Merton College, Oxford. Apparently he would have preferred a fellowship at his undergraduate college, Trinity, but it was not forthcoming. John Meurig Thomas recalled that "From the time he became a Fellow of Peterhouse, Kendrew undertook a number of tasks which, if properly undertaken, make college life so agreeable and charming . . . To all those tasks he brought the same military precision and flair that greatly facilitated his progress as a molecular biologist and director of research."[27] John had Peterhouse accommodations at The Hostel—built in 1926 and situated next to the Master's Lodge—at a cost of £48 (about $192) annually. The average British wage in 1947 was about £278 (about $1,100) annually. The cost of accommodations was reduced to £36 (about $144) when John married Elizabeth Gorvin-Jarvie in 1948.

Thanks to a suggestion by Joseph Barcroft (1872–1947), a physiologist at the Molteno Institute in Cambridge who studied the oxygenation of blood, John's thesis research was to be a comparative study of adult and fetal sheep hemoglobin. In January, just one year after joining the Cavendish, John described his research on hemoglobin and some preliminary work on myoglobin in a report to Paul Cairn Vellacott (1891–1954), a historian and master of Peterhouse (1939–1954):

> The main problem investigated has been the relation between the hemoglobins obtained from the adult and fetal forms of the sheep. This project was chosen both because of its general physiological interest in view of the doubts which have hitherto been expressed as to whether the two hemoglobins are different molecular species, and because it fitted appropriately into the general programme of comparative X-ray studies of hemoglobins initiated by Perutz. The investigation has shown that the adult and fetal hemoglobins must almost certainly be regarded as distinct molecular species, the evidence being obtained both from general morphological examination and from X-ray studies of hemoglobin crystals grown under identical conditions from fetal and adult blood. Furthermore the change-over from the fetal to the adult type of hemoglobin takes place (in sheep) between the 120th day of pregnancy and birth (c. 150 days), the blood at intermediate stages containing a mixture of the two proteins; it is hoped

shortly to investigate the change in more detail by using electrophoresis technique. It has also been shown that the X-ray diffraction patterns of fetal sheep hemoglobin are not inconsistent with a structure similar in general outlines to that deduced for horse hemoglobin by Perutz, and that in particular there seems to be a layered structure in the molecule in both cases.

A new method has been devised for studying the shrinkage of protein crystals on drying, by allowing them to take up equilibrium in atmospheres of known vapour pressure. It has been shown that at any rate in the case of horse hemoglobin the shrinkage definitely takes place in discontinuous steps, as already supposed by Perutz. This phenomenon can most readily be explained by supposing that the water layers interposed between the protein molecules in the crystal have themselves a "fine structure"; that is to say, that the water molecules are arranged in discrete layers. This investigation is at present in a preliminary stage, but it is hoped eventually, by making measurements at different temperatures, to make deductions concerning the thermodynamical relationships involved.

Attempts have recently been in progress to make crystallographic studies of the muscle respiratory pigment myoglobin. Unfortunately the specimens so far available, derived from horse heart, have not yielded crystals large enough for X-ray examination, and the possibility of examining other species is now being explored.

None of the studies outlined above have so far been submitted for publication, but it is hoped shortly to prepare a paper on the relation between adult and fetal hemoglobin.[28]

Myoglobin and Hemoglobin

John had become very interested in the structure of myoglobin, a small and easily crystallized protein closely related to hemoglobin. Myoglobin, a protein used to store oxygen temporarily in tissues, consists of a single polypeptide chain made up of about 150 amino acids and possesses a single heme group that combines with a single oxygen molecule. One reason John gave for his interest in myoglobin was that he wanted his own protein rather than just going along with what Max was doing on hemoglobin.

John's keen interest in myoglobin at the time is revealed in his correspondence with Alessandro Rossi Fanelli (1906–1990) in Italy and Hugo Theorell

(1903–1982; Nobel/Physiology or Medicine, 1955) in Sweden in 1947. He wrote to Fanelli:

> I am very interested to hear that you are beginning some work on the X-ray examination of myoglobin crystals. I have recently begun some work in the same direction, and since I last wrote have succeeded in getting X-ray diffraction pictures from crystals of a horse myoglobin . . . They are still far from adequate, however, and we are continuing the search for better methods of crystallization.[29]

This was followed several months later by a similar letter to Theorell who had crystallized horse myoglobin:

> I am writing to ask your advice on the crystallization of myoglobins. For sometime I have been attempting to prepare large crystals of this protein in order to carry out single-crystal X-ray studies . . . In the case of horse myoglobin I found it easy to prepare crystals by a method closely similar to that developed by yourself, but . . . they were too thin for X-ray purposes. I therefore turned to other species, and . . . have so far examined ox, guinea pig, and whale . . . but so far I have been unable to prepare the myoglobin in crystalline form, except in one single preparation from whale meat, in which crystals were obtained, but far too small for X-ray purposes . . . What I should like to ask, is whether you can give advice on what would be suitable species to try out, both from the point of view of readily crystallizability, and of the suitability of the crystals for X-ray study.[30]

John once again described his ongoing research to the master of Peterhouse in 1948. In the report he pointed out that, as of October 1947, the MRC had established the Unit for the Fine Structure of Biological Systems at the Cavendish with Max and himself as staff members. He went on to describe his work on fetal and adult hemoglobins and, in particular, his promising work on crystals of horse and whale myoglobin:

> The [X-ray] results have been extremely encouraging; the structure of the protein [myoglobin] does appear to be very simple and to be easily related to that of hemoglobin, and a short series of X-ray pictures has yielded nearly as much information as demanded long and tedious experimental

work and computation in the case of hemoglobin; the interpretation was, of course, much facilitated by this analogy between the two.

Clearly it will be worth pursuing this problem much further; I imagine it likely that part of my time for several years will be occupied by horse myoglobin, the next stage being to attempt to prepare even better crystals . . . .

I have also spent some time making a preliminary study of another variety of myoglobin, this time derived from the whale . . . I am doubtful whether it will be so amenable to analysis as horse myoglobin; nevertheless I shall continue to work on it for a time at least.[31]

In the same year John published the article "A Comparative X-ray Study of Fetal and Adult Sheep Hemoglobins" with Max in the *Proceedings of The Royal Society*. He also published some preliminary results on myoglobin in an article, "Preliminary X-ray Data for Horse and Whale Myoglobin," in the journal *Acta Crystallographica*, a portent of what would follow a decade later.

In May 1949, at the age of 32, John submitted his PhD thesis, "X-ray Studies of Certain Crystalline Proteins: The Crystal Structure of Fetal and Adult Sheep Hemoglobins and of Horse Myoglobin" (Thesis Preface, see Appendix A.3). He was awarded a PhD by Cambridge University in 1949 and received an ScD in 1962. John was to remain in Cambridge associated with Max, the MRC, and Peterhouse for the next 25 years.

Indexing Literature

John was always deeply concerned about the organization and retrieval of large amounts of data and information. From the beginning of his graduate career at the Cavendish he was obsessed with devising a method for indexing the vast scientific literature. At the beginning of 1947 he wrote:

For some time a number of colleagues and myself have been thinking about this problem [abstracting and indexing of scientific references], which has become particularly acute in our own borderline field (the X-ray study of crystalline proteins), in which it is necessary to keep abreast of the vast literature of most of biochemistry and much of physiology, chemistry, and physics. We have been thinking both of comprehensive indexing systems suitable for a University Department, and of selective systems suitable for the individual. Our emphasis has mainly been on the latter: and for this it seems important that any scheme should be (a) inexpensive (in time as well as money), and (b) universal—i.e., employing some well reorganized international system of the classification of knowledge.[32]

By spring 1946 John had written several times to Copeland Chatterson Co., London, manufacturer of the Cope Chat Paramount Sorting System, about using Paramount Cards for indexing scientific references. In December John wrote to them with detailed specifications for the cards. The letter reflects John's compulsiveness and attention to minute details:

I have now designed a card which I think will suit my requirements: a copy is enclosed. The size is to be 8" x 6½" approx. so that it can fit standard card index cabinets of this size. There would be no printing on the reverse; standard small Paramount holes would be used—in all 114 holes per card (not including corner holes). The most appropriate stock would be G.70, though I should prefer this weight of paper in white rather than yellow, if possible.

I should be grateful if you could let me have a quotation (including purchase tax) for various quantities of this card e.g. 1,000, 2,000, and 5,000. I should also like to know whether you preserve a block from which the card is printed, so that repeat orders could be executed at the "quantity" price.

We feel it may be desirable to modify the design of the card in the light of our early experience of its use; would it be possible to supply a preliminary order of say 100 cards so that we can test out the system thoroughly before ordering a large quantity? If so, what would be the cost of such a small order?

I should be glad also to have an idea of delivery times.

Have you proceeded further with your scheme for a multisorting device? If so I should be glad to have details at this stage, since the design might have repercussions on the final arrangement of my card. E.G. I have used throughout the 5, 4, 3, 2, 1, 0 code for the sake of simplicity in sorting with a mechanical device the more compressed 7, 4, 2, 1 code might seem preferable . . . .[33]

John's attention to detail is further revealed in a related letter to Maun Industries in which he describes a pair of slot-cutting pliers for use with the Paramount Cards:

Thank you for your letter of 2nd December, and for the sample pair of slot cutting pliers.

The tool appears to be very satisfactory and I have only one small point to raise. This is that at present the working stroke is unnecessarily long and the pliers are therefore rather tiring to use for long periods. It would be better if in the resting position the tool were as sketched—

This could be achieved either by making the cutting head just over 3/15" longer; or alternatively by preventing the jaws from opening fully by fitting a screw as in the Clamping Pliers illustrated in your catalogue, or by modifying the length of the slots at the rear ends of the jaw pieces.

I should be glad of your comments on this, and also perhaps you would let me know whether your original estimate of 30/-applies to the final design.[34]

By summer 1949 John was placing orders for more than 10,000 Paramount Cards, as well as hand nippers and sorting needles, for at least half-a-dozen members of the Department of Physics at the Cavendish. Several years later he published a letter in *Nature*, "Use of a Computing Machine as a Mechanical Dictionary," in which John described a way to index scientific literature on edge-punched cards. He pointed out that with 22 holes and 11 needles one could index the entire Oxford English Dictionary.

MRC Support

By late 1947 salaries for Max and John had become looming problems. John's two-year grant was coming to an end, and it was unclear where his future salary would come from. Max was supported by the Rockefeller Foundation, which had decided that Cambridge University should pay his salary. Bragg recommended Max for a university lectureship, but this did not materialize until nine years later; perhaps because Max was a chemist working on a biological problem in a physics department. Bragg thought it inappropriate to appoint Max and John to staff positions in the Cavendish, a physics laboratory, since neither one was a physicist.

With no biochemical facilities available in the Cavendish, some bench space for Max and John had been provided by David Keilin (1887–1963), Quick professor and director of the Molteno Institute, Cambridge. Keilin had helped Max and John prepare protein crystals and was well aware that their research was in danger of closing down because of a lack of support by the university. At the suggestion and with the strong support of Keilin, in May 1947 Bragg wrote to and then met with Edward Mellanby (1884–1955), secretary of the MRC (1933–1949), at the Athenaeum Club, London. Mellanby was a noted British physician and pharmacologist who had discovered the role of vitamin D in preventing rickets.

The MRC is a government agency that began in Britain in 1913 as the Medical Research Committee and Advisor Council, and became the MRC seven years later. It was assigned the duty of distributing medical research funds in Britain under the terms of the 1911 National Insurance Act. In 1914 their annual budget amounted to £55,000 (about $265,000), whereas 100 years later in 2015/16 the amount had grown considerably to £928 million (about $1.3 billion). Initially established to focus on a cure for tuberculosis and rickets, after WW1 it was decided that funds should not be restricted to specific diseases but should be available for promoting fundamental scientific research generally. Over the coming years research funded by the MRC in the UK would give rise to more than two dozen Nobel laureates and many fellows of the Royal Society.

As a result of the letter and meeting, Mellanby suggested to the MRC council that they provide financial support for a unit at the Cavendish consisting solely of Max and John. He let Bragg know that he was going to bring up the Perutz application to the council at their meeting on Friday, October 17, 1947, for at least a preliminary run. A few days later the council met and concluded that research on molecular structure of biological systems was approved in principle. They intended to fund the unit proposed by Bragg for five years. There would be an annual expenditure of about £2,600 [about $10,480] on salaries and additional non-recurrent expenditures on major items of special equipment, with running costs to be borne by the Cavendish Laboratory.

The good news was communicated to Bragg by Mellanby immediately after the council meeting on Friday, October 17. Bragg was relieved and delighted with the MRC's promise of salary support for Max and John, and in a letter of thanks to Mellanby provided some insightful comments about John and his future career. He described John as a "borderline person" who trained in chemistry, knew a lot of physics, and was learning biology and as an up and comer with great force of character who probably would make a real leader in the future. However, in view of John's background in OR and government service, Bragg was unsure of John's long-term commitment to academic research. Clearly Bragg had a very high opinion of John, but was very perceptive in recognizing that John might not be totally committed to a career as an academician.

In this manner the Unit for Research on the Molecular Structure of Biological Systems was created at the Cavendish with a two-member staff, Max and John, and two technicians. The five-year MRC grant paid the annual salaries of Max, £1,005 (about $4,050) and John, £500 (about $2,015); the average British worker's annual salary was £415 (about $1,672) at the time. Two research assistants were hired. First K.K. Moller from Denmark and then the engineer Tony Broad who was destined to develop the powerful tubes used to generate X-rays for crystallographic studies at the Cavendish.

The MRC funding provided a reasonable period of financial security for Max and John and subsequently the name of the unit was changed to the Molecular Biology Research Unit. It is interesting that Max and John were both trained as chemists, were using physics to solve a biological problem, and had the support of the MRC, a government agency whose goal was to maintain and improve human health. In 1973 John was asked why the MRC was interested in supporting his and Max's research: "I think Bragg and Keilin must have done a terrific piece of sales talk. Roughly speaking, I think they said you can't expect anything to come out of this group for 10 years. But, if and when it does happen, it's going to have long term implications for biochemistry, and through biochemistry to medicine. And that was the basis on which the MRC took us on."[35]

Myoglobin and EDSAC-I

As 1949 began all members of the unit were accommodated in one room on the top floor of the Austin Wing of the Cavendish. Some time later Max and John moved into a small office and Francis Crick and Jim Watson, who had joined the Cavendish in 1949 and 1951, respectively, were put together in another large room so that they could talk to each other without constantly disturbing everyone else.

In the same year a course of lectures for science students, "The Ancient World," was initiated at the Cavendish in order to broaden the outlook of natural science students. Lecturers came from the faculties of archaeology, anthropology, and classics. *Varsity* magazine noted at the time that scientists realized the necessity of keeping a broad outlook, and it was the arts men who were guilty of being confined in their own little world.

In the fall John lectured on "The Narrow-Mindedness in the Humanities" to the Perne Club, an undergraduate science society at Peterhouse. The beginning of John's lecture emphasized his firm belief that non-scientists should know more about science in order to become, as he put it, "a complete man":

> What I want to do this evening is to consider whether it is in fact true that scientists are more interested in what the humanists are doing than vice versa. And if this is true, whether this is a bad thing, how it has come about and what can be done to change it. This a fairly large set of problems, and even if I thought I knew all the answers to them I should not be able to keep within the bounds set by your patience . . . I confine myself to the comment that if a scientist knows nothing of the humanities, he is branded narrow-minded—in my view deservedly; whereas if a humanist knows nothing of the sciences—that is just normal. So I shall stick to the question of whether humanists ought to know more about science; and if you regard that as narrow-minded on my own part I am afraid it is just too bad . . . I may seem a little old-fashioned, but I am a firm believer in the ideal of a complete man.[36]

By 1949 John had completely changed the focus of his research from hemoglobin to myoglobin, the oxygen-carrier abundant in muscle and closely related to hemoglobin. Myoglobin is about one-quarter the size of hemoglobin, and as Max recognized several years earlier, because it was considerably smaller that hemoglobin would probably be much more amenable to X-ray crystallographic analysis.

During his thesis research John had begun to work with myoglobin purified and crystallized from horse hearts. He obtained the hearts from a local slaughterhouse that was preparing horse meat for dining at college high tables. Such preparations of myoglobin produced rather poor crystals and it was obvious to John that he would need to find a more suitable source for the protein. Despite this difficulty, in 1950 his research resulted in a lengthy paper in the *Proceedings of the Royal Society*, titled "Crystal Structure of Horse Meat-Myoglobin: I. General Features: The Arrangement of the Polypeptide Chains." Although a long way from proposing a three-dimensional structure for myoglobin, John reached several very interesting conclusions in this paper. Primary among these was the recognition of striking similarities between the structures of myoglobin and hemoglobin, especially the folding of their polypeptides. Since both proteins function as oxygen carriers, it raised the question of whether their identical function implied identical structures. It would take several more years of structural work to answer this question. It was also in 1950 that John, Max, and Lawrence Bragg published a paper, also in the *Proceedings of the Royal Society*, titled "Polypeptide Chain Configurations in Crystalline Proteins," in which they speculated about the folding of myoglobin and hemoglobin polypeptides; speculations that in time would turn out to be incorrect.

In July John wrote once again to the master of Peterhouse about his research on myoglobin and brought up the novel possibility of using computers in X-ray crystallographic studies:

> My main work during the year has been a theoretical study of the possible configurations of polypeptide chains in crystalline proteins, carried out in collaboration with Sir Lawrence Bragg and Dr. M.F. Perutz. We have recently written a joint paper on that work . . .
>
> I recently embarked on two new projects. The first is to study in detail methods of obtaining large crystals of myoglobin than were hitherto available to me; if, as I am fairly confident, I am successful in preparing them I should be in a position to make further progress next year with the study of this material. The second is to devise methods of using the large Cambridge calculating machine (commonly but inappropriately known as the Electronic Brain) to lighten the tedium of the computations which unfortunately play a prominent part in the X-ray analysis of crystalline proteins.[37]

In January 1951 Bragg wrote to Harold Himsworth (1905–1993), an eminent clinical scientist known for his work on the etiology of diabetes mellitus, who had replaced Edward Mellanby as secretary of the MRC in 1949 (1949–1968). Bragg raised the question of Max's and John's future since the five-year MRC grant would end in October 1952. He pointed out that they should consider the interests of Max, now over 35 with a wife and family, and John, age 33 with a wife, and provide Max with a permanent post and John with a temporary one at the university. Bragg went on to characterize John as a very valuable man, a first-rate scientist with considerable drive and organizing power.

At the time the unit at the Cavendish consisted of John and Max, together with two research students, Francis Crick and Hugh Huxley, with Bragg as its patron. In July, just before leaving for the United States for two months, John wrote a progress report for the master of Peterhouse: "I mentioned last year that I was about to embark on two new projects. The first was to obtain large crystals of myoglobin in order to extend my X-ray work on that protein. In spite of a good deal of biochemical work I have so far been only partly successful. We feel that this protein is one of the most promising and therefore propose to continue these efforts next year."[38]

John's other project was to develop methods for using the Cambridge high-speed computer to shorten the time taken for computations required in protein crystallography. About this project John wrote:

> In this I have collaborated with Mr. J.M. Bennett of the Mathematical Laboratory. We have completed a preliminary program, and have succeeded in reducing the time taken for a typical computation from three days by hand methods to fifteen minutes with the machine: again, a computation which we had done commercially [Scientific Computing Service, London] two years ago at a cost of £600, which took six months [using a punched card machine], can now be done in a matter of 30–40 hours. We have written a paper on this work; it is shortly going to press.[39]

The high-speed computer referred to in John's report was the Electronic Delay Storage Automatic Calculator, known as EDSAC-I. Characterized as a room-sized behemoth, it was used in Cambridge from 1949 to 1958. The computer was built in a room, approximately 2,300 square feet in area, that was at one time the dissecting room of the Anatomy School and on warm days the room was filled with fumes from formalin embedded in the floor boards. John Makepeace Bennett (1921–2010), an Australian research student who was later to become Australia's first professor of computer science, had helped William Renwick (1924–2010) and Maurice Wilkes (1913–2010), Director of the Mathematical Laboratory, build EDSAC-I. Bennett was a friend of Hugh Huxley who in 1948 had become John's first research student. Both Bennett and Huxley were research fellows at Christ's College. Huxley suggested that John and Bennett get together and use EDSAC-I for high-speed computation of myoglobin data. Consequently, in July 1951 John and Bennett were the first to apply an electronic computer to computations for structural biology.

Howard Dintzis described EDSAC-I in his recollections of his time at the Cavendish working with Max in the 1950s:

Three technicians had built it [EDSAC-I] by hand, using dozens of war surplus radio chassis and hundreds of vacuum tubes, almost filling an enormous room with racks of chassis and bundled wires running everywhere. Computer "memory" consisted of an iron pipe 15 feet long filled with mercury, with a loudspeaker at one end of the pipe and a microphone at the other end. The dynamic memory consisted of the sound waves traveling down the pipe, with a cycle time of a millisecond and a total memory capacity of a kilobyte. Computer input was a punched paper tape, and output was an electric typewriter. The contraption usually ran only a few hours before one of the many hundreds of vacuum tubes burned out. Calculation then stopped until one of the three technicians who had built the computer came by to repair it. After 6 p.m., this required waiting until the next morning, or Monday morning if on weekends.

Dintzis further recalled that:

Because our crystallographic computations were among the most complex done at the time, we were only allowed to use the facility after 8 p.m., when other users had finished their work. We could work during the night until we finished our calculations or a vacuum tube burned out, whichever came first....[40]

John and Bennett first described their landmark use of EDSAC-I for computations associated with X-ray crystallographic analysis of myoglobin at the second English computer conference held in Manchester in 1951 (Manchester University Computer Conference Proceedings, 1951). They then published a seminal paper, titled "Computation of Fourier Syntheses with a Digital Electronic Calculating Machine," in the journal *Acta Crystallographica* the following year. In 1958 EDSAC-II, designed and built in the Mathematical Laboratory with a micro-programmed control unit, came into operation and replaced EDSAC-I.

When interviewed by the *Manchester Guardian* in 1959, John admitted that "We are doing things which we couldn't think of tackling at all without high-speed computers. The work we shall be doing in the future will need even bigger and faster machines. For us the high-speed computer is not only a luxury enabling us to work less hard, but an absolute necessity."[41]

During an interview many years later John made the following comments about his work with Bennett on EDSAC-I:

> Well, I rather think John Bennett and I were probably the first people ever to calculate a Fourier synthesis on an electronic computer. You see, the Cambridge EDSAC Mark I had only very recently come into . . . use . . . I had to learn how to program it . . . the total store was 512 words, and I remember the day . . . we got 1024 [words] and it seemed as if infinite horizons were suddenly open. And of course it was very slow by modern standards and the PC on my desk is, of course, much faster and has much more storage capacity . . . It wasn't reliable and you came in at the weekend and if it was working people would bring you relays of sandwiches and beer to keep you going until it broke down, hopefully not before Monday morning. And we did the first calculations on the myoglobin work on that computer.[42]

Insofar as gaining access to computer time on EDSAC-I, John remembered that "there were other people, claimants, like the radio astronomers and people searching for prime numbers and so forth. So you really had to battle for time, but we got it."[43]

In the early 1950s John and Max embraced different approaches toward computations associated with X-ray crystallographic data. John's affinity for high-speed computational methodology to solve problems that were dependent on large data sets is consistent with his wartime experiences as part of OR, his organizational and technological bent, and his fearlessness about novel approaches to solve old problems. On the other hand, Max was unfamiliar and uncomfortable with, and distrustful of, electronic computers. He was too busy mounting crystals, taking pictures and overcoming experimental difficulties to take the time to learn about computers. In referring to Max in a crystallography symposium in 1988, John labeled him technologically conservative since Max apparently didn't believe results from EDSAC-I. For some time Max continued to rely on an army of women to process crystallographic data. In this, as in many other matters over the years, John and Max were at odds. However, with the advent of EDSAC-II and John's obvious success using EDSAC-I, in time Max also turned to the computer to process data from his crystallographic work on hemoglobin.

In 1951 John spent a little more than two months in the United States. He spoke at a Biophysics Symposium in Ann Arbor, Michigan, visited laboratories and gave lectures in Seattle, San Francisco, and Pasadena on the West Coast, attended a Gordon Research Conference in New Hampshire, spoke at an International Chemical Conclave in New York City, and visited Harvard Medical School, Woods Hole Marine Biology Laboratory, and other scientific centers in the northeast.

Suitable Myoglobin Crystals

In his Nobel lecture John commented on his choice of the sperm whale as a source of myoglobin: "First of all it was necessary to find some species whose myoglobin formed crystals suitable, both morphologically and structurally, to the purpose in hand; the search for this took us far and wide, through the world and through the animal kingdom, and eventually led us to the choice of the sperm-whale, *Physeter catodon,* our material coming from Peru or from the Antarctic."[44]

Toward the end of his fifth year as a research fellow at Peterhouse, John prepared another report on his research for the master:

> On the experimental side we are still pretty baffled by the problem of dis-covering how to make crystals which have been eluding us for the last three years—my complaints on this score have, I fear, been a monotonous refrain in these reports. However, . . . I intend to make a major and I hope suc-cessful attack on this problem next year. In the meantime a new kind of crystal turned up which, though not what I really wanted, did enable us to push ahead to some extent. I have a good number of experimental results from these crystals which I have so far found no time to publish.[45]

The myoglobin crystals that John referred to in the report had been prepared from whale muscle. He had found that skeletal muscles of aquatic mammals, who as diving mammals spend much of their time under water, were a very rich source of myoglobin, which was used to store oxygen. Purified sperm whale myoglobin produced crystals so large that they had to be cut with a razor blade and pieces could be placed in quartz capillaries for X-ray analysis.

In December 1952 John wrote to Neil A. Mackintosh (1900–1974), a British biological oceanographer and authority on Antarctic whales:

> We have recently discovered that the meat of whales is extremely rich in this material [myoglobin], and that all the whales we have investigated

(so far blue-, humpback-, finback-, sperm-, sei-, and mink-) contain myoglobin that crystallizes extremely well . . . In particular we are extremely
interested in sperm-whale. This sperm-whale contains myoglobin whose
crystal structure is more favorable than any that we have hitherto encountered, and we think it likely that a detailed study of it will throw important
light on the structure of proteins in general.[46]

John's original source of sperm whale muscle was from the Low
Temperature Research Station in Cambridge, but the amount of raw material available was limited. One of the goals of the Research Station during
WW2 was to determine which kinds of whale meat might be made palatable
for human consumption. Mackintosh had a connection with H.W. Symons
on a whaler in the Antarctic and John wanted frozen specimens of any
species of whales they caught. In January 1953 John received a letter from
Symons confirming that in April he could pick up 25-kilogram blocks of
frozen sperm, blue, finback, and humpback whale meat, as well as a sample
of seal meat. The shipment was picked up in Liverpool from the Pacific
Steam Navigation Company by Michael Bluhm, John's research assistant.
Another shipment of frozen sperm whale, sea lion, tortoise, and seal meat
arrived from Peru on the ship *Reina del Pacifico* in June 1953 and was stored
in garbage cans in a deep-freeze room at the Molteno Institute. John would
point out that "Most of our supplies of meat came from Peru: encouraged
by the cool Humboldt Current, sperm-whales congregate near the western
coast of South America and their meat is accounted a delicacy in the
restaurants of Lima."[47] In 1955 John received 100 pounds of frozen blue and
sperm whale meat and 200 pounds of finback whale meat. He was invited to
come to South America to shoot an Atlantic seal, but he graciously declined
the invitation and decided to wait for the carcass of a recently deceased seal
to be sent to him.

Three New Recruits

In summer 1952 John also let the master of Peterhouse know that he had
successfully recruited three new co-workers during his two-month trip
to the United States. Two were postdoctoral fellows, Robert Parrish and
Jim Watson, and the third a research student, Peter Pauling, son of Linus
Pauling:

during my stay in America I made arrangement for 3 young American scientists to come to the Cavendish Laboratory to collaborate with me in research. Two of these are Merck Fellows; one came last fall, two are due to arrive in six weeks' time . . . In that I was enabled to visit many research laboratories working in the same field as my own, and thus to make contacts and arrange exchanges of information which have been and will be of the greatest value.[48]

Robert Parrish was recruited from Duke University to help John find the most suitable crystals of myoglobin for X-ray analysis. John wrote to Parrish in fall 1951 describing the current state of the myoglobin project:

First of all you are aware that for some years I have been studying the X-ray patterns of horse and whale myoglobin. This work has been held up very much by the lack of suitable crystals. One of our immediate problems is to develop methods of making bigger crystals of myoglobin in particular, and indeed of crystalline proteins in general. In the past the preparation of such crystals has generally been a matter of chance; we are trying to put the preparation of large crystals on a more rational basis. If we are successful in making large crystals of horse myoglobin then a very large field of research lies open. Very much more data have been obtained for hemoglobin than for myoglobin, and if we can really make a more proper start on the latter, there is an enormous amount to be done.[49]

At the Cavendish, Parrish attempted to crystallize myoglobin from a wide variety of sources. He worked mainly with aquatic animals that spend a lot of time under water and are rich in myoglobin, such as whales, seals, and penguins. He settled finally on crystals of sperm whale myoglobin and showed that heavy-atom isomorphous derivatives exhibited significant intensity changes in X-ray diffraction photographs.

Jim Watson was recruited to work on protein structure as the result of a conversation between John and Salvatore Luria (1912–1991; Nobel/Physiology or Medicine, 1969) at a meeting in Ann Arbor, Michigan. Luria pointed out that his very bright graduate student was working with Herman Kalckar (1908–1991) in Copenhagen and was very unhappy there. Watson commented later on that "Most fortunately, Kendrew made a favorable impression on Luria; like Kalckar, he was civilized and in addition supported the Labor Party."[50]

Peter Pauling (1931–2003) was recruited as a research student to work on the structure of myoglobin when John visited Woods Hole in Massachusetts. Anxious to escape from home, Pauling arrived in Cambridge in September 1952, exactly one year after the arrival of Watson. He sat in a room together with Francis Crick, Jim Watson, Jerry Donahue, and Michael Bluhm. More interested in his social and college life in Cambridge than his research, Peter's work proceeded rather slowly at the beginning and worried John a great deal. John hesitated to place any critical portion of the myoglobin structural analysis in Peter's hands since he considered it much too risky. But by October 1954 John was writing quite positively about Peter to his father, Linus Pauling: "I am very happy to be able to tell you that in the last two months I have become much more content with Peter's work. For the first time he strikes me as being really interested in his work, and if only he maintains this enthusiasm I have no doubt that he will succeed in obtaining his PhD."[51]

John and Peter published their work on "The crystal structure of myoglobin II. Finback-whale myoglobin" in the *Proceedings of the Royal Society* in 1956. But Peter's antics continued unabated after more than three years in Cambridge, culminating eventually in the discovery that he was about to become an unwed father. This was the final straw for the new master of Peterhouse, Herbert Butterfield (1900–1979; Master 1955–1968), and the tutors, and Peter was expelled from the college and then from the university. Shortly thereafter Peter married the child's mother in a civil ceremony in Cambridge with John serving as best man and hosting a wedding reception in his flat at Tennis Court Road. In his characteristic supportive manner John came to Peter's rescue, arranging through Bragg for him to transfer as a research student to the Royal Institution in London. Peter went on to receive a PhD and remained in England as lecturer in physical chemistry at University College, London, until 1989. Peter died in 2003 at the age of 71.

Today nearly everyone is familiar with the legendary story of Jim Watson and Francis Crick who discovered the DNA double-helix and were astute enough to understand the profound genetic implications of the structure. Recruited by John to work on protein structure, Watson was a biologist, not a chemist or physicist. However, he was absolutely convinced that DNA was the genetic material and was obsessed with the goal of determining its structure. In Francis Crick, a highly accomplished biophysicist and crystallographer working with Max, he found the perfect partner for the task and in a relatively short time achieved his goal.

Marvelous Myoglobin Crystals

At the end of 1953 John wrote with some excitement to Hugh Huxley, by then a postdoctoral fellow at MIT, about the state of the myoglobin project:

> This is mainly to let you know what is going on in the lab. I am afraid I have owed you a letter for sometime; my failure to write it is largely due to one piece of news, namely that 3 weeks ago we got the most marvelous myo-globin crystals, from sperm-whale of all odd places. They are gigantic crystals which give excellent precession pictures in 12 hours, and, best of all, they are P21 with 2 molecules in the unit cell ... So we are going hell for leather for the 3-dimensional. Besides this we have several other new forms of myoglobin. So altogether there is a great deal of excitement, to such an extent that I am hardly going away at all this Christmas so that we can get on as fast as possible.[52]

In 1954 John and Robert Parrish published a paper in *Nature,* titled "The Species Specificity of Myoglobin." Various crystallographic parameters, such as symmetry, space groups, and unit cell dimensions were reported for myo-globin crystals prepared from whale, seal, and penguin muscle. At this point John had almost everything in hand with which to determine the three-dimensional structure of myoglobin. However, what remained was the unre-solved issue of how to solve the crystallographic "phase problem."

Phase Problem

When a beam of X-rays is focused on a protein crystal the beam is scattered or diffracted in many directions by the atoms that make up the protein. The amplitudes of the waves scattered by atoms are directly proportional to the number of electrons the atoms possess. The diffraction pattern of regularly spaced spots or reflections that results can be collected by a detector, such as photographic film or a radiation counter. From the intensities and spacing of the spots, two components required for solving a structure, the amplitudes of the waves and the volume of the basic repeating unit that makes the crystal can be calculated. However, a third required component, the phases of the waves, that is whether the waves are in step or out of step with one another, are lost during data collection. To calculate a three-dimensional depiction of

the spatial distribution of electron density within a crystal one must approx-
imate the phases.

In the 1930s the isomorphous replacement method was invented by
John M. Robertson (1900–1989) at Glasgow University to solve the phase
problem for crystals of small organic molecules consisting of about 20
atoms. The method was used by the Dutch chemist Johannes M. Bijvoet
(1892–1980) at the University of Utrecht who showed how the three-
dimensional structure of complicated molecules could be derived from X-
ray diffraction patterns of two different heavy-atom variants (strychnine
sulphate and selenate). The method involves attaching an electron-dense
or heavy-atom, such as mercury or gold, to molecules in a crystal without
altering the native crystal dimensions. The native and heavy-atom modi-
fied proteins must be isomorphous; that is, they must have the same crystalline form or shape. By comparing X-ray diffraction data from native crystals
and isomorphous heavy-atom derivatives phases can be approximated. This
method had been used successfully for small molecules, but not for large
molecules like proteins. Although both Bernal and Robertson suggested
that heavy-atoms could be used with proteins to determine phases, the conventional wisdom was that proteins were much too large and complex for
the method to work. Bragg had commented that proteins would take no
more notice of a heavy-atom than a maharajah's elephant would of a gold
star painted on its forehead.

However, in 1953 Max, Vernon Ingram (1924–2006; born Immerwahr),
and David Green applied the isomorphous replacement method using either silver or mercury derivatives of hemoglobin. They showed that the
scattering from a single heavy atom was strong enough to produce repro-
ducible, measurable changes in the intensities of X-ray reflections from
hemoglobin crystals. The diffraction spots appeared in exactly the same
positions as heavy-atom-free hemoglobin, but with slightly altered intensity.
Max was jubilant! He and his colleagues published their results in 1954 in a
paper, titled "The Structure of Hemoglobin. IV. Sign Determination by the
Isomorphous Replacement Method," in the *Proceedings of the Royal Society*.
Bragg called isomorphous replacement the "Rosetta Stone" for interpreting
protein diffraction photos and it became the so-called "Holy Grail" for protein crystallographers dealing with the phase problem.

It was Max's tremendous technical breakthrough that provided John with
the methodology necessary to solve the three-dimensional structure of
myoglobin. Whereas Max, having found a solution for the phase problem,

lamented that he was getting absolutely nowhere with the structure of hemoglobin.

Recruitment to London

In 1953 Bragg was offered the positions of Fullerian professor of chemistry and director of the Davy Faraday Research Laboratory at the Royal Institution in London. He would have an annual salary of £2,000 (about $5,600) and £600 (about $1,680) for expenses, and the normal retirement age of 70 years of age was changed to 75 especially for him. Bragg's father, William, had been director of the same laboratory 30 years earlier and under his leadership it had become a world-class X-ray crystallographic facility. Bragg left the Cavendish for London in January 1954, however, his departure was not without ramifications.

Crystallographic research at the Royal Institution had deteriorated over the decade since William Bragg was director, and his son wanted to restore its reputation. He tried to recruit the entire Cavendish MRC Unit to the Royal Institution, but his attempts were unsuccessful. Max would not leave Cambridge, did not want the unit split between Cambridge and London, and wished to remain completely independent, not under Bragg's direction. The latter issue was of considerable concern to Max since Bragg had suggested openly that protein crystallography at the Royal Institution and Cavendish should all proceed under his direction.

Unable to recruit Max and his group to London, Bragg went after John and the myoglobin group. He courted John excessively during the spring and summer of 1954 sending him several impassioned letters about a potential move to London. Bragg even wrote to Harold Himsworth at the MRC, who in turn wrote to John assuring him that the council regarded him as a free agent and that John should do what is best for his research and for himself. In August, Bragg wrote two more letters to John trying hard to tempt him to move to London. He conceded that John would be giving up college life in Cambridge but insisted that it was the right time for John to establish a show of his own and build up a little empire, all with Bragg's support.

In the end, John, like Max, refused to leave Cambridge and the Cavendish. However, he agreed to accept a part-time appointment as a reader at the Royal Institution, on the condition that Max be offered an identical appointment. John held the part-time reader position for 15 years. It was with a bit of irony that John wrote to Jim Watson about his recruitment to London, confessing that "The only good reason I can think of for going [to London] would be to escape from Max, which would greatly please me, but it is not sufficient to compensate for the loss of Cambridge."[53]

The Hut

In 1954, Nevill Mott, professor of physics and director of the Physical Laboratory at Bristol University, was chosen to succeed Bragg as Cavendish Professor of Physics. The arrival of Mott in Cambridge, into a scientific hierarchy dominated by the chemist Alexander Todd (1907–1997; Nobel/ Chemistry, 1957), led to a terse poem that made the rounds in Cambridge at the time:

"I'm God," said Todd
"You're Not," said Mott

Mott was a solid-state physicist with little or no interest in protein crystallography. Initially, in discussions with Bragg, Mott indicated that the MRC Unit would be welcome to stay at the Cavendish. However, in order to make space for some of his colleagues that accompanied him from Bristol, Mott asked the General Board of the Faculties to evict the crystallographic unit from the Cavendish. A letter sent by Mott to Max Perutz in February 1954 made the situation very unsettling. Mott felt that the Cavendish should not house the unit indefinitely and that he could not commit space to it beyond July 1955. Max replied to Mott immediately stressing that he wanted to keep the members of the unit together since the chance of making significant breakthroughs would be far greater than if they were dispersed to different laboratories. Max also wrote to the General Board of the Faculties pleading with them to let the crystallographic unit remain at the Cavendish. To his relief they agreed that crystallography should remain at the Cavendish until alternative accommodations could be found.

In September, following a recurring illness and a long holiday away from Cambridge, Max wrote to Harold Himsworth at the MRC and proposed that John be appointed deputy director of the unit. Himsworth had suggested this possibility to Max sometime earlier since John had already fulfilled this function for several years. John was appointed deputy director of the MRC Unit at the age of 38, was granted unlimited tenure, and was earning £2,200 (about $6,000) annually; £1,770 from the MRC, £100 from the Royal Institution, and £330 from Peterhouse where he was director of studies and steward. However, by 1956 the size of the myoglobin group had dwindled to such an extent that John wrote to Francis Crick, then at the Brooklyn Polytechnic Institute, about the imminent problem:

> so the net result of all this is that from the New Year the myoglobin group will probably be reduced to one, namely myself. While this will undoubtedly give me a chance to do some experimental work with my own hands, which I will enjoy, I think it would be better to have the group a little larger, particularly at the present critical juncture in the work. So this is really an S.O.S., asking you to keep your ears very wide open for someone who could join me as soon as possible.[54]

In summer of 1957 Mott decided that after October the rooms occupied by the MRC Unit on the top floor of the Austin Wing would no longer be available. As an alternative accommodation Mott offered a prefabricated building known as The Hut that stood in the courtyard/parking lot between the Cavendish and the old Anatomy School on Free School Lane, close to Cambridge's market square. The offer of space was for five years with the period beyond remaining quite uncertain. The Hut was built in 1946 as the Metallurgy Hut for Egon Orowan (1902–1989), a solid-state physicist from Hungary. It was a shabby, brick structure with a corrugated roof. Dark and cramped, it was quite unsuitable insofar as utilities and facilities for research and, in fact, it had been designated for demolition by the university. But Max, with Mott's help, got the demolition order rescinded and The Hut was renovated at a cost of about £7,000. Max wrote to Himsworth saying that while The Hut would provide a substantial increase in space, it was only a temporary solution, and the unit hoped to move into Todd's department when completed in 1960.

About half of The Hut was converted into offices and the other half into laboratory space, and in 1957 members of the unit moved in. There were offices for crystallographers and wet labs, rooms in the basement of the Austin Wing for X-ray and workshop facilities, areas in the old Anatomy building and Greenhouse for Francis Crick and Sydney Brenner to work, and a small wood shed that housed high voltage electrophoresis tanks. John and Max each had their own office, Crick and Brenner shared an office, and there were other small offices for postdoctoral fellows, research assistants, and women involved in processing data. Several visitors to the laboratory, including Renato Dulbecco, Seymour Benzer, George Streisinger, Mahlon Hoagland, and Sewell Champe also occupied the very limited space. Michael Rossmann, who arrived in 1958 to work with Max at the Cavendish, recalled that "Everyone packed tightly into the central corridor of The Hut at 11 o'clock each morning, reaching for the coffee pot brewed by Leslie Barnett, Sydney Brenner's assistant, while debating everything from the relation between church and science in the then new Churchill College to the best way of finding heavy-atom positions in protein crystals."[55]

David Blow, a research student with Max and then staff member at the LMB, recalled the move into The Hut:

> [The] Unit . . . took over a hut in the car park, erected for temporary wartime use. Happily, the X-ray generators remained in the basement of the Cavendish Laboratory's Austin Wing. In the following years, the group of scientists attracted to work with Perutz, Kendrew, Crick and Brenner became far too large to do experimental work in the hut. Every unused room in the vicinity was adopted for use. Kendrew built his model of myoglobin in the recently vacated cyclotron room at the Cavendish, while Brenner and his colleagues used Lord Rutherford's former stable for experiments in molecular genetics.[56]

In September 1957 Max wrote to Himsworth and requested that the name of the unit be changed from "Molecular Structure of Biological Systems" to "Molecular Biology Research Unit." The change in name would coincide with their move into The Hut and with 10th anniversary of the unit's formation in October 1947. The name change was approved by the MRC Council in October 1957.

In an interview with Bernard Dixon of the *Scientist* in 1987, John reflected on the pursuit of scientific knowledge in Cambridge in the 1950s:

> We were lucky in Cambridge after the war. Even within the framework that existed for British science in the 1950s—broadly speaking the old Haldane principle, according to which scientists were funded and then left to their own devices—Max Perutz and I received very favorable support. We were given money to do research for about 10 years, without getting any results at all . . . Perutz worked even longer than I did—about two decades—without

achieving anything. Credit for allowing us to operate in that way goes to a handful of people—Lawrence Bragg, Henry Dale, David Keilin, and Harold Himsworth—who in various capacities backed what we were doing and gave us time. And remember that many pure scientists, including professional crystallographers, believed that our new approach to protein structures was a complete waste of time. Well, they were wrong.... All I can say is that the system at that time allowed our work to proceed without us being asked continually to justify ourselves or provide tangible findings. And we were not the only beneficiaries, by any means. The system also allowed other avenues to be explored and to prosper in the fullness of time.[57]

Various Activities

During the 1950s John was director of studies in natural sciences (1950–1956), college steward (1955–1959), and college lecturer in natural sciences (1956–1967), at Peterhouse. He also was a frequent guest at college feasts and dinners held at Peterhouse and Trinity, including Ascension, All Saints, Audit, Candlemas, and Perne Feasts, and Annual Gathering, Commemoration, Commencement, and Invitation Night Dinners. He also participated regularly in discussions at the Hardy Club, founded in 1949 and named for the biologist William Bate Hardy (1864–1934), attended by a select group of Cambridge biophysicists, biochemists, and structural biologists. John became a member of the Faraday Society and Institute of Biology, a fellow of the Institute of Physics, and lectured to several Cambridge clubs/societies, such as the Carphologists, Space Group, Kapitza, Perne, and Hardy, in the 1950s. As a part-time reader at the Royal Institution he made more than a dozen trips to London annually between 1954 and 1960, and made more than a dozen trips abroad to attend meetings and give lectures throughout the 1950s. After the low- and high-resolution structures of myoglobin were published in *Nature* in 1958 and 1960, respectively, John became somewhat of a celebrity in science and invitations for lectures and other duties poured in from around the world. John had accomplished the impossible!

7
Marriage and Divorce (1948–1956)

[H]is mind usually went in directions other than towards his wife.

Jim Watson

Marriage to Elizabeth

Saturday, May 1, 1948 was a cool spring day in Cambridge. It was the first of a three-day cricket match between Cambridge University and Lancashire scheduled for F.P. Fenner's Ground on Mortimer Road, a 15-minute walk from the city center. The match ended in a draw.

Also scheduled for the same day was the wedding of John Kendrew and Mary Elizabeth Gorvin-Jarvie (1919–2002) at Little Saint Mary's Church on Trumpington Street, at the corner of Little Saint Mary's Lane. A house of worship had existed on this site adjacent to Peterhouse since the 12th century. Elizabeth was one of three daughters born to John Gorvin, a civil servant, and his wife, Winifred Seldon Gorvin. The father of the bride served as assistant secretary at the Ministry of Agriculture and Fisheries and had wide experience in relief work all over the world. At the time of their marriage John was 31 years old, a research student at the Cavendish, and a fellow of Peterhouse. Elizabeth was 29 years old, a medical student at Cambridge and at St. Bartholomew's Hospital in London, where William Harvey (1578–1657) had carried out landmark research on the circulatory system in the 17th century. Elizabeth's class at St. Bartholomew's was only the second intake of women by the hospital.

Joseph N. Sanders, chaplain of Peterhouse and rector of Rampton, served as officiating minister at the wedding; he would later become dean and bursar of Peterhouse. The organ was played by Dennis Mack Smith, a fellow of Peterhouse and later of All Souls. One of John's close friends, John Graham, served as best man. Graham had attended Trinity College, studying classics, at the same time as John and was a civil servant at the Ministry of Food; he eventually became fisheries undersecretary at the Ministry of Agriculture, Fish and Food. Another friend, John Pinkerton, served as a witness at the wedding. Pinkerton had attended Clifton and then Trinity studying natural sciences at the same time as John and did graduate research at the Cavendish. He became an eminent computer designer who designed England's first business computer, the LEO, and for many years was head of research at English Electric Computers.

Graham composed and presented the following epithalamium for John and Elizabeth on the occasion of their marriage:

> Poets enshrine in richest eloquence
> The gods they know by faith and not by sense.
> Thus, the high joys of Heav'n itself we know
> Only from bards who sang while still below;
> And the Church Militant alone acquaints
> Us with the rapture of Triumphant Saints.
> Then, happy couple, pray take no offence
> At my inculpable incompetence
> To hymn the bliss that will descent upon
> Two paragons within a Paragon.
> Such lofty themes are better fitted for
> The lips of an inspired bachelor.
> Myself, a mere prosaic married man,
> Must offer such poor tribute as I can,
> And wish you both a long harmonious life,
> Enlivened with a modicum of strife.[1]

John Graham's wife, Betty, was a sister to Elizabeth's first husband, John Jarvie, who had also attended Trinity, studying classics and economics. Jarvie was a sub lieutenant in the Royal Navy during WW2 and was one of more than 230 men killed on Easter Sunday, April 5, 1942, when their ship, the HMS *Dorsetshire*, was hit by 10 bombs from Imperial Japanese Navy Aichi D3A dive-bombers. The ship sank along with its sister ship, the HMS *Cornwall*, in the Indian Ocean about 200 miles southwest of the coast of Ceylon/Sri Lanka. Consequently, Elizabeth became a widow in spring 1942, only two years after marrying Jarvie. John and Jarvie had been close friends as undergraduates at Cambridge, both had participated in the Cambridge

University Socialist Club, and John had served as best man at Jarvie's marriage to Elizabeth in 1940.

Elizabeth and John spent their honeymoon in Rome and visited John's mother in Florence. Upon returning to England the couple set up housekeeping at Flat 5, The Paragon, Blackheath, while Elizabeth finished her medical training in London at St. Bartholomew's and the Elizabeth Garrett Anderson Hospital for Women. During this period John also had rooms in Cambridge in The Hostel, a wing of Peterhouse situated next to the Master's Lodge, at an annual cost of £36. The Hostel and nearby buildings at Peterhouse were used to house the London School of Economics during WW2. When Elizabeth received her degrees in medicine and surgery in August 1950, she and John moved into a flat owned by Peterhouse at 12 Tennis Court Road, Cambridge, at a cost of £35 annually. Elizabeth became a house physician at Addenbrooke's Hospital in Cambridge and a trainee assistant general practitioner in Huntington.

Divorce from Elizabeth

John's marriage to Elizabeth deteriorated rapidly in the early 1950s. John had a very independent lifestyle as a Cambridge man, spending much of his time on research and participating in the academic, social, and administrative life of Peterhouse, which distanced him from Elizabeth. Jim Watson observed that "The fact that John treated many college functions as more important than being at his Tennis Court Road home had to mean that his mind usually went in directions other than towards his wife."[2] To a significant extent John was married to Peterhouse and bound to his research at the Cavendish. Elizabeth certainly felt neglected and openly expressed her feelings to others from time to time. In addition, John's idiosyncratic nature and his alleged critical view of Elizabeth's intellect and lack of cultural refinement also contributed to their marital difficulties.

As a couple, John and Elizabeth were incompatible, and this was recognized by both parties relatively soon after their marriage. Elizabeth tried to remain loyal to John, but their relationship was severely strained. Perhaps John married Elizabeth as a result of his friendship with her late husband John Jarvie. John valued loyalty and had been very close to Jarvie during their time together as undergraduates at Cambridge. He may have considered his marriage to Elizabeth fulfillment of an obligation owed a close friend who died tragically during WW2. John's unsuccessful attempt to marry Suzy Ambache and have a family several years earlier also may have continued to weigh on him.

By 1950, only two years after marrying Elizabeth, John complained to his mother in Florence about the great difficulties of his marriage and his strong desire to seek a divorce. His mother revealed the fragility of the marriage to John's research student, Hugh Huxley, and a mutual friend, Freddie Gutfreund, during their short visit to Italy in summer 1951. John had arranged for Hugh and Freddie to visit his mother in Florence when he heard of their planned trip to Italy. It is unclear whether or not John had discussed with his mother what she should or should not say to Hugh and Freddie about his marriage.

Although John wanted a divorce, he feared that such an action might jeopardize his Peterhouse fellowship. This was not an unjustified fear in view of the very conservative environment at the college and the stigma associated with divorce in Cambridge at the time. In 1954, following the death of Paul Vellacott, Herbert Butterfield (1900–1979) was appointed master of Peterhouse; he was a deeply religious man concerned with religious matters. Furthermore, it should be remembered that in England in the 1950s marriage was still viewed as a binding legal duty, and divorce was permitted only on the grounds of adultery, cruelty, and desertion for three years or more.

Hugh Huxley had become a research student at the Cavendish in 1948 at the age of 24. This followed four years in the RAF as a radar officer during WW2 and two years back at Cambridge to finish up his undergraduate degree in physics begun in 1941. Hugh joined a small group at the Cavendish that was supported by the MRC and consisted of Max, John, and Francis Crick. Max commented that "What Kendrew, Crick, and Huxley had in common was experience of science applied to war, which made them think harder than the average graduate about their future research and realize that the greatest promise of physics and chemistry lay in their application to the understanding of life."[3]

As John's research student, Hugh got to know John and Elizabeth rather well and thought of them as a very pleasant and well-matched couple, until summer 1951 when he visited John's mother in Florence and was told that their marriage was a failure. Hugh greatly admired John and was very fond of Elizabeth who often joined Hugh, Jim Watson, and John Bennett to play tennis. At John's request, on several occasions Hugh escorted Elizabeth to various social gatherings when John was unavailable. Since Jim Watson had lost his financial support, he was living at the time with John and Elizabeth on Tennis Court Road. He recalled that "John and Elizabeth Kendrew rescued me with the offer, at almost no rent, of a tiny room in their house in Tennis Court Road. It was unbelievably damp and heated only by an aged electric heater."[4]

All of this remained rather innocent until early in 1952 when Elizabeth revealed to Hugh that her marriage was a disaster, confirming the assessment by John's mother the summer before. After her disclosure, Elizabeth and Hugh entered into a closer, more intimate relationship. John was aware of Hugh's involvement with Elizabeth and continued to consider a divorce. He spoke to Hugh about it and the awkward situation caused some tension in the laboratory. But John knew that Hugh had plans to move to the United States for two years after defending his thesis in summer 1952. Hugh did leave for the Massachusetts Institute of Technology with a Commonwealth Fellowship, but his absence between 1952 and 1954 did little to improve John's marriage. In reality John's marital difficulties had very little to do with Hugh.

Despite the disturbing undercurrent at home, John wrote to Hugh about his research on myoglobin and about the possibility of Hugh returning to the Cavendish at the end of his stay in the United States: "It's probably far too early for you to say yet, Hugh; but you might keep me posted about the progress of your thoughts re one-versus-two years in the States. The reason I ask is that we must keep in mind methods of getting you some kind of niche in the Cavendish, that is if you want one, and we ought to get to work fairly soon if anything is to be done by next October."[5] Clearly, John remained strongly supportive of Hugh for a position at the Cavendish. Science superseded any of John's personal feelings regarding his marriage difficulties at the time.

So, with John's strong support, toward the end of 1954 Hugh was invited back to the Cavendish as a staff member. He returned for a short time, but by September 1955 Max was reporting to Himsworth at the MRC that "Unfortunately all this good news is tempered by one sad occurrence. Huxley got into private difficulties and thinks that he must leave Cambridge. This is a great blow to the Unit . . . I shall have to approach you again about this matter as soon as Huxley decides where he wants to go."[6]

By 1956 Hugh had moved to the Biophysics Department at University College, London, as an external staff member of the MRC. However, once again with John's strong support, Hugh rejoined the new LMB on Hills Road in 1962 and remained in the Division of Structural Studies until 1987. In research that began while a research student with John at the Cavendish in the early 1950s, Hugh used light and electron microscopy, as well as X-ray diffraction, to show that muscle is comprised of an array of protein filaments that slide relative to each other during contraction. As a result of his landmark work on the structure and function of muscle, Hugh enjoyed a long and highly distinguished career, marked by numerous prestigious appointments and international honors and awards. He served as joint head of the Structural Studies Division from 1977 to 1987 and as deputy director of the LMB from 1979 to 1987. In 1987 Hugh moved to Brandeis University in Massachusetts and remained there until his death in July 2013 at the age of 89.

Elizabeth separated from John in August 1955 while he was hiking in the Alps with Jim Watson. Jim Watson recalled the incident in his book *Genes, Girls, and Gamow*:

> That day I also learned that John and Elizabeth Kendrew's marriage was on the rocks and most likely irretrievably over. The situation was a ghastly mess with Hugh Huxley somehow involved. With the news, I instantly understood John's uncharacteristic bolting in the midst of our mountain walk in August. The matter was just coming to a head when John left Cambridge to join me in Italy. Apparently, upon reaching Saas Fee he learned that Elizabeth had already cleared out her possessions from their house in Tennis Court Road.[7]

In the end, their marriage was dissolved by an uncontested divorce on November 30, 1956, and made absolute several weeks later on January 14, 1957. Hugh Huxley was named as co-respondent.

Elizabeth became a practicing Quaker and a general practitioner at Seaford on the south coast of England as a member of the medical practice of Sutton, Robinson, Kendrew, Cairns, Thomas, and Marks. She lived first at Sussex Gate, Beacon Road, and then at 3 Chyngton Lane in Seaford. In 1968 in Sussex Elizabeth married a distant cousin, Robert Dexter, who was a judge, and they moved to Transvaal, South Africa. Following her third husband's death, Elizabeth returned to England in the early 1980s and lived for a while in London at 26 Vanburgh Hill. She died in November 2002 in Norfolk at the age of 83, five years after John's death.

Aftermath

During the 40 years following their divorce, whenever Elizabeth's name came up John referred to her as "my wife," and her name and address were always recorded in his diary. John and Elizabeth corresponded with each other when she became ill. In fall 1962, shortly after the announcement that John would be awarded a Nobel Prize, he received a congratulatory note from friends that referred humorously to Elizabeth: " Your ex-wife was with us when the great news [Nobel Prize] came through . . . You will be glad to know that so far as I could see Elizabeth looked more placid etc.—maybe because she is a little plumper. I boldly asked if there was a man in her life to explain it, and with quite a gay giggle she said perhaps she looked happier because there wasn't!"[8]

Divorce from Elizabeth was traumatic for John. He was unaccustomed to failure in anything and although he had wanted a divorce from Elizabeth for some time the failed marriage weighed heavily on him. For 40 years he remained a bachelor, had numerous romantic interests, and even fell in love at times. But John continued to be apprehensive about entering into another marriage, in part for fear of another divorce.

John's dread of another divorce is reflected in excerpts of a letter sent to him by a woman in March 1982 for his 65th birthday: "It is sad that you don't trust me in marriage as we would be so happy together most of the time. . . . I'm not really expecting to influence you. I know quite well you would rather settle for life as it is than risk greater contentment against trauma of another divorce. I'm just hoping for a miracle because of my love for you and a conviction that we are so right for each other."[9] Despite this poignant offer of marriage John remained a bachelor.

8

Structure of Myoglobin (1957–1962)

I always knew a protein would look like this.

J.D. Bernal

New Worlds

In a 1961 *Scientific American* article John revealed his feelings when the low-resolution structure of sperm whale myoglobin was finally completed in 1957:

> When the early explorers of America made their first landfall, they had the unforgettable experience of glimpsing a New World that no European had seen before them. Moments such as this—first visions of new worlds—are one of the main attractions of exploration. From time to time scientists are privileged to share excitements of the same kind. Such a moment arrived for my colleagues and me one Sunday morning in 1957, when we looked at something no one before us had seen: a 3-dimensional picture of a protein molecule in all its complexity.[1]

There were a number of factors that enabled John to be the first to solve the three-dimensional structure of a protein in 1957. Perhaps the most significant of these were the availability of suitable crystals of myoglobin, solution of the crystallographic phase problem by Max Perutz in 1953, development of powerful rotating anode X-ray tubes by Tony Broad in the 1950s, and access to an electronic computer, EDSAC-I, in the 1950s, together with John's brilliance, punctiliousness, and superb organizational skills.

Background

With determination of phases worked out by Max and colleagues in 1953, in the mid-1950s two visiting scientists, Howard Dintzis from the United States and Gerhard Bodo from Austria, succeeded in attaching mercury, gold, or silver to five different sites in sperm whale myoglobin crystals. John decided to use these heavy-atom derivatives to collect X-ray diffraction data at 6 Å or low-resolution.

In X-ray crystallography, resolution is a measure of the level of detail that can be seen in a three-dimensional structure. For proteins, low-resolution structures, 5 to 6 Å resolution, generally reveal only the overall folding or shape of a protein's polypeptide. High-resolution structures, 1.2 to 2 Å resolution, reveal not only the folding of the polypeptide chain but the positions of atoms of the amino acids (carbon, nitrogen, oxygen, and sulfur) that make up a polypeptide. Of course, much more data and work is required for a high- than for a low-resolution protein structure, because the number of reflections collected increases with the inverse cube of the resolution. For myoglobin, a crew of technical assistants, nearly all women, scanned hundreds of X-ray photographs to determine the positions and intensities of reflections. It would require about 25 times more data (number of reflections) and a great deal more effort to determine the 2 Å structure of myoglobin published in 1960 than the 6 Å structure published in 1958. John and several colleagues, including Herman Watson, Charles Coulter, and Carl Brändén, then went on to determine the structure of myoglobin at 1.4 Å resolution which permitted identification of nearly all of the polypeptide's amino acids. A list of John's collaborators and research students and the three stages of X-ray analysis of myoglobin structure are presented in Tables 8.1 and 8.2, respectively.

6 Å Myoglobin Structure

In August 1957 John wrote about the 6 Å resolution myoglobin structure to Michael Bluhm, Gerhard Bodo, Howard Dintzis, Joseph Kraut, Robert Parrish, and Harold Wyckoff, six of his colleagues who had worked on the project:

> To save repeating myself, this is a circular letter to tell you that we finally managed to do a 3-dimensional Fourier synthesis of myoglobin the other

day, with a resolution of 6 Å. The phases were determined using 5 different isomorphous derivatives . . . The calculation was done on EDSAC in 76 minutes . . .

The Fourier shows much more than I personally ever hoped it would. First of all the main parts of the polypeptide chains are quite clear . . . The general shape of the molecule corresponds pretty well with what we have said in earlier papers . . . The arrangement of the chains is complicated . . .

Our plan is to extend the resolution to 3 Å using the same techniques as hitherto . . . Maybe we can do this in 18 months to 2 years . . . This should clear up a lot of the doubtful points. After that atomic resolution! . . . .

I thought you might like to have this advance information about the latest results which, it goes without saying, would not have been possible to achieve without the efforts you made when you were in the Unit . . .

Why not come back for a while and help with the 3 Å stage?[2]

At about the same time Max wrote to the Rockefeller Foundation, which had provided so much financial support to the Unit:

Kendrew completed the first 3-dimensional Fourier of myoglobin a few weeks ago and we are all very thrilled with it. It shows the position of the iron atom with the heme group together with several long stretches of rods of high electron density which clearly represent polypeptide chains. Their configuration cannot be seen directly but judging by the distances between these rods one would guess them to be alpha-helices. The structure of the molecule as a whole is most complex and intricate, and quite different from anything anyone had ever imagined, the heme group being attached to a kind of basket work of polypeptide chain with many different kinds of contact between protein and prosthetic group. The resolution (6 Å) is still too low to make out details but at least the shape of the molecule, general layout of the polypeptide chains and the position of the heme group are clear.

I am immensely encouraged by Kendrew's success and am redoubling my efforts to try and get a 3-dimensional Fourier of hemoglobin at a comparable resolution. Kendrew meanwhile is going ahead with plans for a second 3-dimensional Fourier at twice the resolution of the present one and so far as we can see there are no fundamental obstacles in the way of this being achieved.

I am very pleased that the long continued support of the Foundation for this research is now beginning to bear genuine fruit.[3]

In January 1958 John submitted a paper on the low-resolution structure of myoglobin to Arthur Gale, the joint editor of *Nature* from 1939 to 1961. He wrote saying that

> I was sorry to miss you when you last visited the Unit. Perutz told me that you would be interested to have an article on the recent myoglobin work. I have now written such an article and enclose it herewith. I hope you will not think it impossibly long, but I found it very difficult to tell the story more briefly. Naturally I am anxious to have it in print as soon as possible, but I realize that is length may make that difficult. I am leaving for America, where I shall be for one month. In the meantime I have asked Perutz if he would be kind enough to look after this for me, so perhaps you should write to him if any problems arise.[4]

The paper, titled "A 3-dimensional Model of the Myoglobin Molecule Obtained by X-ray Analysis," was co-authored by Gerhard Bodo, Howard Dintzis, Robert Parrish, and Harold Wyckoff at the Cavendish and David Phillips at the Royal Institution. It was accepted for publication less than two weeks later and appeared in *Nature* in early March 1958. John noted that the paper appeared exactly 100 years after the German scientist F.L. Maschke's 1858 publication of a preparation of protein crystals. John remarked in the paper that "Perhaps the most remarkable features of the molecule are its complexity and its lack of symmetry. The arrangement seems to be almost totally lacking in the kind of regularities which one instinctively anticipates, and it is more complicated than has been predicted by any theory of protein structure. Though the detailed principles of construction do not yet emerge, we may hope that they will do so at a later stage of the analysis."[5]

At 6 Å resolution the structure of myoglobin was, as John put it, "rather blurred" and could "not be recommended on aesthetic grounds." Max commented on the structure by saying that "The first protein molecule beheld by man revealed itself as a long, winding visceral-looking object with a lump like a squashed orange attached to it [the heme group that delivers oxygen to tissues]."[6] On the other hand, upon seeing the 6 Å model of myoglobin with its irregular structure, J.D. Bernal confidently announced that "he always knew a protein would look like this."

It was impossible to deduce the folding of polypeptide chain segments, its secondary structure, or the nature of amino acid side chains at this resolution. On the other hand, it was possible to identify a single myoglobin molecule with dimensions 43 Å x 35 Å x 23 Å, follow the twists and turns of its polypeptide, visualized as long dense rods of high electron density (approximately 5 Å in diameter and 20-40 Å in length, possibly representing alpha-helices), and locate the iron atom present on the heme group to which oxygen binds (fig. 8.1).

There was an overwhelming worldwide response to John's publication of the myoglobin structure. Everyone congratulated John on his magnificent work, but some mentioned that they found the myoglobin structure a horrible object. Shortly after visiting Bragg in London to show him the myoglobin structure, in August 1957 John received a letter from him. Bragg expressed his excitement about the structure by saying

> I feel I must write to say what a thrill it was to see your structure yesterday. It was not only that it represents the realization of so many years of work and optimism that in the end it would be crowned with success, often in the face of real discouragement. It was also so good to see what X-ray analysis can now achieve. It is almost nearly 45 years since I published my first paper on X-ray structure in the Proceedings of the Cambridge Philosophical Society! It has been wonderful to see the subject expand and now it is in its power and its effect on so many branches of science. I did deeply appreciate your coming over so quickly to show it to me.[7]

Following publication of the 6 Å three-dimensional structure of myoglobin John received a great many invitations to give lectures on his research in the UK and abroad. For example, in 1958 John lectured in Austria at the Congress of Biochemistry, in the United States at Brooklyn Polytechnic, California Institute of Technology, Harvard University, National Institutes of Health, Rockefeller Institute, University of Colorado, and Yale University, and in the UK at Cambridge, Oxford, Imperial College, and King's College. In addition, John entertained many visitors to the Cavendish who came along to see the model of the first three-dimensional protein structure and discuss the structure with him. All of the attention heaped on John apparently proved to be quite an annoyance for Max.

2 Å Myoglobin Structure

It was decided that the next stage of the X-ray analysis of sperm whale myoglobin should be at 2 Å resolution since it would permit determination of the

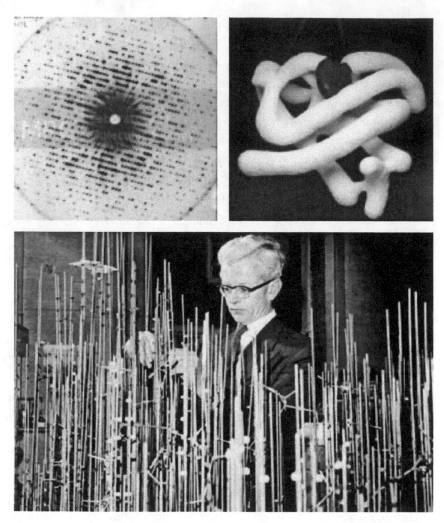

Fig. 8.1 Photos of an X-ray diffraction photo of myoglobin (top left), of a low resolution sausage-style model of myoglobin (top right), and of John Kendrew building a high-resolution model of myoglobin on a forest of metal rods (bottom). Reproduced with permission of the MRC LMB, Cambridge

spatial arrangement of most, but not all of its atoms. To see all of myoglobin's atoms it would be necessary to work at a resolution higher than 1.5 Å, the length of a carbon-carbon covalent bond.

Three collaborators joined John at the LMB to carry out the high-resolution analysis. Richard Dickerson arrived from the United States in fall 1957, Bror Strandberg from Sweden in summer 1958, and David Davies from the United States in spring 1959. In addition, Roger Hart, a newly arrived member of Max's hemoglobin group, assisted in the analysis and David Phillips and Violet Shore collected data for one heavy-atom myoglobin derivative at the Royal Institution in London. A summary of John's collaborators and research students in the 1950s and 1960s is presented in Table 8.1.

Data for the high-resolution structure were processed on EDSAC-II, the successor in 1958 to EDSAC-I that had been used to process data for the low-resolution structure. To process a full set of 10,000 reflections for the 2 Å structure required 10 to 12 hours of computer time on EDSAC-II, as compared with about one hour for a set of 400 reflections for the 6 Å structure on EDSAC-I. Richard Dickerson described access to EDSAC-II:

> There were three grades of access to EDSAC-II: users, partially authorized users, and fully authorized users. Users could only have normal access to the computer during the day while it was manned by computer center staff. Partially authorized users were allowed to run the computer as late as they

Table 8.1 John Kendrew's collaborators and research students

1950s	1960s
Bennett, J.M. (UK)	Banaszak, L.J. (USA)
Bluhm, M.M. (UK)	Brändén, C.-I. (Sweden)
Bodo, G. (Austria)	*Bretscher, P.A. (UK)
Davies, D.R. (UK)	Coulter, C.L (USA)
Dickerson, R.E. (USA)	Creighton, T.E. (USA)
Dintzis, H.M. (USA)	Edmundson, A.B. (USA)
Hart, R.G. (USA)	Kretsinger, R.H. (USA)
*Huxley, H.E. (UK)	Nobbs, C.L. (New Zealand)
Kraut, J. (USA)	Schoenborn, B.P. (Switzerland)
Parrish, R.G. (USA)	Stryer, L. (USA)
*Pauling, P.J. (USA)	Wassarman, P.M. (USA)
Phillips, D.C. (UK)	Watson, H.C. (UK)
Shore, V.C. (UK)	
Strandberg, B.E. (Sweden)	
Watson, J.D. (USA)	
Wyckoff, H.W. (USA)	

Table 8.2 Three stages of myoglobin[a] structure determination

Stage	Resolution	Reflections	Computer	Resolved
I	6 Å (completed 1957)[e]	400[b]	EDSAC-I	Polypeptide chain
II	2 Å (completed 1959)[f]	10,000[c]	EDSAC-II Mercury	Groups of atoms (2/3 aa identified)
III	1.4 Å (begun 1960)[g]	25,000[d]	IBM 7090	Individual atoms (150 of 153 aa identified)

[a] Sperm whale myoglobin consists of 153 amino acids. The chemical formula for myoglobin is $Carbon_{744}Nitrogen_{210}Oxygen_{222}Sulfur_{5}Iron_{1}Hydrogen_{1224}$.

[b] Native crystals and 5 heavy atom derivatives. Reflections per data set.

[c] Native crystals and 4 heavy atom derivatives. Reflection per data set.

[d] High-resolution refinement program.

[e] Kendrew, et al., A 3-dimensional model of the myoglobin molecule obtained by X-ray analysis. *Nature* 181, 662, 1958. First publication of a low-resolution, 3-dimensional structure of a protein.

[f] Kendrew, et al., Structure of myoglobin: 3-dimensional Fourier synthesis at 2 Å resolution. *Nature* 185, 422, 1960. *First publication of a high-resolution, 3-dimensional structure of a protein.*

[g] The 1.4 Å structure of myoglobin was never formally published in a scientific journal. However, the structural coordinates were made available to many investigators worldwide, primarily by H.C. Watson. An abstract, "Progress with the 1.4-Å resolution myoglobin-structure determination," by H.C. Watson, J.C. Kendrew, C.L. Coulter, and C.-I. Brändén, appeared in a supplement to *Acta Crystallographica* 16, in 1963.

chose and to turn it off when finished by pulling a set of wall switches in a defined order. Fully authorized users were permitted to turn the computer on and off. I eventually qualified as a partially authorized user. I do not know of any member of the protein crystal structure group ever rose to the rank of fully authorized user.[8]

Initially it was difficult to identify a majority of the 153 amino acid side-chains in the 2 Å structure of myoglobin. In summer 1960 John wrote to Dickerson saying that

The position about side chains is still rather doubtful. It is true that many of them can be identified quickly and definitely, but there are many others which are much more difficult and for these one gets simply a probable identification. We are now going round the whole molecule in detail and while doing so we try to avoid looking at the Edmundson sequencing data. When we have been right round our plan is to see whether we can in fact correlate our findings with his peptides.[9]

However, by summer 1962 the identities of nearly two-thirds of myoglobin's 153 amino acids had been assigned. The amino acid assignments were extended and strengthened considerably by Allen Edmundson (b. 1932), a postdoctoral fellow with John, who chemically sequenced a large number of myoglobin peptides. Edmundson had initiated his sequencing work on myoglobin while a graduate student with C.H.W. Hirs (1923–2000), Stanford Moore (1913–1982; Nobel/Chemistry, 1972), and William Stein (1911–1980; Nobel/Chemistry, 1972) at the Rockefeller Institute and then with Hirs at the Brookhaven National Laboratory. He began the sequencing of sperm whale myoglobin with 100 milligrams of purified protein delivered to him in New York by Francis Crick.

It was clear that more than three-quarters of the myoglobin molecule consisted of right-handed alpha-helices, designated as helices A–H, consisting of as few as 7 (helix C) to as many as 24 (helix H) amino acids. The eight alpha-helices were joined by short non-helical regions of polypeptide. In this context, Dickerson recalled that when Lawrence Bragg, who had promoted and supported structural biology at the Cavendish for more than 25 years, saw the helices of myoglobin at 2 Å resolution he exclaimed, "Look! It's hollow! It's hollow!"

The high-resolution structure of myoglobin was published by John in *Nature* in February 1960. It was accompanied by a paper from Max describing the low-resolution structure of hemoglobin that, in effect, approximated 4 interacting myoglobin molecules. The 2 Å myoglobin structure revealed a number of important features that were described by John in 1986 when he was asked by the editor of *Current Contents* to comment on the 1960 *Nature* paper that in intervening years had become a "citation classic":

I began this research in 1946 because it seemed to me that the structure of proteins was the most important unsolved problem in molecular biology (this was before the role of DNA was generally recognized). Its culmination was one of the two most exciting moments in my scientific career.

The first had been the earlier determination of myoglobin structure at 6 Å resolution, when after working on the computer for many hours, a colleague and I finally plotted the results (by hand), at about 2 a.m. on a Sunday morning in 1957, we became the first ever to see the actual shape and conformation of a protein. [Kendrew, J.C. et al., *Nature* 181, 662–666, 1958].

The second, described in the present paper, was the determination of the structure at a resolution of 2 Å, which not only demonstrated that more than half the polypeptide chain had the alpha-helix conformation first described by Pauling and Corey and here directly revealed for the first time to me, but also showed the position of the heme group, of the water molecule attached to the iron atom (in the met-form) [heme group with iron in the ferric, Fe^{+3}, not ferrous, Fe^{+2}, state], and of the histidyl residue connecting it to the protein. Furthermore, many of the side-chains could be identified directly from a crude model constructed of steel rods and skeletal atoms, and this made it possible later to establish most of the amino acid sequence, which could be correlated with that determined using the chemical techniques of Edmundson and Hirs. The revelation for the first time that these details of the structure of a protein doubtless explains why our paper has been very frequently cited. [Kendrew, J.C. et al., *Nature* 185, 422–427, 1960].

The principal obstacles encountered in the work were, first, the lack (in the early stages) of sources of X-rays powerful enough to record the weak diffraction pattern of a protein crystal, remedied later by A. Broad's rotating-anode X-ray tube; second, the absence at the beginning of any technique for determining the phases of the X-ray reflexions, an obstacle removed by my colleague Perutz when he showed (in his work with the related protein hemoglobin) that the isomorphous replacement method originally proposed by Bernal before the war could be applied to protein crystals; and the third, the inadequacy of computing techniques then available. Indeed, it was only the timely availability of EDSAC-I, and a year or 2 later of EDSAC-II, that made the work feasible as early as the late 1950s.

The work was carried out at the Cavendish Laboratory in Cambridge, and its success absolutely depended on my many collaborators; 11 of them are listed as authors of the papers cited here, and there were a number of others. It resulted in the award to me of the Nobel Prize in Chemistry for 1962 with my colleague and friend Max Perutz. Since the publication of the structure of myoglobin, that of many other proteins, by now numbered in the hundreds, has been reported; a comprehensive review has been published by Richardson.[10]

The A. Broad that John referred to was D.A.G. (Tony) Broad (1922–2015), the engineer hired by Bragg to build X-ray tubes, together with instrument mechanic L.G. (Len) Hayward, for crystallographers at the Cavendish. In an interview in 1973 John described Broad's extraordinary accomplishment, the design of rotating anode X-ray tubes:

one of the problems about taking X-ray photographs of protein crystals is that they diffract weakly, and with conventional X-ray tubes the exposure times are impossibly long. Back in the fifties we hired an extremely good engineer [Tony Broad] who designed a rotating anode X-ray tube. In an ordinary X-ray tube the electrons [from a cathode] hit a target [tungsten anode] and X-rays come off, the target heats up and that limits what you can do, even if you water cool the target. In a rotating anode tube you have the target spinning [3,000 to 9,000 revolutions per minute] so that the bit that's been overheated is moved away and the next bit is hit by the X-rays. By spinning, you can increase the power of the tube by a factor of 20 or 30. Well, at the time we got into this, rotating anode tubes were not commercially available. It was quite a major development because there were technical problems about the high vacuum, the need to spin the anode at high speed, drawn through a seal, and so on and so on, and it took 5 or 6 years to develop. For us it was absolutely essential for progress to have that tube, it wasn't available commercially. Our design was taken up by industry afterwards and now is commercially sold. But the point was that we had to do it ourselves otherwise we would have been delayed perhaps 5 years.[11]

Between 1959 and 1969 John received several hundred requests for photographs of the sperm whale myoglobin structure that would subsequently be reproduced in innumerable articles and textbooks and used in teaching worldwide. As a result of his extraordinary accomplishments John was elected a fellow of the Royal Society in 1960, awarded a share of the Nobel Prize in Chemistry in 1962, made Commander of the Order of the British Empire in 1963, and received the Royal Medal from the Royal Society in 1965.

1.4 Å Myoglobin Structure

The third stage of John's analysis of sperm whale myoglobin was at 1.4 Å resolution in an attempt to identify all atoms in its polypeptide chain. Having solved the low-resolution structure of myoglobin, at the time John looked to the future and concluded that he was "Fortunate that we can expect further improvements in techniques, otherwise on the basis that I took 1 year to determine the phases of the first 400 reflections it will take 50 years before I can present to you the same structure at a resolution of 1.4 Å."[12] Such was not the case, and in 1963 John described in the journal *Science* some conclusions reached in the previous two years about the 1.4 Å myoglobin structure (see appendix, A.6). He pointed out that "this extension of the analysis was made possible by the availability of automatic data-collecting equipment with proportional counters, and of still larger computers, such as the IBM 7090."[13] Although the 1.4 Å myoglobin structure was never formally published in a scientific journal, a significant omission by John that was criticized by Max in a letter to John Edsall at Harvard in 1965, the structural coordinates were made available to many investigators worldwide primarily by Herman Watson. John's interests had veered away from research and he became intensely involved in other activities, although he continued to host several postdoctoral fellows at the LMB between 1960 and 1967 and was involved in refining the myoglobin structure. A summary of the three stages of myoglobin structure determination by John and colleagues is presented in Table 8.2.

Myoglobin Models

Models are used by crystallographers to visualize the spatial relationships of atoms in molecules. With the 2.0 Å or high-resolution structure of myoglobin solved in the late 1950s, John needed to address the issue of how to construct a suitable model of the protein in the LMB workshop. Building the 2 Å structure of myoglobin was not an easy task due to the size and complexity of the protein's polypeptide chain, consisting of 1,260 atoms, excluding hydrogen atoms. Despite the difficulty, and inspired by Meccano toy construction sets produced in Liverpool, John and members of the workshop constructed a model made up of a forest of about 2,500 steel rods, each about 6 feet tall, inserted into a wooden platform. To this vertical grid of steel rods, colored clips were attached representing electron density; the brighter the color the higher the electron density. Short metal rods were used to represent covalent bonds. Using a plumb-line, John determined atomic coordinates from the model which aided considerably in refining the myoglobin structure.

Richard Dickerson heard about John's model building efforts and wrote to him in October 1959 commenting on the size of the myoglobin model. He had heard from Bror Strandberg that John had three miles of steel rods (6 feet tall x 2,500 rods equals 15,000 feet; only slightly less than 3 miles which equals 15,840 feet) and was building a myoglobin model that "you could walk through." John responded saying that "You were quite right in thinking that we are building models on a large scale—actually 5 cm to one Å corresponding with the scale of our standard atomic models."[14] This scale was used to permit relatively easy access to the steel rods so colored clips and other pieces of the model could be attached.

With the virtual completion of the 1.4 Å structure, in 1965 John approached A.A. Barker, the owner-operator of a small model-making business. Barker was making ball-and-stick type models of DNA, vitamin B12, insulin, alpha-helices, and at least a dozen other molecules at a scale of 2.5 cm to 1 Å. John, on the other hand, planned to use small perspex balls (polymethyl methacrylate) 6.9 mm in diameter, provided by Professor B.A. Beevers at the University of Edinburgh, to build myoglobin models on a scale of 1 cm to 1 Å. They started to take orders for myoglobin models in May 1966 with the price set at £210 (about $600). The models were produced at the rate of one per month and were supplied to customers by John's colleague, Herman Watson. Nearly 30 orders for the myoglobin model were received over a two-year period, 1966–1968.

Any discussion of myoglobin models must include mention of the extraordinary drawings/paintings by the molecular artist Irving Geis (1908–1997). In 1961 John published an article in *Scientific American*, titled "The 3-dimensional Structure of a Protein Molecule," illustrated in part by Geis. At a time before computer graphics, Geis's colored portrait of myoglobin is considered an icon of scientific illustration.

Other Activities

In November 1960 John gave a lecture in the Division of Biology at the California Institute of Technology in Pasadena. Shortly after his lecture John received a very flattering letter from George Beadle (1903–1989; Nobel/ Physiology or Medicine, 1958): "Thanks again for coming out and doing such a fine job. The favorable comments on your lecture were many and enthusiastic. One: 'His work is so magnificent it would have been a good lecture even if presented badly—and it was not presented badly—quite the contrary, it was a fine presentation.' Our best to you and all your good colleagues."[15]

A few months later, in March 1961, John gave a series of 12 lectures in a biophysics course organized by Arthur Solomon (1912–2002), professor of Biophysics at Harvard Medical School in Boston. His first lecture was titled

"The New Approach to Biology" and John began by saying, "So I and many others like me, though not trained as biologists, decided it would be enormous fun to spend our time doing research in biology, using as tools the trades we had learnt—the method of physics and chemistry."[16] In his twelfth and final lecture on "The Future of Biology," John concluded that

> We are indeed only at the beginning. Even though the rate of scientific progress is faster than ever before in the history of man, it will take years to understand some of these things—and when we do it probably won't be called molecular biology . . . What will the next revolution [in science] be? That I can't tell you—scientists can't predict the future any better than anyone else, even about their own field—meanwhile our present revolution is enormously exciting and there is plenty of life in it still.[17]

During the early to mid-1960s John was offered various leadership positions at several institutions in England and the United States. Among these were the following: Birkbeck College (chair of crystallography), Florida State University (professor of biophysics), Harvard Medical School (professor of biological chemistry), Imperial College of Science and Technology, Oxford University (professor of biophysics), University of California at Los Angeles (professor of biophysics or director of the Institute of Molecular Biology), University of California at San Francisco (professor of biophysics), University of Michigan (director of the Biophysics Division), University of Newcastle upon Tyne (vice chancellor), University of Sheffield (chair of physics), Yale University (chair of molecular biology), Glaxo Labs Ltd., and UNESCO, among others.

In July 1965 at a Gordon Conference on proteins John quipped that "You may say things have gone fast—but from now on they will go at a gallop. Many protein structures will undoubtedly be solved in the next few years even if the time has not quite arrived when the determination of one will be considered inadequate as a qualification for a PhD."[18]

9

Journal of Molecular Biology (1957–1987)

[T]he subject and, for that matter, the Journal, were doomed to success from the start.

Sydney Brenner

Academic Press

In 1942 Academic Press was founded by Walter Johnson (1908–1996; b. Jolowicz) and his brother-in-law, Kurt Jacoby (1892–1968), in New York City. Walter's father, Leo Jolowicz (1867–1941), and Jacoby had been publishers of scientific books and journals in Leipzig, Germany. In the late 1930s Johnson and Jacoby fled to America when the Nazi government seized their publishing business, *Akademische Verlagsgesellschaft*. Both men spent a short time in a concentration camp in 1938, were released, and traveled through Russia, Japan, Scandinavia, Australia, and Cuba before arriving in New York in the early 1940s. Leo Jolowicz was forced to resign from *Akademische Verlagsgesellschaft* in 1937, but remained in Germany. He was refused an exit visa and died, perhaps by his own hand, in 1941 at the age of 73.

Invitation

In June 1957 John was invited by Kurt Jacoby, vice president of Academic Press, to become editor-in-chief of a new journal to be called the *Journal of Molecular Biophysics* (*JMB*). John would receive £250 annually plus travelling and personal expenses. Prior to contacting John, Jacoby had asked Paul Doty (1920–2011) at Harvard University whether he had any interest in editing a new journal in molecular biophysics, but Doty declined the offer. Apparently Doty suggested to Jacoby that John would make a most suitable editor-in-chief for such a journal.

John sought advice about the proposed journal from four eminent scientists, Paul Doty, John Edsall, and Jim Watson at Harvard and Linus Pauling at the California Institute of Technology. Doty responded immediately to John's inquiry about the proposed journal and stressed that such a journal was needed and it should avoid publishing results of research on intermediary metabolism and the like. In his response to Doty, John agreed that the title *Journal of Molecular Biology* was a better title for the journal, and he had thought of the same title himself. John also sought the advice of Max Perutz who was against the proposed journal at first but subsequently decided it would serve a useful purpose.

John was certainly interested in becoming editor-in-chief of the *JMB*. However, he felt it necessary and appropriate to receive permission from Harold Himsworth, secretary of the MRC, before considering the position further. In October 1957 John wrote to Himsworth about Jacoby's offer, saying,

> I am being pressed to accept the Editorship of this journal . . . I am not anxious to edit a journal for its own sake, and as you know I have only too much on my plate just now in any case, both of research and administration . . . The case for doing it, as put by Bragg for example, is that in England we are ahead in this field and ought therefore to take a leading part in publishing it. I should of course prefer that the journal should emanate from a British learned society, rather than from an American publisher; but the fact is that the latter is determined to start it, and if I don't take on the job it is likely to revert to an American."[1]

In view of John's significant commitments to research and administration at the Cavendish and Peterhouse, Himsworth recommended that John turn down the editorship of *JMB*, but assured him that he would be a better editor than anyone else likely to be chosen by the Americans.

Shortly thereafter three issues were resolved that made accepting the editorship of *JMB* more appealing to John. Peterhouse would release him from his extensive duties as steward of the college, Academic Press would provide him with a part-time secretary, and he would be allowed to appoint an assistant editor. John contacted Himsworth once again saying that "Assuming that some or all of these proposals were put into effect I should like to do the job, though of course I shall bow to your opinion if you still feel that I should not."[2] This time Himsworth was completely supportive of John's wish to become editor-in-chief of *JMB* and explained that his prior recommendation to refuse the editorship was to provide John with the means of refusing if he felt doubtful about it and was being importuned.

In December 1958 Academic Press issued a partially revised contract to John after he had sought legal advice from Fairchild, Greig, and Company about the initial contract. He signed the contract and became the editor-in-chief of *JMB* on December 26, 1958. In August 1959, with the structure of myoglobin solved at 2 Å resolution, John took on the task of getting *JMB* under control.

Editorial and Advisory Boards

JMB was the first journal that the London branch of Academic Press published after it was established in 1959. To begin, the company intended to publish six issues of the journal annually, totaling about 600 pages, and to address the nature, production, and replication of biological structure at the molecular level and its relation to function. By spring 1959 John had assembled a distinguished six-member Anglo-American editorial board for *JMB* consisting of Paul Doty at Harvard, Andrew Huxley at Cambridge, Robert Sinsheimer at the California Institute of Technology, Jim Watson at Harvard, and Maurice Wilkins at King's College, London. The choice of board members was left completely up to John by Academic Press. At about the same time, an international advisory board comprised of 20 distinguished scientists was also assembled (table 9.1).

Table 9.1 *JMB* advisory board members in 1959

Seymour Benzer	Peter Medawar
Melvin Calvin	Leslie Orgel
Francis Crick	Max Perutz
James Danielli	Reginald Preston
Alfred Hershey	G.N. Ramachadran
Francois Jacob	Alexander Rich
Arthur Kornberg	Gerhard Schramm
Cyrus Leventhal	Frederick Sjöstrand
Salvador Luria	Itaru Watanabe
Daniel Mazia	Robley Williams

Sydney Brenner was appointed assistant editor in 1961, John and Sydney became joint editors-in-chief in 1985, and in 1987, by which time John was 70 years old and had served the journal for nearly 30 years, Sydney took over as sole editor-in-chief. In 1990 Peter Wright at the Scripps Research Institute in California took over as editor-in-chief of *JMB* and continues in that role today. A list of 15 landmark *JMB* publications that appeared between 1959 and 1980 is provided in Table 9.2.

Nascent Journal

From 1958 through 1959 approximately 80 manuscripts were submitted to *JMB* for publication, and its first issue was published in April 1959, only 4 months after John became editor-in-chief. Volume 1 consisted of five issues published in April, June, September, and November/December, priced at £5. The first issue was a rather thin volume of less than 90 pages bound in an unpretentious cover with articles from Melvin Calvin, Paul Doty, and Robert Sinsheimer. The subsequent four issues that year included articles from Sydney Brenner, Norman Davidson, Paul Doty, Alfred Hershey, Niels Jerne, Jacques Monod, Robert Sinsheimer, Gunther Stent, Alfred Tissierès, and Jim Watson. This was a truly outstanding group of scientists, including five future Nobel laureates, publishing some of their best research in a fledgling journal focusing on molecular biology, still an emerging area of research.

For more than a decade the technical editing of manuscripts accepted for publication in *JMB* was carried out in John's rented rooms in Peterhouse. The editorial office moved out of Peterhouse to All Saints' Passage, Cambridge, when John left for Heidelberg as DG of EMBL. Everything was done by hand on index cards and in ledgers. From the very beginning John was intimately involved in every aspect of *JMB* production, from the choice of cover design and color, paper, and typeface, size and layout of the pages, form of references, handling of proofs, to establishment of submission guidelines for authors. John also corresponded extensively with members of the editorial board about the quality and appropriateness of papers for the journal. In 1959 John hired Leslie Barnett (1920–2002), a microbiologist who worked with Francis Crick and Sydney Brenner, as technical editor for the journal. Leslie took on major responsibilities for establishing *JMB*'s style and many authors came to dread the green pen she used to point out so-called editorial sins.

Table 9.2 Fifteen landmark *JMB* publications (1959–1980)*

Pardee, A.B., et al. (1959)
The genetic control and cytoplasmic
expression of "Inducibility" in the
synthesis of β-galactosidase by *E. coli*.
JMB 1, 165–178.

Brenner, S., et al. (1961)
The theory of mutagenesis.
JMB 3, 121–124.

Jacob, F., Monod, J. (1961)
Genetic regulatory mechanisms in the
synthesis of proteins.
JMB 3, 318–356.

Cairns, J. (1961)
An estimate of the length of the DNA
molecule of T2 bacteriophage by
autoradiography.
JMB 3, 756–761.

Marcker, K., and Sanger, F. (1964)
N-Formyl-methionyl-S-RNA.
JMB 8, 835–840.

Crick, F.H.C. (1966)
Codon-anticodon pairing: the wobble
hypothesis.
JMB 19, 548–555.

Klug, A., and Finch, J.T. (1968)
Structure of viruses of the papilloma-
polyoma type IV. Analysis of tilting
experiments in the electron
microscope.
JMB 31, 1–12.

Abelson, J.N., et al. (1970)
Mutant tyrosine transfer ribonucleic
acids.
JMB 47, 15–28.

Mandel, M., and Higa, A. (1970)
Calcium-dependent bacteriophage DNA
infection.
JMB 53, 159–162.

Flavell, R.A., et al. (1974)
Site-directed mutagenesis: generation of an
extra- cistronic mutation in bacteriophage
Qβ RNA.
JMB 89, 255–272.

Sanger, F., and Coulson, A.R. (1975)
A rapid method for determining sequences
in DNA by primed synthesis with DNA
polymerase.
JMB 94, 441–448.

Southern, E.M. (1975)
Detection of specific sequences among DNA
fragments separated by gel electrophoresis.
JMB 98, 503–517.

Bernstein, F.C., et al. (1977)
The protein data bank: a computer-based
archival file for macromolecular structures.
JMB 112, 535–542.

Rigby, P.W.J., et al. (1977)
Labeling deoxyribonucleic acid to high
specific activity in vitro by nick translation
with DNA polymerase I.
JMB 113, 237–251.

Sanger, F., et al. (1980)
Cloning in single-stranded bacteriophage as
an aid to rapid DNA sequencing.
JMB 143, 161–178.

*Adapted from figure 5 of T. Picknett, Journal of Molecular Biology: A Publishers Perspective, *JMB* 293, 165–171 (1999).

Initially John became very pessimistic about the prospects for *JMB* and not only because of the time consuming and often tedious work: from early October until the end of November 1959 no manuscripts were submitted to *JMB* and numerous unanticipated printing problems were encountered. In this context John wrote to Jacoby in December 1959:

> You may remember that before the decision to publish the journal was taken you asked Dr. Wilkins to make a survey of the literature and to give you an estimate of the amount of material we might expect. His opinion was that the amount of material available was marginal, but that during the next year or two it was probable the field would expand enough to fill a journal of this size. As you will see his prognostications were very accurate. I have no real doubt that the flow of papers will increase and that we shall be able to achieve our target eventually. On the other hand I am bound to say that I am somewhat unhappy for the immediate future, and we think it only fair to warn you that it may not be possible to achieve our target in 1961. All members of the editorial board are aware of the position, and we unanimously feel that it is more important to maintain the standards of the journal at this early stage in its history than to try to pad it out with second rate papers. I believe that in fact the standard of the paper which has so far appeared in the journal is very much higher than the average of any comparable journal. We shall certainly do our best to achieve 600 pages in 1960 but I hope you will agree that in the long run high standards are more important than bulk.[3]

Jacoby responded promptly and in a very supportive fashion, pointing out that the subscription list for *JMB* had been growing steadily and the latest issue had been mailed to more than 800 subscribers.

Right from the beginning John had many headaches as editor-in-chief of *JMB*. Since new issues of the journal were sent to the United States from Britain by regular post, not airmail, they took weeks, or even months, to arrive. Paul Doty complained bitterly about the June 1959 issue of *JMB* not arriving until November and concluded that this failing had turned what might have been a labor of love into a dismal chore. Because of the long delays Jim Watson threatened to send his manuscripts elsewhere. As a result of many complaints, Jacoby flew to Boston and visited Doty and Watson at Harvard to humble himself before them. Despite Jacoby's visit Watson continued his rantings, berating the atrocious binding, horrid layouts, and slow speed of publication of *JMB*, and Hugh Huxley complained bitterly about the poor reproduction of micrographs.

JMB became an administrative nightmare for John due in no small part to the outstanding editorial board: each member had very strong opinions and was unafraid to express them. A great deal of correspondence about problems associated with *JMB* went back and forth between John, Jacoby, and Charles Hutt, a managing director of Academic Press London. John was spending an inordinate amount of time handling *JMB* affairs during this period and worried constantly that the major names in molecular biology would send their papers elsewhere for more rapid publication.

After a rather tumultuous start, prospects for *JMB* improved significantly between 1960 and 1962. The number of manuscripts submitted doubled from about 100 in 1960, to 200 in 1962 and increased to 250 in 1963, and Academic Press was very pleased with the number of subscribers. Things were definitely looking up for the journal. The number of submissions to *JMB* continued to increase significantly in subsequent years, reaching highs of 550 in 1965 and nearly 700 in 1968. Naturally, the considerable increase in submissions and acceptances meant that John had to devote more and more time to editing the journal. The field of molecular biology had taken root, and *JMB* was definitely the place to publish exciting papers.

On the occasion of Kurt Jacoby's 70th birthday in December 1962, John wrote to him on behalf of himself and the editorial board of *JMB*:

> I have the greatest pleasure in sending you felicitations on your birthday, not only on my own behalf but also in the name of the Editors of the Journal of Molecular Biology. I would add my congratulations—not only on your having attained your seventieth birthday with such vigor that it is hard to believe there is not some mistake in the reckoning, but also on your part in funding and guiding the fortunes of the best scientific publishing house in the country. You could not have made a greater contribution to the progress of science if you had spent your entire life in a laboratory; *si monumentum requiris, circumspice* [if you seek his monument, look around]. I am indeed happy to have had the opportunity of so many connections with you and Academic Press, and I look forward to a long period of future associations, as pleasant and successful as those in the past. With warm personal regards, and many happy returns.[4]

By 1964 *JMB* was doubling in size annually, and John was looking for a way to deal with an ever increasing number of submissions. Between 1958 and 1965 the number of pages of *JMB* increased nearly 10-fold, from about 400 to about 3,500 pages per year. In 1965 John proposed to the *JMB* Board much tighter refereeing and an increase in the number of rejected submissions without becoming arbitrary and unfair. Typical of responses John received back was one from Seymour Benzer who felt that "the standard maintained by the *Journal of Molecular Biology* has been quite good, and that tighter refereeing would not stem the tide of contributions from this field, which is simply growing . . . Frankly, I feel that molecular biology has gotten out of hand altogether. What is a shame is the amount of duplication of activity, and that is something that the Journal can hardly control."[5] This was followed by Jim Watson's latest complaint to John about having too many manuscripts to review for the journal and requesting that absolutely no more manuscripts be forwarded to him from England until October 10th. He expressed concern that the *Journal of Molecular Biology* was almost out of control and would soon have to be published twice a month.

Early in 1966 John pointed out once again to the advisory board that *JMB* was growing very rapidly such that some sort of action needed to be taken: "It is clear that something has to be done, if only because our monthly issues are now becoming so fat that the binding will soon collapse under the strain. Also Max Perutz tells me that he likes to read *J. Mol. Biol.* in the train going to London, but that the issues are now too heavy to carry. It seems to me that several courses are open to us . . . I should be very grateful for your opinions of these as well as any other ideas you may have . . . ".[6] After receiving feedback from the editorial board John concluded that

> I am bound to say, having read them [suggestions] and having talked to many people, I have come to the conclusion that at present there is no radical solution to our problems which would really be workable. I think we have to accept the fact that the Journal will continue to grow, and take what steps we can to control growth, to maintain standards, to secure uniformity of treatment as between Editors, to make the Journal as convenient as possible to read, and to follow the extending field of molecular biology.[7]

John joined the board of directors of Academic Press, London in 1968, the same year that Kurt Jacoby died at the age of 75.

Established Journal

By the mid-1970s *JMB* was one of several journals publishing papers in molecular biology. Competition among journals for the most exciting papers in the field was getting quite fierce, and scientists were becoming more and more concerned about the speed of refereeing and publication. Molecular biology, like biochemistry, had become an experimental approach applied to virtually all areas of biological research. The novelty of *JMB* in the 1960s had worn rather thin by the 1970s, and its stature declined as both old and new journals competed successfully for submissions.

In 1976 John became concerned about his tenure as editor-in-chief of *JMB* and wrote to a director of Academic Press:

> About my agreement. I never felt the need of one while you were around, nor indeed do I during the reign of your immediate successor who will have your briefing. But personalities change and memories are short—and I would be happier to have an agreement for that reason, if you could find time to do it. I would say that the natural duration would be until my 65th birthday, namely 24 March 1982, and it would be even better if it could be a little open-ended giving a possibility of further extension if by that time I am active and non-senile: in retirement I might well be able to give it more time than I have been doing in the last year or two. But you may have rules that would make this difficult.[8]

The director responded to John's concerns, saying that John should continue as editor-in-chief until his 65th birthday in March 1982 and thereafter would continue on a year-to-year basis. During his time at the EMBL in Heidelberg, 1975 to 1982, John returned to the UK about 3 times a month, in part to take care of *JMB* matters.

During the mid to late 1970s the number of manuscripts submitted to *JMB* declined, perhaps in large part because of the slowness of publication. Whereas between 1968 and 1975 nearly 700 manuscripts were submitted to *JMB* each year, beginning in 1976 there was a steady decline in manuscript submissions, falling below 500 in 1980. That year a director of Academic Press addressed the problem, saying that "JMB had become an institution, but an institution potentially subject to decay." Academic Press continued to look into ways to alter *JMB* and thereby prevent its decay.

At about the same time John addressed some of the issues affecting *JMB* in an announcement from the editors:

In 1980 the *Journal of Molecular Biology* is twenty-one years old. For journals as for human beings, coming of age is a good time to take stock and to make plans for the future. Many changes have taken place since the first issue appeared in April 1959. The field called molecular biology has ramified through so many areas of biological research that some have questioned whether it still exists as a separate discipline. We believe that it does, though even if we use the conventional criterion of "studies at the molecular level" there is an element of arbitrariness in laying out its boundaries. We believe that the Journal has in the past played a significant part in establishing and defining those boundaries, and we are confident that by extending and modifying its coverage in the future it will continue to do so. Scientific publication has changed too, with the proliferation of "quickies" which whatever benefits they may have brought, have never been imitated by the *Journal of Molecular Biology*: nevertheless they have raised the expectations of rapid publication among authors and readers so that a journal like this one, resolved to maintain high standards by careful refereeing and editing and by paying close attention to typography and quality of reproduction, must nevertheless respond to such expectations as far as it can without sacrificing its standards . . .

The Editorial Board and the Publishers have devoted much thought to these problems, and have consulted many advisers, during recent months. We plan, from the beginning of 1981, a substantial reorganization . . .

We hope that the new policies will find favour with you, and we are confident that the widening of our subject areas will attract important new material as well as new readers. During a vigorous childhood and adolescence the *Journal of Molecular Biology* established for itself a central position in the literature of the subject. This reorganization is not necessarily final, rather it is a continuing process of growth and development. The *Journal* enters the years of maturity with the intention not only of maintaining this position, but of increasing its value to the biological community. With your support and suggestions it will do so.[9]

However, serious criticisms of *JMB* continued into the early 1980s. There were complaints that the journal was strictly a Cambridge product, partial and parochial, its rejections ill-advised, and that it was dull and missing exciting issues in molecular biology.

The year 1983 was a time of turmoil for Academic Press London, a subsidiary of the New York office. It was decided that the London site would continue to publish journals, but would no longer publish books, and there would be major reorganization of the London staff, including the removal of the longtime managing director and redundancy of more than 100 staff to save the company from going under. After a very chaotic year that saw the departure of the editorial and personnel directors, Peter Jovanovich was appointed managing director of Academic Press London. He was the son of William Jovanovich, president and chief executive of Harcourt Brace Jovanovich which had bought both Academic Press New York and London in the late 1960s.

John was now 66 years old and had been editor-in-chief of *JMB* for nearly 25 years. By 1983 a large number of highly significant papers from the foremost laboratories in the world had appeared in *JMB*. However, between 1976 and 1983 *JMB* lost nearly one-third of its subscribers, declining from nearly 3,000 subscribers in 1976 to about 2,000 in 1983; this coincided with a substantial drop in manuscript submissions. Academic Press was extremely concerned and emphasized that they could not just sit back and let the trend continue. Although past his 65th birthday and currently president of St. John's College, Oxford, John was very reluctant to hand over the reins of *JMB* to anyone else. Quite characteristically he wished to remain in charge of what he considered to be "his journal." However, John's long and pleasant relationship with Academic Press began to slowly unravel.

In 1984 John received a letter outlining some of the publisher's concerns about the current and future management of *JMB*. Among the changes to be implemented, John's name would appear on the title page as founding editor, Sydney Benner would be promoted to editor and he would appoint a deputy editor; John's salary and expenses would continue for three years and he would remain a director of Academic Press. After intense discussions between John and the management of *JMB* in spring 1984, John received another letter in which it was acknowledged that the publisher was pleased that he would continue as editor of *JMB* at least until he reached the age of 70 in March 1987, with Brenner joining him immediately as co-editor.

Unpleasant End

In summer 1984 the managing and editorial directors at Academic Press London announced that there would be major changes in the format and publishing frequency of *JMB*. They felt that the appearance of *JMB* had become old fashioned, so the format would change to a larger page size with double columns, and it would be published 24 rather than 30 times a year, to modernize the journal, reverse the decline in subscribers, and reduce production costs. The changes would be made just as Sydney Brenner became co-editor.

John was very unhappy with the proposed changes for *JMB* and felt that "this is much too short notice for major changes of this character to be made, and that whatever the final outcome much more time must be given for discussion—so far there has been virtually none, at least involving the Editor-in-Chief."[10] Particularly galling to John was that a number of discussions about various issues had taken place with Brenner and others before they had been discussed with him. In response the managing director apologized for some of that conduct, especially for discussing possible changes in *JMB* with Brenner without first communicating with John—behavior that John characterized as reprehensible and inconsiderate. On the other hand, it was pointed out to John that the managing director was directly responsible to the company for everything published and they alone were responsible for any changes in the format or appearance of *JMB*. A short time later it was announced that Brenner had been appointed joint editor-in-chief of *JMB* and that a change in both format and frequency of publication of *JMB* would occur from the first issue of 1985. John continued to be openly negative about the planned changes in format and cover design, but despite his complaints the changes were implemented at the beginning of 1985.

By summer 1987 management at Academic Press had concluded that *JMB* needed some new blood and that there was a dire need to bring younger molecular biologists onto the editorial board. Although John expressed a desire to remain in charge of the journal until such a time as a successor had been appointed, management felt that Sydney Brenner was extremely competent to take over as the sole editor-in-chief of *JMB* on September 1, 1987. They planned to groom a new editor-in-chief over a two-year period to replace Brenner no later than 1990.

It was made clear to John that he would not be able to maintain his current office in Cambridge, but that Academic Press would find an alternative office for him with appropriate secretarial support. John felt betrayed and was extremely upset by the terms laid out and responded angrily in summer 1987: "I am replying to your letter of 12 June, and I will not at this time comment on its first two paragraphs (though I may do so later) except to say that in the world I move in it would not be considered reasonable or decent (perhaps even legal?—but I have not yet consulted my lawyer) to give someone the sack with less than 3 months' notice even if he had been employed a much shorter time than 28 years. But perhaps your world has different standards."[11]

Shortly thereafter the managing director responded to John's letter, emphasizing how grateful they were to him for his many years of service to *JMB* and hoping that their relationship could remain cordial. However, John remained adamant and maintained that his use of the words dismissal and sacking was accurate and he would use them when asked by anyone about the circumstances of his dismissal by Academic Press.

In fall 1987, when John was 70 years old, Sydney Brenner replaced him as editor-in-chief of *JMB* (at about the same time John was replaced by William Hayes, director of the Clarendon Laboratory and principal bursar at St. John's, as 34th president of St. John's). A very long and fruitful partnership between John and Academic Press ended on a very sour note. In March 1988 John was also removed as a director of Academic Press and in April of the same year the *JMB* editorial office moved from All Saints Passage into new quarters at St. Edward's Passage, Cambridge. In 2000 the Dutch publisher Reed Elsevier purchased Harcourt Brace Jovanovich and Academic Press became an imprint of Elsevier. Today, 60 years after John founded *JMB* and began as its editor-in-chief, the journal is published weekly and continues to cover original, groundbreaking research on all aspects of contemporary molecular biology.

10
Nobel Prize in Chemistry (1962)

The achievement is so great that one is outside the ordinary range of one's vocabulary.

Harold Himsworth

Alfred Nobel

Alfred Bernhard Nobel (1833–1896), the Swedish chemist, engineer, and inventor, in his third and final will in 1895 left nearly all of his vast fortune, 31 million Swedish Crowns, equivalent at the time to £1.7 million (about $8.3 million), to establish five Nobel Prizes in Physics, Chemistry, Physiology or Medicine, Literature, and Peace. Alfred Nobel's philanthropy has been attributed to a mistaken French newspaper report of his death in 1888; it was his younger brother, Ludwig, who had died. The report characterized Alfred as the merchant of death who had made his fortune by manufacturing dynamite and weapons of war. Troubled by the damaging report and very concerned with his legacy, Nobel decided to use his fortune to fund a prize that recognizes those who during the previous year had conferred the greatest benefit to mankind. Ragnar Sohlman (1870–1948), Alfred Nobel's assistant and an executor of his last will, created the Nobel Foundation and served as its Executive Director from 1929 to 1946. Today the Nobel endowment has grown more than 200-fold to about $1.7 billion.

The following excerpt is from Alfred Nobel's final will in 1895:

The whole of my remaining realizable estate shall be dealt with in the following way: the capital, invested in safe securities by my executors, shall constitute a fund, the interest on which shall be annually distributed in the form of prizes to those who, during the preceding year, shall have conferred the greatest benefit to mankind. The said interest shall be divided into five equal parts, which shall be apportioned as follows: one part to the person who shall have made the most important discovery or invention within the field of physics; one part to the person who shall have made the most important chemical discovery or improvement; one part to the person who shall have made the most important discovery within the domain of physiology or medicine; one part to the person who shall have produced in the field of literature the most outstanding work in an ideal direction; and one part to the person who shall have done the most or the best work for fraternity between nations, for the abolition or reduction of standing armies and for the holding and promotion of peace congresses. The prizes for physics and chemistry shall be awarded by the Swedish Academy of Sciences; that for physiology or medical works by the Karolinska Institute in Stockholm; that for literature by the Academy in Stockholm, and that for champions of peace by a committee of five persons to be elected by the Norwegian Storting. It is my express wish that in awarding the prizes no consideration be given to the nationality of the candidates, but that the most worthy shall receive the prize, whether he be Scandinavian or not.[1]

Nominations

Nomination forms for Nobel Prizes are sent out in September to more than 2,000 people worldwide for submission at the end of January of the following year. A list of 200–500 candidates, many of whom had been nominated in previous years, is assessed by the Nobel Committee in consultation with experts in February and March. The list is shortened, special evaluations are carried out from April to August, and final recommendations are made in late September and submitted to the Royal Swedish Academy of Sciences. In October the Academy selects the Nobel laureates in chemistry from among candidates recommended in September by the five-member-plus Nobel Committee for Chemistry. The names of the laureates are made public in early October. The Nobel lectures and associated festivities take place over six days in Stockholm, December 6–11, with the awards ceremony always held on December 10, the anniversary of Alfred Nobel's death in 1896.

With completion of the high-resolution structure of myglobin in 1959, Bragg decided to nominate John and Max for a Nobel Prize in Physics in 1960 and wrote to Linus Pauling to gain his support. This, however, was not forthcoming as Pauling felt that it was premature to make an award in protein crystallography because the most significant papers had not yet been published. Nevertheless, Bragg went ahead and nominated John, Max, and Dorothy Hodgkin for the 1960 physics prize. But the physics and chemistry prizes in 1960 went instead to Donald Glaser (1926–2013) at the University of California, Berkeley for the invention of the bubble chamber and Willard Libby (1908–1980) at the University of California, Los Angeles for devising a method to use carbon-14 for age determination in archaeology, geology, geophysics, and other branches of science.

In 1960 Bragg once again campaigned for John, Max, and Dorothy Hodgkin. But the physics prize was awarded jointly to Robert Hofstadter (1915–1990) at Stanford University for his studies of electron scattering in atomic nuclei and on the structure of nucleons and Rudolf Mössbauer (1929–2011) at the California Institute of Technology for his research on the resonance absorption of gamma radiation and his discovery of the effect that bears his name. The chemistry prize was awarded to Melvin Calvin (1911–1997) at the University of California, Berkeley, for his research on carbon dioxide assimilation in plants.

By summer 1961 Bragg had come around to the conclusion that he would give first priority to awarding a prize to John and Max jointly and nominated them for the 1962 Nobel Chemistry, not the Physics, prize. In addition to Bragg's support, John and Max received support over the years from several other eminent scientists, including Adolph Butenandt (1903–1995; Nobel/Chemistry, 1939), Cyril Hinshelwood (1897–1967; Nobel/Chemistry, 1956), Jean Roche (1901–1992), Fred Sanger, Harold Scheraga (b. 1921), and Hugo Theorell. Some nominators recommended that they receive the award in physics or physiology or medicine rather than in chemistry.

Announcements

This time around the academy selected John and Max for the Nobel Prize in Chemistry, noting that "John Cowdery Kendrew's and Max Ferdinand Perutz's contributions are of the highest class and they are extraordinarily worthy of receiving a Nobel Prize." They went on to say that their contributions represented the most important results obtained by X-ray diffraction methods since they were introduced by Max von Laue and William and Lawrence Bragg at the beginning of the 20th century.

On Thursday, November 1, 1962 John received a telegram from Erik Rüdberg, Permanent Secretary of the Nobel Foundation, with marvelous news that would significantly affect his future (see fig. 10.1). The telegram informed John that the Royal Swedish Academy of Sciences had decided to award him and Max one-half each of the 1962 Nobel Prize in Chemistry for their research on the structure of globular proteins. John replied immediately to Rüdberg, saying "I thank you for your telegram and would ask you to convey to the Royal Swedish Academy of Sciences my profound sense of the honor it has done me in conferring the greatest distinction it is possible for a scientist to receive."[2]

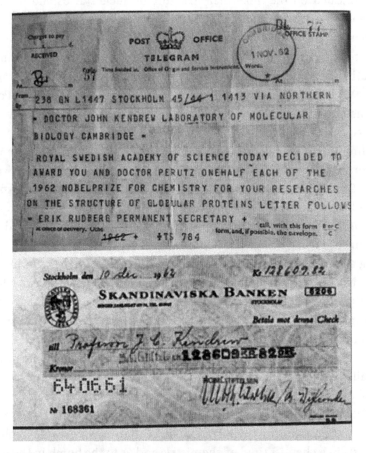

Fig. 10.1 Photo of a telegram sent to John Kendrew by Erik Rüdberg, Permanent Secretary of the Royal Swedish Academy of Science, announcing that John and Max Perutz would share the 1962 Nobel Prize in Chemistry (top). Photo of a check from the Skandinaviska Banken, Stockholm, issued to Professor J.C. Kendrew for 128,609 Swedish Crowns (bottom). Reproduced from items in the Kendrew Archives, Weston Library, Oxford

In 1962 the Nobel Prizes in Chemistry and in Physiology or Medicine would go to five long-time colleagues and friends. John and Max would share the Chemistry prize for their studies of the structures of globular proteins, which included a cash award to each of a little more than 128,000 Swedish Crowns, approximately $25,000 at the time. They were the 41st and 42nd British winners of the Nobel Prize for science. Max would use the cash award to pay off the mortgage on his house and buy a car. Francis Crick, Jim Watson, and Maurice Wilkins would share the prize in Physiology or Medicine for their research on the molecular structure of nucleic acids and its significance for information transfer. The Physics prize would go to Lev Landau, a Russian, for his theories for condensed matter, especially liquid helium and the Literature prize to John Steinbeck, an American, for his realistic and imaginative writings exhibiting sympathetic humor and keen social perception. The 1962 Peace prize was awarded one year later in 1963 to Linus Pauling who had campaigned ceaselessly against all warfare as a means of solving international conflicts.

John, Max, Francis Crick, and Jim Watson all performed their prize-winning research at the Cavendish Laboratory, and Maurice Wilkins worked only 55 miles away at King's College, London. It was completely unexpected that all five friends would be awarded Nobel Prizes in the same year, *annus mirabilis*, the miraculous year, and consequently the awards raised more excitement in Cambridge than ever. Bragg, who was in the hospital undergoing an operation for prostate cancer, was elated by the news of John and Max's Nobel Prize; after his operation the doctor told Bragg's wife that he was over the worst but he might die of excitement. Bragg confessed that news of the awards in medicine and chemistry gave him the deepest pleasure of any event in his scientific life. It is likely that Bragg's satisfaction about the 1962 Nobel Prizes was related in part to the fact that he had finally managed to prevail over Linus Pauling, his long-time US rival.

Following announcement of his Nobel Prize John wrote fondly to J.D. Bernal acknowledging his crucial role in John's career: "I'm always grateful that it was you, during the war, who more than anyone awoke biological interests in a chemist tired of chemistry: and as to the bomb in the jungle, I've been dining out ever since on the story . . . Anyway, after being a renegade chemist all these years I now seem to be a respectable one once more and, as to you, it seems to me you've fathered five Nobel Prizes this year alone."[3]

Ceremonies

The laureates were to receive their Nobel Prizes at a ceremony on Monday, December 10, the anniversary of Alfred Nobel's death at the age of 63 in 1896 in San Remo, Italy. John arrived in Stockholm on Saturday afternoon, December 8, on an Air France flight from London. He and the other laureates were met at the airport and accompanied to all official functions by junior members of the foreign ministry. John was escorted by Hans Andén from the Royal Ministry for Foreign Affairs and stayed with the other prizewinners at the Grand Hotel, Stockholm. During the time leading up to the awards ceremony and banquet the laureates gave public lectures on a topic related to the subject of their prize. Two days of parties and receptions together with the Swedish Scientific Society and people associated with the Nobel Foundation preceded the awards ceremony.

On Saturday a dinner was held in the evening at the home of Erik Rüdberg on the Science Academy's grounds. Max provided some details of the dinner in his diary:

> The dinner, like most of the meals that were to follow, was opened by a solemn speech on the part of the host, in praise of John and me . . . The meals themselves were good and comparatively simple, consisting of no more than 3 courses; even at the Royal Palace there were only 4 . . . The company was mostly older than ourselves; people were remarkably good looking, extremely well bred and spoke immaculate English. They were a little formal, perhaps, but full of a genuine warmth and friendliness that made us feel welcome and happy.[4]

Other events took place on Sunday. These included morning tours of a Venetian paintings exhibition at the National Museum of Fine Arts on Blasieholmen and a visit to the Swedish battleship Wasa built between 1626 and 1628 and sunk in Stockholm soon after. In late afternoon an informal reception sponsored by the Board of Directors of the Nobel Foundation was held at the Library of the Swedish Academy, and in the evening a dinner was hosted by Arne Engström (1920–1996) and his wife in the Naval Officers' Mess Rooms. The dinner was attended by John, Jim Watson, Max and his wife Gisela (1915–2005), and Francis Crick and his wife Odile (1920–2007), as well as by the UK and US ambassadors.

On Monday there was a rehearsal in the morning for all Nobel laureates in the Stockholm Concert Hall, Stockholms Konserthus. The Nobel Prize awards ceremony took place in late afternoon in the Grand Auditorium of the Concert Hall where it had been held since 1926. The 80-year-old Swedish King, Gustaf VI Adolf (1882–1973), accompanied by Queen Louise (1889–1965), a sister of Louis Mountbatten, and their family arrived with a fanfare of trumpets followed by the playing of the Swedish National Anthem "Du Gamla Du Fria," "Thou ancient, Thou free." They sat in the front row just beneath the stage. The laureates emerged through doors at the back of the stage and took their seats as the entire audience stood and the orchestra played. Behind the 1962 Nobel laureates sat four previous Nobel Prize winners: Hans von Euler (1873–1964; Nobel/Chemistry, 1929), Manne Siegbahn (1886–1978; Nobel/Physics, 1924), Hans Krebs (1900–1981; Nobel/Physiology or Medicine, 1953), and Harold Urey (1893–1981; Nobel/Chemistry, 1934).

Each recipient was introduced by a short speech and then descended from the stage to receive their prize from the king, all to a fanfare of trumpets, taking of photographs, and grand applause. The order of recipients was Max, John, Francis Crick, Jim Watson, Maurice Wilkins, and John Steinbeck (fig. 10.2). After an opening speech by the President of the Nobel Foundation, Max and John were presented by Gunnar Hägg of the Royal Swedish Academy of Sciences (appendix, A.9). The king presented each recipient with a 23-carat gold Nobel medal bearing a profile of Alfred Nobel, a diploma decorated with artwork on special handmade paper and mounted in a blue goatskin cover, and a confirmation of the prize money. The recipients gave a slight bow to the king, queen, and audience, and then returned to the stage. The awards ceremony ended with the singing of the Swedish National Anthem. Regrettably, Lev Davidovich Landau (1908–1968), the Physics Laureate, was not present at the Nobel ceremony since he had not recovered from a serious automobile accident that occurred in early 1962 and left him with brain damage. The Nobel medal and diploma were handed to him in Moscow by the Swedish Ambassador.

Following the awards ceremony laureates were driven to the City Hall of Stockholm, Stadthuset, on Kungsholmen in the early evening for the Nobel Banquet. More than 800 guests, including the Swedish royal family and students, were seated in the Great Golden Hall, site of the banquet since 1934. It was a strictly formal affair with gentlemen in white tie and tails and ladies in evening gowns. Prior to entering the hall laureates and their families

Fig. 10.2 Photo of the 1962 Nobel Prize winners at the ceremony in Stockholm. Shown left-to-right are Maurice Wilkins, Max Perutz, Francis Crick, John Steinbeck, Jim Watson, and John Kendrew (ca. 45 yoa). Lev Landau, the Nobel laureate in Physics, was absent. Reproduced from an item in the Kendrew Archives, Weston Library, Oxford

were introduced to the king and his court. John was seated for dinner between the Swedish Queen and 28-year-old Princess Margaretha (b. 1934), the eldest sister of King Carl XVI Gustaf of Sweden and a first cousin of Queen Margrethe II of Denmark (fig. 10.3). The banquet began with a toast by the president of the Nobel Foundation to their majesties, the king and queen, followed by a toast by his majesty the king to the memory of Alfred Nobel.

The laureates and royal family sat at a long, slightly elevated table running down the center of the hall. The three-course menu for the evening included smoked trout, Parisian style, roasted chicken, foie gras with Madeira sauce, golden potatoes, seasonal salad, peach with crème Chantilly; to drink there was Grand Marnier; wine, Chateau Bellevue, 1955, St. Emilion; champagne, Pommery Greno Brut; fruit liqueur, Marie Brizard; cognac, Courvoisier; and coffee. After coffee was served, Rickard Sandler (1884–1964), a member of the Royal Swedish Academy of Sciences and former prime minister of Sweden (1925–1926) and minister for foreign affairs, gave a speech.

Fig. 10.3 Photo of Princess Margaretha of Sweden and John Kendrew sitting next to each other at the 1962 Nobel banquet in Stockholm. Reproduced from an item in the Kendrew Archives, Weston Library, Oxford

John's Speech to Audience

At the conclusion of dinner, university students paid tribute to the laureates in the Blue Hall below the banquet hall. The student choir sang and the president of the central committee of the Stockholm Student Organization gave a short speech. John followed this with a speech of thanks that he had been requested to deliver to the audience:

> Your Majesties, your Royal Highnesses, Ladies and Gentlemen and—perhaps I may add—Citizens of Sweden: today you have raised your flags over the buildings of Stockholm to do us honor. As I was walking through the city this morning there came over me, not for the first time since I received a telegram from the Secretary of the Swedish Academy of Sciences on the first of November, a feeling of unreality—could it really be true that it is I who am here as one of the principal actors in your annual ceremonial, or am I perhaps a tourist, who has somehow by accident strayed into Stockholm on Nobel Day, and has been given a ticket for the presentation of

the prizes? But a little later, when I looked at my four colleagues, it seemed to me that they looked solid enough, and I thought I could not be dreaming and that it must after all be true.

I think anyone in my position must have some thoughts along these lines, when he remembers the unique tradition which you in Sweden have built around the will of Alfred Nobel, and when he contemplates the fact that he has been called to become a member of that distinguished company who have received an honor greater than any other, civil or academic, to which it is possible for a scientist to aspire.

The truth is, I think, that when one is actually doing one's work in the laboratory one may have a variety of motives; one does it because one finds it interesting, because one is trying to solve a problem which seems important, or perhaps because one is paid to do it. And then when an extraordinary event like this happens one is astonished to find that the work has been accorded a recognition which, at an earlier stage, one would not have dreamed of.

Also at this solemn and happy time I find myself thinking much of others—of those whose work has gone before, of Bragg, whose name has already been mentioned this evening, of Bernal and Astbury and Pauling; those who laid the foundation of our work, and before them a line of others disappearing into the past and representing the continuity of science. Indeed, as has been said in another connection, we stand on the shoulders of giants. I think too of my colleagues for, in the modern fashion, my work has been altogether a team affair. I could, if I had the time, mention perhaps twenty names of those who have made essential contributions to a result which has won for me alone a share of the Prize. It was indeed a long road, and many hands were employed. Reading the Statutes governing your awards and your Nobel Institutes, and finding that one of the functions of the latter is to investigate the discoveries of potential Laureates—and perhaps if necessary to repeat their work in order to make sure that it was correctly done?—I cannot help thinking that in my own case perhaps the Nobel Institute may have interpreted its instructions rather liberally.

I have described this as a happy as well as a solemn time, and I think it is so, not only for the obvious reason but also because somehow you succeed in combining informality with the formality of a great occasion in a most delightful way. This year there is the added circumstance that you have invited here tonight five of us who are all old friends. I suppose it is something unique in the history of the Nobel Foundation, and for all I know it may never happen again, that in one year five men who have worked together, drunk together and talked together for the last fifteen years should all be coming simultaneously to Stockholm, and on the same errand. Thus we have many reasons to feel happy, and I thank you for the warmth with which you welcomed us in Stockholm and with which you are receiving us tonight.[5]

Max Perutz recalled that John spoke very well so that everyone could hear him and at the conclusion of John's speech the grand ball began with refreshments and dancing in the Great Golden Hall until the early hours of the morning.

John handed in his Nobel Lecture at the Nobel Foundation on Tuesday morning and then delivered his lecture in late afternoon to an audience of scientists at the Royal Institute of Technology. In the evening he dined with the king and queen at the Royal Palace. It was an informal dinner with approximately 100 guests and was followed by a party during which the guests mingled and conversed with their royal hosts.

On Wednesday the laureates visited the University of Uppsala and its Chemical Institution, a visit organized by Professors Hägg, Tiselius, and Olovsson, where they gave short lectures and had lunch as arranged by the rector of the University, T. Segerstedt. On Thursday afternoon the laureates were invited by Bengt Feldreich, science editor at the Swedish Broadcasting Company, for a radio discussion in mid-afternoon on Sveriges Radio that was broadcast early the same evening.

John's Speech to Students

On Thursday evening the laureates attended a Lucia Festival or "Festival of Lights" organized by the Union of the Students of Science at Stockholm University. There John was invited by Anders Bergendorff, president of the union, to be knighted into the "order of the smiling and jumping frog." As part of the Lucia Festival ceremony, John was requested to deliver a speech to the students:

Students of the University of Stockholm—I have been asked to express thanks to you for your very kind greeting on behalf not only of myself but also of the other Nobel Laureates, among whom I should like to include those who are present this evening, as well as our so unfortunately absent Laureate, Professor Landau of Moscow. This has been a wonderful occasion for us all. You have sung songs to us—I do not recall ever being serenaded by the students of my own University at Cambridge—and I am told that one of the songs you sang is one which you traditionally sing when you have passed your examinations; it seemed to me that this was a particularly appropriate song for us, in the circumstances. In advance I was warned of the solemnities of this occasion and of the complexities of its ceremonial; but as it has turned out you have made it all very easy for us. I assure you that the ceremony for taking degrees at Cambridge is much less simple than this one; I have not had to listen to any Latin speeches this evening.

However, I understand that we have now reached a point at which solemnity should give way to a very different spirit. As a matter of fact, it has already occurred to me, as I looked round during the ceremonies, that always in the background there were students to be seen. I have no doubt that the Nobel Foundation and the Academies think that they are running this show, but I suspect that really it is the students who run it—and I believe that from now on this will become even more obviously true. Your celebrations seem to me quite different from anything I have seen before; for example, I have never in Cambridge seen so many students gathered in one place, and there the ratio of the sexes is not at all the same. It all looks extraordinarily agreeable to me: why I have been chosen to address you I do not know, but perhaps it is because I am one of the two unmarried and presumably more vulnerable members of this year's team of Laureates. I can only say that we are very much looking forward to meeting the Santa Lucia ladies of 1962—not only the unmarried Laureates but, I think, the married ones as well. On behalf of us all I thank you for your welcome.[6]

John was invited to give a lecture in Lund on Friday evening, arranged by Einer Bladh and the Chemical Society, and flew from Bromma outside Stockholm to Malmö that day. He stayed overnight at the Grand Hotel, Lund. Shortly thereafter John travelled to CERN in Geneva with Jim Watson to begin discussions with Victor Weisskopf and Leó Szilárd about creating an international laboratory of molecular biology in Europe.

11

The Old Guildhall, Linton (1964–1997)

Everyone knew his home, the Guildhall, from the outside.

St. Mary's Church Newsletter

Home in Linton

For much of his adult life in Cambridge John lived in quarters owned and rented to him by Peterhouse. In 1964, using money awarded with his share of the Nobel Prize and inherited following the death of his father, John purchased The Old Guildhall at 4 Church Lane, Linton, about 12 miles southeast of Cambridge (fig. 11.1). The name Linton is derived from its original name dating back to 1008, Lintune, meaning flax farm. At the age of 47 John finally had a home of his own, and in August 1965, after completion of some extensive renovations of the house, he left 12 Tennis Court Road, Cambridge, and moved to Linton.

The Old Guildhall is considered by some as the most noteworthy building standing in Linton. John purchased the house for £12,000 (about $32,000) from Gordon Long and Company, Saffron Walden, who represented the owners, Alan Wells and John Mitchell. Wells was a structural engineer who had lived there with his wife Rosemary (née Mitchell) since 1958, but moved to Belfast, Ireland, in 1964 where he became professor of structural science.

The Old Guildhall had only a few owners between 1912 and 1958. Three members of the Hallidie family, Andrew, followed by Alice, Andrew's widow, followed by Margaret, Andrew's daughter, owned it from 1926 to 1958. John owned the house for more than 30 years and retired there in 1987 after serving as president of St. John's. Following John's death, the house was sold to David and Judith White who owned it from 1997 to 2008. The present owners, Bill and Frances Bickerstaff, purchased the house in 2008, carried out some necessary renovations, and currently live there with their children.

Fig. 11.1 Photo of an advertisement by Gordon Long and Company for The Old Guildhall, Linton. The property is described and offered for sale at £12-thousand. Reproduced from an item in the Kendrew Archives, Weston Library, Oxford

The Old Guildhall is unique. It is a 16th century Tudor residence with a Grade II* listing (National Heritage List for England) indicating that it is a particularly important building of special architectural or historic interest and therefore worth protecting. It was built in the mid-1520s as the Hall of the Holy Trinity Guild on land purchased in 1484 and belonged for some time to Pembroke College, Cambridge. Following suppression of the guilds in 1547 it was used as a town meeting house, "Town House," for about 150 years and then became a private house in 1697. The timber-framed, two-storied house stands on a little more than one acre of land, of which about nine-tenths are mature wooded gardens that back onto the River Granta, a tributary of the River Cam. The site is directly opposite St. Mary's Church and an infant's school and is close to a detached house, Church Cottage, just up the road.

The ground floor of The Old Guildhall is comprised of three reception rooms, living room, drawing room, lounge, and a kitchen. The first floor is comprised of five bedrooms. Adjacent to and belonging to the house is a 16th-century detached barn that is timber-framed and weather boarded with a hipped thatch roof and rubble flint base. In addition to the barn, which provides garage space for three cars, there is a greenhouse, fruit store, workshop, and stable. The estate agents, Gordon Long and Company, described the approximately nine-tenths-acre garden that extends to the north and west as an attractive garden "laid to lawns, flower beds, herbaceous borders, flowering shrubs, bushes and trees, including conifers, ash, laburnum, and the side entrance and walls are covered with Montana Clematis plants. There is a productive kitchen garden, also apple, pear and plum trees, asparagus beds, and two soft fruit cages." The garden borders the River Granta on the west side.

John dearly loved The Old Guildhall with its extensive gardens and he lived there on many weekends even while working in Heidelberg as DG of EMBL and in Oxford as president of St. John's. It was well known to visitors that John stored his ample fine wine collection in the house's priest hole. Upon purchasing the house John immediately arranged for substantial changes to it. These included alterations and refitting of bathrooms and kitchen, stripping the extensive oak work throughout back to natural wood, exposing oak studs and fireplace brickwork, replacing various interior doors, and installing period wood paneling on several interior walls. He even had plans drawn up for modification of the gardens, which included installation of a large swimming pool, a project that never took place. John applied for and succeeded in having the house listed as an English Heritage Building (Identification Number 51909) in November 1967.

Since The Old Guildhall sits so close to the banks of the River Granta it has been subjected to floods that have plagued Linton for centuries. There were disastrous floods in 1879, 1918, 1968, and 2001. In 1968, the year of the so-called "great flood of Linton," John awoke to the tapping sound of his sofa, records, books, and other possessions banging against his bedroom floor. John's library included a large number of his father's books. The water had risen several feet and was at ceiling level. The flood did considerable damage to the ground floor of the house and some treasured possessions, including his Nobel diploma, were damaged. However, his solid oak furniture dried out quickly once the waters subsided and his secretarial assistant spent long hours using a hair dryer to dry out many of John's books caught in the flood.

After John's death in 1997 the following description of the 2001 flood in Linton appeared in the November issue of the *Linton News*: "The worst flooding for over 30 years badly damaged the historic heart of Linton last month [October], with properties around St. Mary's Church flooded . . . The ancient Guildhall in Church Lane was badly affected as the water rose to a depth of several feet . . . Judith and David White [the owners] from the Guildhall had to be rescued from an upper window."

Linton Neighbors

Although he resided in Linton for more than 30 years, quite characteristically John remained fairly aloof from the community. While he was known by a few shopkeepers, a doctor, and some others, John did not make any real effort to befriend his neighbors. Exceptions were the late Charles Anderson and Jeremy Bray (1930–2002), the latter a member of the Labor Party and parliament for 31 years who launched Labor's manifesto for science in 1987. John simply was not interested in the village, did not walk around it very much, and kept the window blinds down most of the time on the village side of his house. He was detached from the people who lived in Linton, but did employ a local woman as a housekeeper and a local man as a gardener. John was extremely fastidious about his garden (fig 11.2).

Fig. 11.2 Photo of John Kendrew in retirement in his garden at The Old Guildhall, Linton. Reproduced from an item in the Kendrew Archives, Weston Library, Oxford

John's behavior toward the inhabitants of Linton is alluded to in the following brief note that appeared on the occasion of his death in the October/ November 1997 newsletter of St. Mary's Church, Linton:

> With the death in August of Sir John Kendrew, Linton loses its most distinguished citizen. Everyone knew his home, the Guildhall, from the outside. Not many penetrated it, to enjoy his gentle hospitality and his beloved garden, though Sir John lived in Linton for over thirty years. Sir John was an agnostic, and only once entered St. Mary's—and that was for the funeral of the wife of his dear friend Charles Anderson. May he rest in peace.

Although John was famously known for not visiting St. Mary's, he did contribute generously to the cost of restoring cradles for the church's old bells.

12

Laboratory of Molecular Biology
(1962–1974)

In many ways, the lab was a kind of paradise for dedicated research scientists.

Hugh Huxley

Proposal

By early 1956 the Research Unit for the Study of the Molecular Structure of Biological Systems at the Cavendish had grown so large that Max spent his time scrounging around for a little bench space here and there and toyed with the idea of asking the MRC to build the unit their own laboratory. Finally, in view of a serious shortage of suitable space for the unit at the Cavendish, in 1957 Max approached the MRC about the possibility of providing support for a building for the newly named Molecular Biology Research Unit. It was auspicious timing since John had just solved the low-resolution three-dimensional structure of myoglobin, the double-stranded nature of DNA proposed by Watson and Crick had been confirmed by biochemists, and Fred Sanger, who was invited to join the unit, had finished the complete amino acid sequence of insulin.

At the time the unit consisted of Max, John, Leslie Barnett, Gerhard Bodo, Sydney Brenner, Francis Crick, Anne Cullis, Joan Hunt, Vernon Ingram, and Tony Stretton. There were also several notable visitors from abroad from time to time, including Seymour Benzer, Paul Doty, Mahlon Hoagland, Alex Rich, George Streisinger, and Bror Strandberg. However, Max only approached the MRC after Fred Sanger in the Biochemistry Department in Cambridge agreed to join the unit following long discussions with Francis Crick. This new alliance was particularly significant since Max recognized that, as it stood, the unit was weak in chemistry. Sanger brought tremendous chemical expertise to the alliance, was already supported by the MRC, and required expanded laboratory facilities. In addition, he was about to be awarded a Nobel Prize in Chemistry in 1958 for his pioneering work on protein sequencing.

Prior to receiving Max's proposal, Harold Himsworth, secretary of the MRC, had heard a rumor about the plan from Charles Harington (1897–1972), director of the National Institute for Medical Research, following Sanger's discussions with him. In July 1957 Himsworth sent a confidential note to Max and John confirming his knowledge of their proposal for a new building and his intention to speak with Sanger and three other Cambridge professors, Nevill Mott (Cavendish professor of physics), Alexander Todd (1907–1997; professor and 1702 chair of chemistry; Nobel/Chemistry, 1957), and Frank Young (1908–1988; Sir William Dunn professor of biochemistry) about it. But he assured Max that no undertakings or commitments of any kind would occur until they had been thoroughly discussed with both Max and John. Max responded promptly to Himsworth apologizing for the rumors and saying that Young disliked the idea altogether and Todd and Mott approved, although Todd preferred to have some members of the unit in the Chemistry Department's new building on Lensfield Road. Himsworth also discussed the proposal with Edgar Adrian (1889–1977; Nobel/Physiology or Medicine, 1932), master of Trinity and vice chancellor of the university, but Adrian had little enthusiasm for the proposal in view of other ongoing departmental needs and fiscal concerns at the university.

Negotiations

At the beginning of August 1957, Himsworth, Richard Fort (1907–1959), a lay member of the MRC, and Quintin Hogg (1907–2001), a member of the House of Lords and liaison with the MRC, met in Cambridge with Mott, Todd, Young, and others. This was followed by a meeting with Max, John, Francis Crick, and Fred Sanger. Mott, Young, and Todd, all influential Cambridge professors, espoused a variety of reasons for being mildly to strongly unenthusiastic about the proposal for the unit to leave the Cavendish. Mott felt it would be detrimental for the Cavendish if they lost contact with the unit and that it should retain a foothold in the physics department. Todd was inclined toward having a part of the unit housed with him in chemistry. Young was strongly opposed and felt that members of the unit should be in university departments; if they did not want to do this then they ought to be moved to

Mill Hill. He objected to an institute being put up for the unit in Cambridge because it would compete with university departments. Young prophesized that a non-university laboratory that was divorced from university teaching would soon fossilize.

Insofar as the feelings of members of the unit, Max, Francis Crick, and Fred Sanger and the people working with them were all for the proposal, but John was rather against it. Max was anxious to keep the chemists, biophysicists, and biologists in the unit together, Crick wanted to collaborate with Sanger and remain in close contact with Max and John, and Sanger was particularly desirous of more laboratory space for his research. On the other hand, John was hesitant about the unit's possible move away from the center of Cambridge since he felt the university would cut off their supply of PhD students and in other ways impede their progress. John wished the unit to remain as part of the university, not as a separate entity located some distance away, so that they could retain the ability to recruit research students who would receive their degrees from the university. Since John worked at the Cavendish, a very short walk from both Peterhouse and his flat on Tennis Court Road, it is likely that he considered this proximity important and did not wish to be miles away from the center of Cambridge, from his flat, and from Peterhouse. John talked further about the proposal and his reservations with Harold Himsworth and in a short time realized that progress in molecular biology in Cambridge would be accelerated by bringing relevant scientists together at one site. He understood that a unit consisting of Brenner, Crick, Huxley, Klug, Perutz, Sanger, himself, and others would be an ideal setting for research. As a result of the conversation, John's reservations about the move from the Cavendish were alleviated.

Approximately one year later the proposal to move the unit away from the Cavendish was still being considered by the MRC Council, by various departments at Cambridge, and by a university-appointed committee. For a brief time the unit even considered an offer from Hans Krebs (1900–1981; Nobel/Physiology or Medicine, 1953), professor of biochemistry in Oxford, to occupy a floor of his new building or to move to an adjacent site on South Parks Road. Max was very impressed by Kreb's attitude toward future scientific developments in Oxford, as well as by the progressive attitude of Oxford university. He found it all quite refreshing compared to the policy of containment in Cambridge.

Soon after the MRC approved the unit's plan to move to a new site enthusiastically, but the university's approval was needed and this proved to be far more difficult. For self-serving reasons university administrators were strongly opposed to the establishment of a non-university laboratory in Cambridge. However, Nevill Mott was soon elected to the general board of the university and he assembled a sub-committee consisting of Todd, Young, and himself to look into the unit's plans. As a result, outvoted by Mott and Todd, Young reluctantly signed a report recommending approval of the unit's plan and the general board had to agree.

Addenbrooke's Site

In the end, after two years of bitter negotiating between the university and the MRC the unit accepted an offer of a building site by Joseph Mitchell (1909–1987), Regius Professor of Physic at Cambridge. Mitchell was a member of the council and, consequently, privy to the many discussions about a building site for the LMB. The site offered was at the Postgraduate Medical School on Hills Road, about two miles south of the city center. In addition to the Cambridge scientists, Max, John, Francis Crick, and Sydney Brenner from the Cavendish and Fred Sanger from the Biochemistry Department, Aaron Klug and his virus group, including Ken Holmes and John Finch, from Birkbeck College in London, and Hugh Huxley from the Biophysics Department at University College, London, joined the LMB in 1962. Both Klug's and Huxley's research already had been supported for some time by the MRC.

At the end of 1961 members of the Molecular Biology Unit were beginning to be relocated to a new four-story building of about 22,000 square feet on a 66-acre site off Hills Road that today is known as the Cambridge Biomedical Campus. The land had been purchased in 1951 for £4,350 (about $12,000) and subsequently became the site for the new Addenbrooke's Hospital. The hospital was named after John Addenbrooke (1680–1790), a physician who left £4,500 in his will in 1719 to establish a small hospital for poor people. The LMB building and equipment costs, about £310,000 (about $840,000) and £90,000 (about $243,000), respectively, were financed primarily by the MRC with additional support coming from the Wellcome Trust and University Grants Committee. A portion of the MRC's contribution came from a benefaction, about £160,000, left to the council in 1957 by Lady Julia Wadia in memory of her late husband.

The new LMB was opened in May 1962 by Queen Elizabeth II with all staff present except for Francis Crick and Sydney Brenner who disapproved of the monarchy and absented themselves from the celebration (fig. 12.1). The principles by which the LMB operated right from the beginning included hiring outstanding scientists and support staff, giving colleagues intellectual freedom and complete credit for their work, showing interest in everyone's research, and facilitating the interchange of ideas. The laboratory initially had a governing board consisting of five members, John, Max, Sydney Brenner, Francis Crick, and Fred Sanger, but in 1965 a sixth member, Hugh Huxley, was added. Max persuaded the MRC to appoint himself as chairman of the governing board rather than as director of the LMB. This arrangement reserved major decisions of scientific policy to the board which never directed the laboratory's research, but tried to attract or retain talented young people and give them a free hand in their research. The board considered

Fig. 12.1 Photo of John Kendrew (ca. 45 yoa) showing a low-resolution model of myoglobin to Queen Elizabeth II (seated) at the opening of the MRC LMB on Hills Road in May 1962. Reproduced with permission of the MRC LMB, Cambridge

the budgets prepared by each division and submitted a single statement of fiscal requirements, including a section for common services, to the MRC. If it proved difficult or impossible to resolve a fiscal matter between a division and the board, the heads of division reserved the right to present their case to the secretary of the MRC.

In 1965 Francis Crick and Sydney Brenner became joint heads of the Division of Molecular Genetics and the name of the Division of Protein Chemistry, headed by Fred Sanger, was changed to Protein and Nucleic Acid Chemistry. Each division had its own floor of the building, with the stores, workshop, X-ray equipment, and electron microscopes on the ground floor, and the library and canteen/tearoom on the fourth floor. Michael Fuller, hired as a technician by Max in 1952, took charge of the stores and general purchasing and in time became an LMB legend. The canteen, run by Max's wife Gisela, became a focal point of the laboratory, not just for morning coffee, lunch, and tea, but for daily discussions and exchange of ideas between senior staff, technicians, students, postdoctoral fellows, and guests. It was ensured that there was absolutely no class distinction at the LMB canteen as there had been at the Cavendish where the scientists and technicians, the latter group even including engineers, had separate canteens.

In 1962 there were approximately 55 members of staff at the LMB, which included all visitors, postdoctoral fellows, students, and attached workers. Between 1963 and 1969 the number of staff nearly doubled to about 95 and between 1969 and 1973, with the addition of a new extension to the LMB building that nearly doubled the amount of research space, the number of staff tripled to about 150. In addition to the six members of the governing board, John, Max, Sydney Brenner, Francis Crick, Hugh Huxley, and Fred Sanger, in the 1960s the LMB was fortunate to have a number of other outstanding researchers on its staff. Among these were David Blow and Aaron Klug in the Division of Structural Studies; Mark Bretscher (b. 1940), Brian F.C. Clark (1936–2014), Robin Monro (b. 1932), and John Smith (1924–2003) in the Division of Molecular Genetics; and Ieuan Harris (1925–1978), Brian Hartley (b. 1926), Kjeld Marcker (1932–2018), César Milstein (1927–2002; Nobel/Physiology or Medicine, 1984), and Alan Weeds (b. 1939) in the Division of Protein and Nucleic Acid Chemistry. During the 1960s the LMB attracted an unusually large number of outstanding American postdoctoral fellows with their salary support provided by individual fellowships from the United States, as well as many very clever Cambridge university research students, a few of which also were American.

In 2013 Hugh Huxley commented on the environment at the LMB in its early days by saying:

In many ways, the lab was a kind of paradise for dedicated research scientists. No one had to do any teaching, or to write grant applications, or even to publish frequently. Expert practitioners in most of the relevant fields were readily at hand. Within reason, all supplies and equipment were either immediately available or rapidly obtainable, via Michael Fuller [Laboratory Steward]. The lab had sizeable mechanical and electronic workshops, where any novel equipment that the experimental groups needed and designed could be skillfully constructed and maintained [including a service engineer to keep instruments in running order]. There were also good facilities for do-it-yourself construction work. Such arrangements followed the traditions of the Cavendish Laboratory. There was a democratically organized cafeteria, where everyone went for morning coffee, lunch and tea, and much scientific conversation and advice was always available. This was of great importance. People worked extremely hard and collaborated with each other extensively, because of the excitement and satisfaction of getting important scientific experiments to work properly, and to find out more and more fascinating new facts about nature.[1]

Several members of the LMB were to receive Nobel Prizes after 1962, the *annus mirabilis*, miraculous year. These included Fred Sanger (Chemistry, 1980; Sanger's second Nobel Prize in Chemistry), Aaron Klug (Nobel/Chemistry, 1982), César Milstein and Georges Köhler (Nobel/Physiology or Medicine, 1984), John Walker (Nobel/Chemistry, 1997), Sydney Brenner, Bob Horvitz, and John Sulston (Nobel/Physiology or Medicine, 2002), Venki Ramakrishnan (Nobel/Chemistry, 2009), Michael Levitt (Nobel/Chemistry, 2013), Richard Henderson (Nobel/Chemistry, 2017), and Greg Winter (Nobel/Chemistry, 2018). In addition, 45 members of the LMB were made fellows of the Royal Society, 10 received the society's Royal Medal, and nine received the Copley Medal between 1965 and 2017. More than 70 members of the LMB were granted membership in EMBO, five awarded the EMBO Gold Medal, and 23 made fellows or honorary fellows of the Academy of Medical Sciences. A truly remarkable track record for a single research institution.

Outside Activities

By the time the LMB opened in 1962 John was immersed in so many activities away from the laboratory that he had little time to be directly
involved in any bench research. He continued to sponsor several postdoctoral fellows in the Division of Structural Studies, including Len Banaszak,
Tom Creighton, Bob Kretsinger, Chris Nobbs, Lubert Stryer, and others.
However, from 1962 to 1968 the day-to-day supervision of research was
in the hands of his colleague Herman Watson since John was away much
of the time. Between 1964 and 1968 John published seven primary research papers, five in *JMB* and two in *Nature*, all with Herman Watson as
a co-author. Thereafter, following Watson's departure for the University
of Bristol in summer 1968, John was not at all involved in bench research
and did not publish any new primary research papers. At the age of 51,
after about 20 years as a practicing scientist in Cambridge, John's career as a
principal investigator was essentially over.

Much of the time between 1962 and 1974 John was frequently away from
the LMB and Cambridge. However, he continued to serve as deputy chair
of the LMB and head of the Division of Structural Studies. In response to
John's frequent absences, David Blow, Hugh Huxley, and Aaron Klug set up a
Structural Studies Committee that would be chaired by John when he was in
Cambridge, but could meet without him and make decisions in his absence.
The 589th meeting of the committee took place in January 2018.

In addition to continuing as an active fellow at Peterhouse, editor-in-chief
of *JMB*, and part-time reader at the Royal Institution, from 1961 to 1963
John was also a part-time member of the MoD and thereafter a member
of the Defense Research Committee, Defense Scientific Advisory Council,
and Council for Scientific Policy (1965–1972). He was also a member of the
Council of the Royal Society (1965–1967), the British Biophysical Society
(1960–1965), the BBC Science Consultative Group (1964–1967), and
served as an advisor to the World Health Organization (1963–1966) and
on the board of governors of the Weizmann Institute (1963–1997). John
was appointed honorary secretary of the steering committee of the British
Biophysical Society (1962–1965), vice president, president, and honorary
president of the IUPAB (1964–1972), chairman of the British National
Committee for Biophysics, governor of the Weizmann Institute, trustee of
the British Museum, governor of Clifton College, and member or chairman
of various working groups, committees, and sub-committees.

During this same period, 1962–1974, John's role in the founding of EMBO, EMBC, and EMBL was particularly time consuming, especially following the death of Leó Szilárd in 1964. John became the frontman for those working to establish an international European laboratory for molecular biology. He was appointed secretary general of EMBO and EMBC in 1969 and 1970, respectively, and project leader of EMBL in 1971. Beginning in 1962 and continuing until 1974 much of John's time and effort was devoted to expanding and improving Europe's role in molecular biology. At the same time, as a Nobel laureate, he was also in constant demand at home and abroad for lectures and participation in symposia on protein structure and various other aspects of molecular biology.

Ministry of Defense

During WW2 Prime Minister Winston Churchill (1874–1965) also held the post of minister of defense, in order to exercise control over the chiefs of staff and coordinate defense matters. At the end of WW2, Clement Attlee's (1883–1967) government established the MoD and First Lord of the Admiralty Albert Alexander (1885–1965) was appointed minister, a cabinet post that represented the military. In 1964 the position minister of defense was abolished and replaced by the position secretary of defense with the appointment of Peter Thorneycroft (1909–1994) who had been minister since 1962, followed in late 1964 by the appointment of Denis Healey (1917–2015), a Labour Party member of parliament.

In May 1960 John received a letter from Chief Scientific Advisor to the Secretary of State for Defense Solly Zuckerman, in which he was asked to serve on an ad hoc panel to consider a particular aspect of defense policy. The relationship between John and Zuckerman went back to WW2 when Zuckerman served as a scientific advisor to Combined Operations Headquarters, including Supreme Headquarters Allied Expeditionary Force.

The first letter was followed by another a month later asking John to become a part-time scientific advisor to the ministry with a retainer of £1000 annually and suggesting he be available two days each week, possibly with portions of a third day. John questioned "whether it might not be better in the first instance to arrange something more flexible, in the nature of a consultancy" and that "he would be quite content for the moment [if he] was simply paid [his] expenses, and any other financial arrangement could be left over until we see how things work out."[2] Zuckerman agreed that there would be no definite commitment about actual hours. He felt that it would be more important to talk intimately with John for a short time on matters that concerned the MoD than to have John committed to two to three days on unimportant issues.

The 1960s was the height of the Cold War, which lasted for more than 35 years after WW2. It was characterized by extreme tension between capitalism in the West and communism in the East and by the rapid growth of nuclear weapons. Zuckerman, as the chief scientific advisor to the MoD, was relying on so-called war games to evaluate the consequences of potential nuclear attacks launched by the Soviet Union against Britain and by Britain against the Soviet Union. In addition, he was setting up panels on anti-submarine warfare, anti-ballistic missiles, chemical warfare, biological warfare, and other issues at the ministry. Some panels were formed to decide the role of strategic nuclear weapons in supporting national policy.

By the end of July 1960 John was working part-time at the MoD as a member, later chairman, of a study group dealing with the Joint Inter-Service Group for the Study of All-Out Warfare (JIGSAW) project. As part of an overall study of the effects of nuclear weapons and the nature of deterrence, the aim of JIGSAW was to determine the cumulative effect of nuclear damage that would lead to a collapse of national structure in the UK, United States, and Soviet Union. The group met twice a month and analyzed such subjects as mutual deterrence, nuclear sufficiency, use of tactical nuclear weapons, the possibility of limited war between East and West and means of preventing its escalation. Its meetings were attended by military officers, scientists, and representatives of MI5 (internal/domestic military intelligence reporting to the home secretary) and MI6 (external/foreign military intelligence reporting to the foreign secretary). The group was asked to select 40 Soviet Union cities on which attack with 1-megaton weapons would achieve the maximum effect on the population from blast and fallout, to select target cities whose destruction would pose the maximum threat to the greatest number of Russian people. It was assumed that any government, East or West, would be swayed by this potential for a huge loss of life in the event of a nuclear war, but no one knew what would be considered an acceptable level of damage.

One of the first domestic model exercises the study group addressed was to consider the consequences of a 1-megaton bomb, 50 times the size of the bomb dropped on Hiroshima, exploded over Birmingham. The studies showed that one-third of Birmingham's million inhabitants would be killed at once by blast, fire, or a lethal dose of radiation, with total destruction within a radius of a about 2 miles from the center of the explosion and few, if any, survivors. The results of the Birmingham study were presented to Britain's senior military officers, including Earl Mountbatten of Burma, chief of the defense staff, and Secretary of State for War John Profumo (1915–2006). John remembered that the politicians and at least some of the military men were appalled by the Birmingham study. In fact, John became a contentious figure when the *Daily Telegraph* published a scathing review of the activities of the JIGSAW study group more than 30 years later in 1996.

In spring 1961 John was spending considerable time at the ministry in London, two to two and a half days a week. The ministry was very pleased with John's participation and suggested the possibility of rooms, an office, a dictating machine, secretarial support, clerical and typing support, and a remuneration for him. John was quite happy with the arrangement and wrote:

> I do indeed hope to spend rather more time with you from now on, and it is kind of you to think of arrangements from my work in the Ministry. I wish for my own sake that I could settle on two regular days each week, but I fear that this will not be possible having regard to the unpredictable way in which committees with which I am associated arrange their meetings . . . it would certainly be very useful to me to have some pied-a-terre as close to Sir Solly's office as possible. As it is, I generally make use of his office when he is not there, but obviously this is not always possible . . . .[3]

By late summer John had been helping Zuckerman for about one year and the only monetary compensation he had received was for travelling expenses. He became concerned about the amount of time he was spending on ministry matters and suggested that they pay a significant portion of his MRC salary. However, the treasury concluded that as a member of the MRC staff he was already a government employee and, consequently, was not entitled to any salary, only to expenses. John argued the case, but unsuccessfully.

In 1963 the ministry was prepared to offer John a three-year contract, but by that time he had serious reservations about continuing on as deputy scientific advisor. In October John wrote to Edward Playfair (1909–1999), permanent secretary at the ministry, about some of his reservations:

> I am anxious to do all I can to help Solly and the Ministry, but I still have serious doubts whether in acting as Solly's Deputy on the basis he proposes

I would not be putting myself in a position which would make it impossible to reconcile these responsibilities with my own research and teaching which, as I have said from the beginning, I feel must claim the major part of my time. I also feel that the longer term future is so open that it would be unfortunate if anyone were to have the impression that I would automatically, or even probably, want full-time employment with the Ministry later. I am not even certain whether full-time or part-time employment carrying executive responsibility would be in the best interests of the Ministry, since the whole object of the exercise, as I understand it, is to get independent advice from fresh minds; how long could one's mind remain either independent or fresh?

My feeling is that if I were to accept a three years' contract immediately, on the basis you proposed last Thursday, I would do so with a sensation of strain which would not be in the best interests of anyone ...

Finally, I should like to reiterate my desire to be of assistance to the Ministry in every possible way. I only want to avoid getting myself into a position of moral commitment to undertake more than I can in fact do.[4]

Soon after John resigned his part-time appointment with the ministry saying that "I find I still do not want to come right in . . . The alternative is to get right out, and I honestly believe this is the right course for me."[5] However, he continued to serve for several years on a number of committees and panels dealing with defense related matters, such as the Defense Research Committee. In one instance John headed a committee looking into how troops could be used to help police in the event of race riots or other outbreaks of violence in the UK. They assessed both the extent of troop involvement and types of weapons that could be employed in such situations.

Council for Scientific Policy

John joined the Council for Scientific Policy at its inception in 1964, was reappointed twice, and continued as a member until 1972 when its work ended. At that juncture its functions were taken up by the Advisory Board for the Research Councils and then in 1992 by the newly founded Office of Science and Technology.

The Council for Scientific Policy was established in January 1965 by the Labour government's Department of Education and Science. It was set up to advise their secretary of state for education and science in the exercise of their responsibilities for the formulation and execution of government scientific policy. It began with 14 members recruited from universities and industry, and included assessors from the Medical and Agricultural Research Councils, Department of Scientific and Industrial Research, University Grants Committee, Cabinet Office, and Ministry of Technology.

John served as deputy chairman of the council from 1969 to 1972, first under Harrie Massey (1908–1983; Chairman 1965–1970), a mathematical physicist, and subsequently under Frederick Dainton (1914–1997; Chairman 1970–1972), a physical chemist who authored a report in 1968 on the relatively low number of British students entering science and engineering at university (the "Dainton Report"). John also served as chairman of the Standing Committee on International Scientific Relations, which held tripartite meetings between France, Germany, and Britain to evaluate areas of research suitable for collaborative efforts, such as the brain and behavior research program, the high flux reactor project at Grenoble, and astronomy research.

Between 1967 and 1971 John was a chairman or member of several of the council's working groups, including one initiated by John that was to evaluate plans for teaching, recruitment, and research in molecular biology in the UK. This group included John as chairman, Michael Swann (1920–1990), Maurice Wilkins, Hans Kornberg, James Gowans (b. 1924), William Hayes (1913–1994), Albert Neuberger (1908–1996), and Michael Stoker (1918–2013); the latter four directors of MRC units at the time. After more than a dozen meetings over a nearly two-year period a report was issued in 1969 that became known as the "Kendrew Report."

The immediate task of the group was to look into the current status of and future plans for teaching, recruitment, and research in molecular biology in the UK. However, it was recognized early on that molecular biology was often carried out under names of different disciplines, such as biochemistry, genetics, or virology. So the assignment became an inquiry into the status of and future for biology at the molecular level. In the end, the group addressed both the size and distribution of research in biology at the molecular level in the UK.

Overall, the "Kendrew Report" highlighted the relatively small number of UK scientists engaged in molecular biology and the clustering of the best of them in only a few institutions. This was considered to be in marked contrast to the situation in the United States. The group made several recommendations that were intended to reform the teaching and practice of molecular biology at universities, relax traditional departmental boundaries to allow the formation of large interdisciplinary research groups, and allow universities to participate more closely in MRC staff appointments.

Various aspects of the proposal met immediately with criticism by the MRC, biochemists at UK universities, and the scientific and popular press. The Biochemical Society set up a subcommittee that produced the "Krebs Report" to counter the Kendrew Report since they saw molecular biology as simply a sub-division of biochemistry. De Chadarevian concluded that, "While molecular biologists saw a need for "national planning and central direction," biochemists expressed themselves against "administrative measures" imposed from above and preferred to rely on independent peer review as the approved mechanism to maintain high standards of research."[6] The council's working group had initiated a scientific and political movement for molecular biology that would be continued into the 1970s and beyond, and John would continue to be a central figure in the movement.

John served on other working groups that dealt with biological research in or near Cambridge and Edinburgh, the organization and support of pure and applied scientific research and postgraduate training, and the science budget for high-flux neutron beam research. In addition to attending frequent meetings and preparing reports, John's council service involved travel to France, Germany, Italy, the Netherlands, North Africa, Sweden, and other European countries. John also acted as an assessor for a working group set up in 1968 that was to consider whether the UK should support an EMBO proposal for establishment of an international laboratory. In 1972 the council recommended to the secretary of state that the UK should participate in the establishment of the EMBL in Heidelberg and its outstations in Hamburg and Grenoble.

BBC Television and Books

On Saturday, May 3, 1960, as part of the series *Science on Saturday*, BBC-TV broadcast a program entitled "Eye on Research—Shapes of Life" with John, Max, and J.D. Bernal participating. The series was rebroadcast on Saturday, November 4, 1961. The program featured what the BBC called one of the great triumphs of British science, the unravelling of the structure of a protein molecule, the basis of life, about which John and Max conversed with moderator Raymond Baxter (1922–2006).

Subsequently in 1963, John was invited to deliver 10 lectures, under the title *The Thread of Life*, on BBC Television. He received 75 guineas (about £78) per program plus 25 guineas (about £26.5) per program for preliminary work, consultation with and advice for the producer. The programs were recorded at the Royal College of Surgeons of England between November 25 and December 4, 1963 and were transmitted weekly from January 4 to March 17, 1964, and repeated in July 1964. A German version, "Der Faden des Lebens," was recorded by John in Munich in August 1964 for Bayerische Rundfunk.

On a Thursday evening in 1964, in a Cambridge pub, John eavesdropped on the following conversation between two locals talking about the BBC's *The Thread of Life*:

Local 1: "If I could be born all over again I think I would like to be a Russian."

Local 2: "Why?"

Local 1: "Oh they are so much more go ahead and advanced in their research."

Local 2: "Well have you seen this programme on television, "The Thread of Life," I think it's called, comes on on Saturdays and Mondays."

Local 1: "No, what is it about?"

Local 2: "Heredity and viruses—I think it's the best programme on television at the moment. It's so interesting and do you realize that almost all this work is being done in Cambridge."[7]

John was most amused by the conversation, but noted that they trailed off before they had a chance to discuss the lecturer.

At about the same time, in January 1964, John was approached by S.L. Dennis at G. Bell and Sons, publishers in London, about the possibility of producing an adaptation of the BBC program *The Thread of Life* in book form. John accepted, received some editorial assistance from a writer at the *Observer*, and the book appeared in June 1966. A US version, from Harvard University Press, also appeared in the same year and a German version was published by Nymphenburger Verlagshandlung in 1967. The book proved to be very successful and was translated into several languages, including Japanese, Hungarian, Danish, Norwegian, Polish, Russian, Spanish, and Braille.

In October 1994, retired and living in Linton, John wrote to Richard Roberts (b. 1943) who, together with Phil Sharp (b. 1944), had received the

Nobel Prize in Physiology or Medicine in 1993 for their discovery of split genes. In his Nobel autobiography Roberts had commented very favorably on John's book, *The Thread of Life*, in that it was his first exposure to molecular biology and got him hooked. John was flattered by Robert's comments and wrote to him:

> I have just got the 1993 volume of Nobel Lectures, and found in your auto-biographical sketch the reference to my book The Thread of Life.
>
> It is much the nicest—and the most important—thing that was ever written about the book and I feel quite overcome by the thought that it was this which turned you from being a chemist to a molecular biologist. It makes the book worth writing!
>
> I also began as a chemist and underwent the same transition. In my case the important book was Schrodinger's What is Life (though there were personal influences too—Bernal, Waddington and Pauling).
>
> You have what may be a unique distinction—to become a Nobel Prizewinner before becoming F.R.S. Perhaps they didn't know your nationality [British]? At any rate I am confident that this will shortly be rectified.[8]

As John predicted, Roberts was elected a Fellow of the Royal Society shortly thereafter in 1995 and was made Knight Bachelor of the British Empire in 2008.

13

European Molecular Biology Laboratory (1962–1982)

> I still think that my time at EMBL was the most exciting, interesting, and enjoyable period of my whole life.
>
> John Kendrew

International Competition

By the early 1960s the term molecular biology had firmly taken hold and there was even a journal, *JMB*, devoted to it. John recognized that in most European countries there was not enough money for building or equipping laboratories, creating positions to fill them, and for travel. On the other hand, during the post-Sputnik-era the United States was financing all aspects of scientific education and research and as a result threatened to recruit the best European scientists and corner the market in molecular biology. John saw that young Americans who recognized the promise of molecular biology had easy access to fellowships, but young Europeans had no such possibilities. In the United States there were summer schools in molecular biology, but no money was available for such activities in Europe.

In a letter to Solly Zuckerman in summer 1962, John pointed out the evolving disparity between science policy in England and the United States:

> In my own field there are almost no openings for good young PhDs in England, owing to the reactionary policy of British universities toward Biology, to shortage of money and to the rigidity of the traditional departmental framework; in America there is a tremendous boom and no self-respecting university can now face the world without a Molecular Biology Department—and the (literally) dozens of such departments now springing up will absorb all the above mentioned British PhDs unless

something drastic is done, and soon. It seems very sad that a new and important field largely started over here should now be flourishing on so large a scale in America.[1]

John confessed in 1960 that he was content to remain working in the UK since the MRC had been extremely good to the unit financially and it was soon to have a new building. But he was perpetually amazed that more scientists didn't go abroad and concluded that they must stay out of pure sentiment. The situation rankled John, a dyed-in-the-wool European, and he committed much of his future efforts to pursuing the goal of establishing an international European laboratory—what eventually became the EMBL in Heidelberg.

Meeting at CERN

Following the Nobel Prize ceremonies in Stockholm in December 1962, John traveled to Geneva to meet with Victor (Viki) Weisskopf (1908–2002), Leó Szilárd (1898–1964), and Jim Watson at the Centre Européen de Recherche Nucléaire (CERN), or European Council for Nuclear Research. They intended to get together there so the four of them could discuss the need for a European laboratory of molecular biology. CERN was established in 1954 on the Franco-Swiss border by 12 member states to provide particle accelerators and other infrastructure needed for high energy physics research. Currently the annual budget for CERN is about 1.2 billion Swiss Francs (close to $1.2 billion) paid for by 22 member states.

Both Weisskopf and Szilárd were nuclear physicists, the former born in Austria and the latter in Hungary. Weisskopf and Szilárd became naturalized American citizens in 1943 and 1962, respectively. Szilárd had worked on the atomic bomb, was appalled by the destruction it caused at Hiroshima, and moved from the United States to Switzerland during the Cuban missile crisis in October 1962. Weisskopf, on leave from the Massachusetts Institute of Technology, held a five-year appointment as DG of CERN and insisted that everyone there be regarded as European and not as citizens of particular nations. He had been looking for a way to strengthen the growing influence of physics on biology, and Szilárd suggested the formation of an international laboratory of molecular biology modeled after CERN.

The idea of a European laboratory for molecular biology was subsequently hatched in Geneva by Weisskopf, Szilárd, and Italian physicist

Gilberto Bernardini (1906–1995). Without hesitation Szilárd telephoned Francis Crick in Cambridge to invite him to form the laboratory, but Crick refused saying that "it just was not his glass of champagne." Szilárd then spoke to Sydney Brenner who identified John Kendrew as just the man he needed. Apparently Szilárd had not yet heard of him, so Brenner explained that John had just been awarded a Nobel Prize in Chemistry. Years later John recalled that Leó Szilárd's original idea was that the DG of EMBL should be Francis Crick, however, it was only when Brenner explained to Szilárd that Francis Crick was a hopeless administrator that John got landed with the job.

Weisskopf, Szilárd, Bernardini, and John were all concerned with the decline of European science relative to science in the United States since the latter had undergone tremendous expansion following the launching of Sputnik I, the world's first artificial satellite, by the Soviet Union in early October 1957. With the recent loss of some of the best European scientists to the United States, John worried that

> the growth of molecular biology in the United States has been spectacularly rapid and the general quality of the work done there is extremely high. This transatlantic development has quite outpaced our progress here in Europe. Yet many (though not all) of the most fruitful parts of the subject developed from European roots . . . only in America is it possible to muster the necessary resources of money and talent on a national basis. This is the reason why almost all promising young men in the field want to go to America as postdoctoral fellows, and why many European countries are suffering a serious permanent loss of more senior workers to America.[2]

Unlike the situation at many universities in the United States in the 1960s, most departments at European universities were autonomous and insular, and this did not encourage enthusiasm for interdisciplinary research. Both financial restraints and archaic attitudes seriously threatened the progress of molecular biology in Europe. To many it seemed quite clear that for Europe to compete successfully in molecular biology with the United States, an international European laboratory had to be established. The idea began to take shape that Europe needed a Centre Européen de Recherche Biologique (CERB) by analogy with CERN. After a dozen grueling years this idea would eventually lead to the founding of EMBL in Heidelberg with John as its first DG.

Significant Events Establishing EMBO, EMBC, and EMBL

1962—Discussion about an international laboratory when John Kendrew, James Watson, Victor Weisskopf, and Leó Szilárd meet at CERN.

1963—Meeting in Ravello, Italy, to discuss a proposal for a European Organization of Fundamental Biology.

1964—Establishment of EMBO after meeting in Ravello, with Max Perutz, chairman, and John Kendrew, secretary general; EMBO was to ensure creation of a central laboratory, EMBL, and establish networking to enhance interactions between European scientists; death of Leó Szilárd.

1968—Establishment of EMBC by 14 member states; most of EMBO's activities are funded by member states that together form the intergovernmental organization, EMBC, that monitors EMBO, but does not interfere with its activities.

1969—Meeting in Konstanz to discuss the proposed European laboratory; appointment of John Kendrew as head of Laboratory Committee.

1971—Appointment of John Kendrew as EMBL project leader; set up of Building and Provisional Scientific Advisory Committees.

1972—EMBC decision to establish EMBL in Heidelberg, Germany.

1973—EMBL agreement signed in Geneva; German architectural firm Lange, Mitzlaff, Böhm, and Müller chosen.

1974—EMBL agreement ratified by 10 member states; John Kendrew appointed first DG of EMBL.

1975—Construction of EMBL in Heidelberg; establishment of EMBL Outstation in Hamburg, Germany.

1976—Establishment of EMBL Outstation in Grenoble, France.

1978—EMBL building completed; inauguration ceremony in May.

1982—John Kendrew replaced by Lennart Philipson as DG.

Beginnings of EMBL

About one month after the meeting in Geneva, John was already on the lookout for well-known scientists for the proposed laboratory in Europe. He described the plan to a potential recruit as

a scheme put up by Weisskopf, Director of CERN, and Szilárd. Their proposal is that there might be established in Geneva an international laboratory of Molecular Biology which we provisionally call CERB to show that it
would be analogous and parallel to CERN. A number of people are enthusiastic about this plan and I am going to do some canvassing of it during the
next few months, but it may be a slow job since all the governments concerned will have to be persuaded separately.[3]

At the end of January 1963, after John met with Szilárd in Washington,
Szilárd wrote to Weisskopf that it would be awkward for John to be a candidate for DG of the proposed international laboratory and a prime mover in
England for its establishment. He felt it would be wise to invite one or two
people from England, France, and Germany who might be helpful with the
project. Soon after John wrote to Weisskopf addressing his role in the project: "I feel that my personal position is a rather difficult one. If I am to be
thought of as a possible Director, it is really not for me to take the leading part
in pressurizing the countries of Western Europe. Nevertheless, I feel strongly
that some single person must be found to undertake this task."[4]

In March and again in June, meetings took place in Geneva to discuss
establishment of a European laboratory of molecular biology. A dozen
or more scientists, predominantly molecular biologists, including John,
Jacques Monod (1910–1976; Nobel/Physiology or Medicine, 1965) from
France, Conrad Waddington (1905–1975) from the UK, Max Delbrück
(1906–1981; Nobel/Physiology or Medicine, 1969) from the United States,
and Arne Engström from Sweden, participated in the meetings. Then in
September 1963, 25 scientists from Europe and the United States met in
Ravello, Italy, at a summer school sponsored by the Italian Physical Society.
There the participants discussed the possibility of founding a European laboratory of fundamental biology and forming a federal organization of existing European laboratories. For the meeting John and Waddington drafted
a "Proposal for a European Organization of Fundamental Biology" based on
previous discussions at the Geneva meetings. The proposal was largely John's
effort and was considered an eloquent and historic document. However,
whereas John strongly supported the formation of a new European laboratory to be called CERB and resembling CERN, Waddington preferred
supporting a federal organization of about a dozen existing European laboratories. At the time some scientists were not at all comfortable with John's

proposal for a new European laboratory since some saw it strictly as an attempt by the elite of European biologists to solve problems peculiar to them.

Founding of EMBO and EMBC

At Ravello a majority of participants concluded that a European laboratory of fundamental biology should be created and several names were considered for the laboratory. It was decided that the enterprise should be European rather than worldwide, a plan was developed for fellowships, travel funds, and teaching of advanced courses, and there was agreement that a formal body, called EMBO, should be created. EMBO would promote the founding of a European laboratory and a federal organization of existing European laboratories would act as a central agency for discussion of international initiatives in molecular biology and undertake the establishment of a European fund for research in fundamental biology. Max and John were elected chairman and secretary general, respectively, of EMBO council, and it was proposed that EMBO membership should consist of 100–200 leading biologists, all from European countries. The founding of EMBO satisfied both John and Waddington; however, the Ravello report recognized the difficulties associated with establishment of a European laboratory and the problems inherent in dealing with so many different governments.

Early on, EMBO expenses were covered by modest sums from Israel and Interpharma, an association of research-based pharmaceutical companies in Switzerland. However, in 1965 a more substantial sum of $600,000 was provided EMBO for three years by the Volkswagen Foundation, which was founded in 1961. The grant was awarded to support international postdoctoral fellowships, accounting for nearly three-quarters of the funds, and meetings, travel, and advanced courses. At about the same time three committees were set up—an Executive Committee, Laboratory Subcommittee, and Federal Organization Subcommittee, chaired by Max, John, and Adriano Buzzati-Traverso, respectively.

Initially EMBO was to be registered in Belgium and René Goffin, a lawyer in Brussels, prepared a first draft of its constitution. However, there was resistance to registering the organization in Belgium, especially from Max Perutz and Jeffries Wyman (1901–1995), so within a short time it was decided to register the organization in Switzerland. A Swiss law firm, Helg, Grandjean, Lalive, and Picot, prepared its constitution and EMBO was incorporated in

Geneva as a non-profit organization under Swiss law. Just prior to incorporation, 140 leading scientists were chosen as EMBO's first members. In spring 1966 a booklet was prepared by Ray Appleyard (1922-2017) summarizing the history, current status, and objectives of EMBO. Appleyard, a radiation biologist at the European Atomic Energy Community in Brussels, had been appointed executive secretary of EMBO in 1965. In 1973 Appleyard was replaced by John Tooze (b. 1938), a highly respected and experienced molecular virologist who retained the position for more than 20 years.

The EMBC was founded in 1968 to serve as an intergovernmental organization that would fund EMBO and provide a support structure for EMBL. The scale of contributions from 14 countries in 1970 (Austria, Belgium, Denmark, France, Germany, Greece, Israel, Italy, the Netherlands, Norway, Spain, Sweden, Switzerland, and the UK) was based on their average national incomes for 1962 through 1964, with four countries, France, Germany, Italy, and the UK providing about 75 percent of the EMBC budget. EMBC would provide for training, teaching, and research scholarships, assist universities and other institutions of higher learning that wished to receive visiting professors, and establish programs of courses and meetings, with execution of these purposes entrusted to EMBO. In addition, EMBC would provide a framework to establish the EMBL.

John was elected secretary general of EMBC in 1970 and project leader for EMBL a year later. Ken Holmes (b. 1934) pointed out in his memoir of John for the Royal Society that "The incorporation of three posts in one person was administratively efficient but raised issues of omnipotence. John was persuaded to relinquish his EMBO and EMBC posts on becoming DG of EMBL at the end of 1974."[5] However, John was very unhappy about relinquishing these two secretary general positions since it left him only his position with EMBL.

Founding of EMBL

In May 1964 Léo Szilárd died of a heart attack in his sleep at the age of 66 and overnight John became the primary driving force for establishing a European laboratory. He worked together with a prestigious committee of scientists from several European countries, including Sydney Brenner from the UK, Adriano Buzzati-Traverso (1913-1983) from Italy, Arne Engström, Francois Jacob (1920-2013; Nobel/Physiology or Medicine, 1965) from

France, A.M. Liquori (1926–2000) from Italy, Ole Maaløe (1915–1988) from Denmark, Max Perutz from the UK, and Jeffries Wyman from Italy and the United States. John was encouraged by recent events and wrote to a friend about his long-term vision for European science:

> I have a feeling that in a generation from now we may find we have all sorts of European Laboratories, and perhaps a European Research Council with its own funds; and that indeed is what European cooperation will mean in a scientific context. Speaking for myself I should welcome such a development, and should be quite ready to see EMBO merged into some wider scheme. Perhaps naively, I am an enthusiast for international laboratories; properly set up (like CERN) they seem to me the most exciting places to work in, and to be good politically as well as scientifically.[6]

In November 1969 a three-day EMBO meeting was held in Konstanz, a city located on Lake Constance (Bodensee) in southern Germany, attended by 20 leading molecular biologists, cell biologists, and neurobiologists. At the meeting three working groups were set up, for cell genetics, subcellular structures, and instrumentation, chaired by Sydney Brenner, Vittorio Luzzati, and Hugh Huxley, respectively. The group decided that the European laboratory should consist of about 60 scientists, with only 5–10 having long-term contracts of six years or more, while the remainder would be on short-term contracts, senior visitors, or postdoctoral fellows. They estimated that the buildings should cost about £750,000 (about $1.8 million), equipment £1.25 million (about $3 million), and the annual budget £850,000 (about $2 million).

It was also decided in Konstanz that the new laboratory should focus its efforts on technology that was too large or too expensive for national laboratories. This ultimately resulted in setting up the first two outstations of EMBL: the Deutsches Elektronen-Synchrotron (DESY) in Hamburg, Germany in 1974, due in large part to the efforts of Ken Holmes and Gerd Rosenbaum; and the EMBL Outstation at the Institut Max von Laue-Paul Langevin (ILL) for neutron diffraction in Grenoble, France in 1975. Both outstations would provide synchrotron X-radiation and neutron diffraction for determination of high-resolution structures. These outstations became part of the proposal and helped convince European governments that EMBL could provide facilities not readily available to national laboratories. Ken Holmes has pointed out that by the 1980s,

Somewhat ironically, the real strength of synchrotron radiation was to be in protein crystallography. Synchrotron radiation transformed protein crystallography from an important but somewhat esoteric method into a very powerful general technique, later to take its place alongside DNA sequencing as one of the pillars of the new biology. The EMBL outstation at DESY Hamburg thrived and became very popular. In particular, for protein crystallography it made itself indispensable. Moreover, at critical moments the DESY outstation provided the EMBL with a raison d'être.[7]

The Konstanz meeting also resulted in adoption of several goals for EMBL. It was decided that the laboratory should focus on cell genetics and subcellular structures, development of advanced instrumentation and specialized computer applications, production of rare enzymes and other biological macromolecules, and provide advanced training of postdoctoral fellows, summer schools, advanced courses, and sufficient space for visitors to use the instrumentation at EMBL.

In the late 1960s it was unclear whether the UK would support the proposed European laboratory. John realized that "The UK is by a long way the least enthusiastic country over the EMBO laboratory . . . I find I differ from many of my colleagues who are against it. In most other countries the opinion is more favorable—as is shown by the offer of sites by Austria, France, and Italy. Several other countries are considering offers which shows their enthusiasm, but opinion in the UK is still quite cool."[8] There was a meeting at the Royal Society to discuss the proposed European laboratory and decide whether the UK should commit to supporting it. John provided background information for the meeting and, in the end, there was considerable support for the laboratory from several participants, but a number of others were dead set against it. The reasons for the lack of support varied, but in some cases the negative response reflected chauvinistic attitudes toward the rest of Europe, as well as the worry that it would reduce remaining funding for institutions in each of the funding countries, such as the UK.

Between 1969 and 1972, 14 governments—Austria, Belgium, Denmark, France, Germany, Greece, Israel, Italy, the Netherlands, Norway, Spain, Sweden, Switzerland, and the UK—participated in an intergovernmental action to ensure support for EMBO. Their financial contributions ranged from about 1 to 26 percent of EMBO's overall budget, with the largest contributions coming from France, Germany, Italy, and the UK, accounting for more than 80 percent of EMBO's budget. The number of governments supporting EMBL has more than doubled since 1972 and currently about 15 of the member states provide financial support.

Heidelberg Site

The proposed European laboratory was taken seriously enough in 1968 to elicit offers from Belgium, France, Greece, Italy, and Switzerland for a laboratory site in or near Brussels, Paris, Athens, Rome, or Geneva, respectively. In France scientists lobbied to build the laboratory in Nice, whereas Weisskopf strongly preferred a site in Geneva next to CERN on the Swiss-French border. Germany entered the competition in 1970, offering not only several potential sites for the laboratory, one near Munich, but sweetening the potential deal with a promise of 12 million DM (about $3.3 million) for capital expenditures. A Heidelberg site was promoted primarily by two scientists, Hermann Bujard at Heidelberg University and Rheinhold von Sengbusch at the Max Planck Institute for Medical Research in Heidelberg. Following extensive discussions, in 1971 Germany agreed in principle on Heidelberg as the site for the European laboratory. The site was located in Odenwald, east of the village of Rohrbach, on a hillside overlooking the Upper Rhine Plain. It was only 5 kilometers from Heidelberg University, while the Max Planck Institute for Nuclear Physics was next door.

The EMBC agreed to establish EMBL in Heidelberg in June 1972; an agreement was signed by 10 participating countries in May 1973 and certified by 10 member states in July 1974. John delivered an EMBL "Signature of Agreement" speech in Geneva on November 10, 1973 (appendix A.10). John was delighted that the laboratory was no longer just an item on the agenda of an intergovernmental meeting and expressed satisfaction in the choice of the site, especially since it was an ancient university city located at the center of Europe. He also liked the proximity of Heidelberg to Frankfurt airport and to Strasbourg, which was less than 60 miles away and there he could enjoy an excellent meal.

In January 1972 the 8-member building committee for the European laboratory met for the first time in Heidelberg; it was chaired by Arthur Rörsch from the Netherlands and John was a member. As a result of this meeting it was decided to choose an architect that had experience constructing biochemical or physical institutes, spoke German fluently, and had experience building in Germany. Further meetings were held in Göttingen in July and

September and resulted in a draft, titled "Information on the Proposed EMB Laboratory in Heidelberg, Germany, Concerning Construction of the Building—Provisional Programme of Requirements," and extensive discussions about potential architects. By June 1973 John had set up a project office in Heidelberg.

Architects from Denmark, England, Germany, Italy, the Netherlands, Sweden, and Switzerland were considered and, in the end, four were invited to take part in a competition: Henn and Petersen from Germany; Lange from Germany; Guillisen from Belgium; and Burckhardt from Switzerland. Following consultations with city authorities, the committee decided to appoint a jury for the competition consisting of three architects, Barcon from Spain, Hansen from Denmark, and Wilerval from France; three scientists, Brenner from the UK, Rörsch from the Netherlands, and Wyman from Italy and the United States; and one representative of the city of Heidelberg. The jury was assisted by a technical committee consisting of some members of the building committee and representatives of the building control office at Heidelberg University. The jury was to recommend the winner of the competition to John on the basis of a simple majority vote. A deadline of October 15, 1973, was set for the architects to submit their proposals.

In November John wrote to Lange in Mannheim, the architectural firm that had built the Max Planck Institute for Nuclear Physics in Heidelberg, to tell him that his firm had won the competition. The contract with Lange, Mitzloff, Böhm, and Müller was signed and the project finally presented to the finance committee of EMBC in fall 1974. Because of expected increases in building costs, a small working group consisting of representatives from Germany, the UK, and CERN was set up to examine financial aspects of the project as presented by the architects. They recommended that the budget should be increased to about 22 million DM (about $7.8 million), including 1 million DM (about $360,000) for the heating plant for which Germany assumed responsibility. EMBO's council gave its final consent for the project in December 1974.

Director-General

In 1974 John was appointed the first DG of EMBL. Initially the offer was a three-year appointment with two possible extensions of two years each, but was changed to a five-year appointment with a possible extension to seven years. With the two-year extension John would be 65 years old when the appointment ended. The appointment, which became effective in January 1975, carried an annual salary of 130,000 DM (about $53,000). EMBL also would provide travel expenses for John for up to a dozen journeys annually between Heidelberg and London. At about the same time, John applied for a secondment from the MRC, meaning that he would be detached from the MRC for temporary assignment to EMBL. John's annual salary from the MRC at the time, about £9,000 (about $22,000), would be paid into his bank account while he remained in Heidelberg and EMBL would reimburse the MRC for his salary and other related contributions. EMBL would then pay John the difference between his MRC salary and his salary as DG.

The MRC sent a letter to John at the end of January 1975 confirming that he would remain an MRC employee while DG of EMBL, and he remained a member of the MRC staff until he moved to St. John's in 1982. However, in the end he complained forcefully about his treatment by the council with respect to his salary and pension during the secondment. He felt strongly that his salary had not been handled in a way that was commensurate with salaries of others of equal or lesser rank at the LMB in Cambridge and that it was truly an injustice. The council felt otherwise and believed they had honored the spirit of the letter of agreement governing John's secondment to EMBL. As with the termination of his editorship at *JMB* and his directorship at EMBL, John's final detachment from the MRC took place on a very sour note.

In 1974 John expressed his plans for establishing EMBL:

We've passed beyond the stage in which a Watson and Crick can sit down and in a short time produce a dramatic result. So it's difficult to create the right conditions. But my policy is not to have hierarchies; to have a lot of small groups of youngish people, to give bright people the freedom to do what they want to do, and to interact with each other. I want to avoid a pyramid with a big boss on the top. One thing I'm happy about is that the governments, for their own reasons, have insisted that most people will be there on five-year contracts, not in tenured posts. It means that the lab is bound to have a continuous flow of people. This contrasts with the MRC's new tenure policy which allows more people to get tenure at a much earlier age. That's good for career structure, but I'm not sure it's the way to get really good work done in the lab.[9]

John's intention was to model the EMBL along the lines of the Cavendish and LMB in Cambridge since members of both laboratories had managed to accomplish so much outstanding research for such a long time.

Despite John's many years of diplomatic engagement in founding EMBL, he was extremely apprehensive about signing on to serve as DG. He spent a tumultuous night awake, pacing the floor, and was extremely agitated before finally signing the contract. John was not convinced that he really wanted the job. After more than a decade of extremely complex negotiations with scientists, administrators, and governments, John had very serious concerns and reservations about becoming DG of EMBL. He recognized the enormity of the job, especially the need for great interpersonal skills, and privately had serious self-doubts as to whether or not he was up to it.

Plans and Construction

John and his staff would need temporary space in Heidelberg while EMBL was in the planning (1972–1974) and building (1975–1977) stages. Administration space for EMBL was provided at the German Cancer Research Center, initially with Daniel Guggenbühl as head of administration, but followed shortly thereafter by Jack Embling and then by Bernard Bach. John wished to have two-thirds of the EMBL staff in place, about 175 people, by the time the building was completed in 1977. Prior to occupation of the new building, workshops, instrumentation, computing, and electron microscopy were located in temporary wooden huts adjacent to the Max Planck Institute for Nuclear Physics, close to the EMBL building site. Some research laboratories and a library were located downtown at the University of Heidelberg Biochemical Laboratory where Robert Bunsen (1811–1899), the famous German chemist, had worked. John anticipated that four offices and one laboratory would be needed in 1973 and that up to four additional offices and five additional laboratories would be needed through 1976. In the mid-1970s up to 50 percent of John's time was spent in Heidelberg where he had an office initially in the German Cancer Research Center and later in one of the two large wooden huts near the Max Planck Institute. The huts remained on site until 2006 when they were demolished to make way for a car park. Since John had spent some time in The Hut in Cambridge it must have been a déjà vu experience for him in Heidelberg.

For his Heidelberg home address, John finally settled on an apartment at 79 Panoramastrasse.

The architect's initial plans for EMBL were reviewed by a committee consisting of construction experts from Switzerland, the UK, and Germany, and several building elements were found to be too expensive. The architects and EMBL representatives produced a revised plan at the end of 1974 which included a cost reduction from 26 million DM (about $10 million) to 22 million DM (about $8.5 million) and in April 1975 documents for a call for bids on the project were issued. To ensure a timely start of construction an access road and supply lines for necessary services were installed under the supervision of the city of Heidelberg and paid for by Germany, and in September 1975 construction of the new building began. Internal construction and technical installations began in fall 1976 when the building's skeleton and roof were completed. Research groups started to move into the new building in fall 1977 and the move was completed by spring 1978. The buildings included the main laboratory, about 8,000 square meters, workshop, Szilárd Library headed by Mary Holmes, and administration, canteen, animal house, and a containment block.

John was deeply concerned with all aspects of EMBL construction, design, and fitting-out of the building. This is made abundantly clear in a letter sent by him to laboratory administrators in spring 1976:

> As you know, I attach great importance to problems of aesthetics and design in the Laboratory. Although there is no such thing as an absolute judgement in this area I am anxious that we should do what we can to prevent irreversible steps being taken without adequate consideration having been given to the aesthetic and design aspects. It is not easy to devise a mechanism to avoid this, because virtually any purchase or contract made by any branch of the Laboratory can involve aesthetic judgement. Since I happen to be very much interested in these questions I would like to be personally involved in all such decisions that have to be made.
>
> You are aware that I have asked Sir Misha Black, a distinguished designer, to act as my consultant in discussions with the architect about aesthetic and design aspects of the building . . . In addition there are other items, not directly architectural, which I shall be discussing with Sir Misha (and the architect) . . . In many cases purchases are being made of items which we currently use in our temporary accommodation, but which will be transferred to the new building when it is completed, and which, therefore,

require the same scrutiny as purchases which will only be made for use in the building . . . I am copying this minute to the members of the staff who are most likely to be directly concerned so that they may be aware of my wish to be personally consulted when problems of this kind arise in any of the areas listed above, or others that may turn up.[10]

Also in spring 1976, as Weisskopf had done at CERN, John addressed the international nature of EMBL in a note to administrators: "Referring to a recent conversation we had, I notice that the labels on the doors in the Administration building are still in German. As I said before, this laboratory is international and not German, and visitors should not gain a wrong impression."[11] In this context, apparently John never spoke German during his time as DG of EMBL, although he was fluent in the language.

Inauguration

An inauguration ceremony for the EMBL took place on Friday, May 5, 1978, that was attended by nearly twenty ambassadors or representatives of member states. Among the guests present at the inauguration were Federal Minister for Science and Research Hertha Firnberg of Austria, Federal Minister for Research and Technology Volker Hauff of Germany, Minister of Educational and Cultural Affairs J.-E. Wikström of Sweden, Minister of Culture Wilhelm Hahn of Baden-Württemberg, Lord Mayor Reinhold Zundel of Heidelberg, and many other officials and scientists from the member states. At the last minute President Walter Scheel of Germany was unable to attend the inauguration ceremony due to a visit from General Secretary of the Central Committee of the Communist Party of the Soviet Union Leonid Brezhnev (1964–1982), but he sent a message of congratulations to John.

John delivered an inauguration speech (fig. 13.1) and a scientific symposium was held to celebrate the opening. The symposium included lectures by Sydney Brenner from the UK (1927–2019; Nobel/Physiology or Medicine, 2002) on "Biological Complexity"; Manfred Eigen from Germany (1927–2019; Nobel/Chemistry, 1967) on "Molecular Biology and Molecular Logic of Life"; François Gros (b. 1925) from France on "Present Trends in Developmental Biology"; and Niels Jerne from Switzerland (1911–1994; Nobel/Physiology or Medicine, 1984) on "Molecules and Cells of the Immune System." Although Vicki Weisskopf was unable to attend the inauguration

Fig. 13.1 Photo of John Kendrew (ca. 61 yoa) giving a speech at the opening of EMBL, Heidelberg, May 1978. Reproduced from an item in the Kendrew Archives, Weston Library, Oxford

ceremony, he sent a letter to John marking the occasion, saying that he hoped that EMBL would be a symbol of everything great and worthwhile.

New Staff

By the time the EMBL building was occupied in 1978 it consisted of four divisions: Biological Structures, Cell Biology, Instrumentation, and Recombinant DNA. When it came to recruitment of group leaders, John depended heavily on his multi-national Scientific Advisory Committee (SAC; appendix A.11), first appointed in 1971 and consisting of 16 members, for recommendations and advice. John was not very successful in his attempts to recruit established senior scientists who would have provided immediate recognition and maturity for EMBL. Ken Holmes pointed out that "A division of neurosciences under Ricardo Miledi (1927–2017) never happened; neither did a nuclear magnetic resonance group under Ray Freeman (b. 1932); or a cell biology division under Lennart Philipson (1929–2011)."[12] On the other hand, John was quite successful in recruiting relatively young scientists who were interviewed either by himself or by one of his advisers, principally Sydney Brenner and Henry Harris (1925–2014) in the UK. This permitted scientists to be hired on three-year contracts, not to be extended beyond nine years, rather than as tenured appointments. When he addressed the issue of tenured appointments in 1978, John made it clear that it was something that he would restrict and resist. This seemed to assure reasonable turnover of staff and the real possibility of remaining at the cutting-edge with respect to new avenues of research and new technology. For above all else John felt that all work at EMBL, useful or useless, had to be excellent.

Between 1975 and 1978 about 26 group leaders set up independent laboratories within the four divisions at EMBL. They comprised about one-half of the 60 scientific staff positions planned. During these early years there were occasional disagreements between John and the EMBL council's finance committee. The latter felt that each year they funded salaries for scientists that had not been hired by John and consequently a significant monetary surplus accumulated at EMBL. The finance committee insisted strongly that this surplus should be deducted from the annual budget. John disagreed strongly and felt that by decreasing the budget the council was interfering with the running of EMBL. This sharp difference of opinion occasionally led to considerable friction between John and the council.

Among the first hires in 1975 were Arthur Jones, Kevin Leonard, Chica Schaller, Kai Simons, Nick Strausfeld, and Hans Weiss who brought expertise in electron microscopy, developmental biology, and membrane biochemistry to EMBL. More than 20 additional scientists joined EMBL between 1976 and 1978, significantly increasing the total number of group leaders hired over four years and greatly expanding the scientific expertise (table 13.1). By the end of 1978 there were about 200 people at EMBL, representing about 75 percent of the funded posts.

Since nearly all of the junior group leaders were young, on three-year contracts, and striving to establish their own independent research careers, there was an air of competitiveness at EMBL. At the same time there was a genuine camaraderie among most of the group leaders and a strong desire for EMBL to succeed. Many of John's early hires had very successful research careers at EMBL and elsewhere and went on to become leaders in their chosen areas of research. Over the years several of them were elected to membership in the Royal Society (UK), National Academy of Sciences (USA), and EMBO, and a few became Nobel laureates.

Among John's early hires for the Division of Cell Biology were Christiane (Janni) Nüsslein-Volhard (b. 1942) and Eric Wieschaus (b. 1947), two highly accomplished *Drosophila* geneticists. Both had worked in Walter Gehring's (1939–2014) laboratory in Basel, Switzerland, and in 1978 were hired as junior group leaders to carry out research on the genetic control of early embryonic development in *Drosophila*. They shared a tiny office, a tiny lab, and one technician. They worked more like independent postdoctoral fellows than group leaders, and apparently felt uncomfortable and unappreciated throughout their time at EMBL. Some of their colleagues felt that John should not have brought *Drosophila* into EMBL and that Janni and Eric were not doing the "right kind" of genetics. John spoke to Eric, but rarely to Janni.

While at EMBL, in October 1980, Janni and Eric published an article in the journal *Nature*, titled "Mutations affecting segment number and polarity in *Drosophila*," that over time became a classic in developmental biology. However, in the same year Eric accepted a job offer from Princeton

Table 13.1 Initial EMBL group leaders (1975–1978)

	Date Hired		Date Hired
Biological Structures		**Cell Biology**	
Hajo Delius	1976	Bernhard Dobberstein	1977
Jacques Dubochet	1978	Eric Jost	-
Brigitte Jockusch	1978	Daniel Louvard	1978
Reuben Leberman	1976	C. Nüsslein-Volhard	1978
*Kevin Leonard	1975	*Chica Schaller	1975
Don Marvin	1976	*Kai Simons	1975
Akira Tsugita	1978	*Nick Strausfeld	1975
Graham Warren	1977	Eric Wieschaus	1978
*Hans Weiss	1975	Marc Zabeau	1978
Instrumentation		**Recombinant DNA**	
Richard Herzog	1978	Hotse Bartlema	-
*Arthur Jones	1975	Hans Lehrach	1978
S.M. Provencher	-	Vince Pirrotta	-
Roelof van Resandt	1978		

*First hires for EMBL in 1975

University and Janni moved to the University of Tübingen, both without any counter-offer from John. Janni recalled that it would have been good had she been able to stay and establish a group with students at EMBL since at the time it was very difficult to find a suitable job in *Drosophila* developmental genetics.

In 1995 Janni and Eric, together with Ed Lewis (1918–2004) at the California Institute of Technology, were awarded the Nobel Prize in Physiology or Medicine for their discoveries concerning the genetic control of early embryonic development. Janni was the first German woman ever to receive a Nobel Prize. It is likely that John came to deeply regret not providing more substantial support and enthusiasm for Janni and Eric's research at EMBL since it was ultimately deemed worthy of a Nobel Prize.

Jacques Dubochet (b. 1942), a Swiss biophysicist who had used electron microscopy (EM) to study viruses and nucleic acids as a research student, was hired by John in 1978 as a group leader in the Division of Biological Structures. This was part of John's effort to develop cryo-EM for determination of the structure of biomolecules in solution. Jacques realized that to eliminate problems associated with the presence of liquid water in biological samples prepared for EM, the samples would have to be imaged in a frozen state. Together with his colleague Alasdair McDowall, Jacques devised a procedure for creating a film of non-crystalline solid water, so-called vitrified or amorphous water, on specimen grids used for EM. The procedure was an essential step that enabled the application of cryo-EM to a variety of biological samples and led to a revolution in structural biology.

In 2017 the Nobel Prize in Chemistry was awarded to Jacques Dubochet and Joachim Frank (b. 1940) at Columbia University, and to Richard Henderson at the MRC LMB for developing cryo-EM for the high-resolution structure determination of biomolecules in solution. At the time Jacques was an Honorary Professor of Biophysics at the University of Lausanne. John's instinct and vision in the 1970s had led to a Nobel Prize for research carried out by an early member of the EMBL staff.

Replacement

John's contract was due to expire at the end of December 1981, three months prior to his 65th birthday in March 1982. This term (1974–1981) included his five-year appointment (1974–1979) and a two-year extension (1979–1981).

EMBL rules and regulations stipulated that the DG should retire at the end of the month in which they turned 65. But by the end of 1977 John was already being queried by the SAC (table 13.2) about how long he intended to serve as DG. John responded that the probability of him remaining DG was high, but he could not exclude the possibility that he might wish to leave at the end of 1979. At about the same time it was decided that the SAC would be responsible for putting together a short list of potential candidates to replace John as DG. The matter would not be raised again until spring 1978.

At a meeting of the SAC in November 1979 a procedure for selecting the next DG was adopted. A Search Committee consisting of seven SAC members, including E.C. "Bill" Slater (1917–2016; Netherlands) chairman of the EMBL council; Michael Sela (b. 1924; Israel) chairman of the SAC; Hans Kornberg (b. 1928; UK) vice president of EMBO; Maurizio Brunori (b. 1937; Italy); Pierre Chambon (b. 1931; France); Niels Kjeldgaard (1926–2006; Denmark); and Peter Starlinger (1931–2017; Germany) was assembled. The Search Committee was asked to obtain names of potential candidates by informal discussion with delegates of the member states, SAC, and EMBO council, and other appropriate sources. Initially the committee had a list of nearly 30 names proposed, but by March 1980 had settled on four of six potential candidates: Roger Monier (1924–2008) from France, Lennart Philipson from Sweden, and Gottfried "Jeff" Schatz (1936–2015) and Charles Weissmann (b. 1931) from Switzerland. Two additional candidates withdrew almost immediately.

It had been decided that the new DG should not come from the UK or Germany, should be of an age that would permit continuation in office, and should be an authority on molecular biology at the Nobel level. Weissmann withdrew as a candidate soon after receiving an inquiry from the Search Committee and Schatz withdrew following a visit from Slater and a visit to EMBL. Consequently, Lennart Philipson, professor and director of the Institute for Microbiology at Uppsala University became the lead candidate. In the late 1970s Philipson had been a candidate for a group leader position at EMBL when John was actively recruiting new staff members, but it was decided not to hire him because of his overall research interests and his desire to have an unusually large research group.

In May 1980 a report was issued by the Search Committee. Several experts recommended that John's appointment be extended a few years beyond 1981, but the Search Committee felt it would be creating an unfortunate precedent should the first DG's appointment be prolonged appreciably

Table 13.2 Initial EMBL SAC members (1971–1985)

1971–1975	Sydney Brenner (UK)
1971–1982	Maurizio Brunori (Italy)
1971–1976	Manfred Eigen (Germany)
1971–1977	Paolo Fasella (Italy)
1971–1976	M. Grunberg-Manago (France)
1971–1976	Henry Harris (UK)
1971–1975	Niels Jerne (Switzerland)
1971–1975	Aaron Klug (UK)
1971–1975	Vittorio Luzzati (France)
1971–1978	Ole Maaløe (Denmark)
1971–1977	Peter Reichard (Sweden)
1971–1977	Werner Reichardt (Germany)
1971–1976	Arthur Rörsch (Netherlands)
1971–1975	René Thomas (Belgium)
1971–1974	Victor Weisskopf (USA)
1971–1978	Charles Weissmann (Switzerland)
1975–1976	Max Birnstiel (Switzerland)
1975–1977	Hans Tuppy (Austria)
1976–1981	Donald Caspar (USA)
1976–1981	Pierre Chambon (France)
1976–1981	Hugh Huxley (UK)
1976–1981	Michael Sela (Israel)
1976–1978	L. van Deenen (Netherlands)
1977–1978	Jean-Pierre Changeux (France)
1977–1982	E. Kellenberger (Switzerland)
1977–1979	Klaus Rajewsky (Germany)
1977–1982	Rudolf Rigler (Sweden)
1977–1982	Peter Walker (UK)
1978–1983	Alfred Gierer (Germany)
1978–1983	Ole Maaløe (Denmark)
1978–1983	Alberto Monroy (Italy)
1978–1983	Manfred Schweiger (Austria)
1979–1984	Henri Buc (France)
1979–1984	Jan Drenth (Netherlands)
1980–1985	Ernest Helmreich (Germany)

beyond December 31, 1981. While the Search Committee conceded that it might be impossible to appoint someone who measured up to John, they concluded that a time must come in any organization when someone else with their own qualities should take over. The report went on to say that it would be best during the transition if John and the future DG were in office simultaneously.

In July 1980 John received a letter from Kjeldgaard, now chairman of the EMBL council, informing him that they were happy to extend his contract to the end of September 1982, but not beyond. John had support from German, Israeli, and Swiss council members for an appointment extended beyond September 1982, but the majority of members were against it. In August, Kjeldgaard tried to arrange a meeting between John, Philipson, and himself, but John declined. He felt that his involvement in such a meeting would be improper since the council had not formally appointed Philipson the new DG. At about the same time John wrote to Philipson to let him know that he would not be in Heidelberg during his upcoming September visit.

In November 1980 the EMBL council appointed Philipson DG for a five-year period beginning in April 1982. John's adverse feelings about the appointment of his successor are blatantly revealed in a letter sent to a colleague:

> I hope you will excuse this personal note sent to you at rather short notice. Without going into details, I want to inform you that during the meeting of our laboratory Council earlier this week that matter of appointing my successor (after I leave Heidelberg at the end of March 1982) was handled in a manner extremely disagreeable to me, and as a result my own personal relations with our Council and its Chairman, as well as with my successor, are now extremely difficult. Under the circumstances I hope you will understand me if I say that I would find it impossible to give a lecture to your Academy in January about the development and prospects of EMBL; psychologically the only talk I could give at the moment would be one so negative in tone that it would be wiser to keep it unsaid.[13]

Several months later John sent a letter to Philipson in which he expressed his thoughts on the DG appointment procedure:

> I have been intending for some time to write you a personal letter in order to offer you some words of explanation, from my side, of the difficulties we have been going through.

I should say for a start that I never had the slightest wish to continue as DG after about the middle of 1982, and at no time requested, or even considered requesting, any extension beyond that period; though I did say that if there were a real difficulty in appointing my successor I would of course be ready to continue for a limited period in order not to leave the Laboratory without a head.

Well, as you know, my successor has been appointed!—and the difficulty for me has been, not with the substance of that appointment but with the manner in which it was carried out. For a start I think the whole process began about 12 months too soon; but that is not important. The main point for me is that in my opinion I was treated by Council, and in particular by certain members of it, in a quite intolerable and abominable manner. I do not forget or forgive this, and my whole object from then on has been and still is to have the minimum possible connexion or collaboration with Council. This is my point of view, and things may be different; but I feel deeply injured in ways that I do not believe it would be right for me to set down on paper.

Thus my quarrel is with Council, and by no means with you. Unfortunately, but perhaps inevitably, however, the virtual breakdown of diplomatic relations between me and Council has affected you, and I would not be surprised if you are upset by this. I do assure you that none of this has been directed in any way at you personally, nor at the choice of you as my successor, and I am indeed sorry if you have been the innocent sacrifice pinned between two opposing forces. I would add (as I have mentioned before) that I am perfectly satisfied with your assurances as to the future program of the laboratory; and furthermore that I wish you well in every way in taking up the job that I am laying down.

It is true, I think, that in England we have a very different tradition from the Continental one; with us it is normal that hand-over periods are minimized, and the jobs I have taken in the past I have usually picked up with literally only an afternoon's overlap—if that. Perhaps in England we carry this too far; on the other hand I think that an overlap which is too long has great disadvantages too, for both the outgoing and the incoming occupant of the position. But I would not wish to maintain any particular philosophical standpoint about this.

Really the object of this letter is to say that of course there are many matters that I believe we could usefully discuss together either sooner or later, and I am anxious to do this provided only that it is clearly understood that this would be a matter between you and me only, and that I intend to minimize all dealings with Council, and particularly with certain members of it whose behaviour towards me has in my opinion been unforgiveable. I also believe that, having by now long experience in dealing with Councils and with national delegates, I might be able to give you some useful hints and tips in this area; as the Laboratory becomes better established the forces of bureaucracy become even stronger, and an important part of the function of any DG will be to maintain a very strong stand vis-à-vis the Council, largely made up as it is of professional (and some amateur!) bureaucrats of only middle or lower level.

In summary, I have written this letter, which is intended only for your personal eye, to say that I am indeed sorry that you have been the innocent victim of a quarrel that was none of your making, and that I am very ready to discuss any problems that you would like to discuss—indeed there are many that, sooner or later, I shall wish to raise with you. None of them are urgent from my side, but I shall look forward to your suggestions and then perhaps we can arrange to meet (I think you will understand, in view of the remarks I have made above, that I would prefer that any such meeting in Heidelberg should not take place at the time of a Council meeting).[14]

Soon after John wrote to Kjeldgaard concerning plans for future meetings and other engagements prior to John's departure from EMBL:

It has come to my notice that various plans have been under consideration involving the Laboratory Council, its Chairman, and my successor, in connection with my departure from the Laboratory, including for example special meetings, celebrations and press conferences.

I am writing to make it quite clear, in case I have not already done so, that any such arrangements involving the Council or my successor would be absolutely contrary to my wishes, and that I certainly would not be present at, or take part in, any of them if they were to take place.

I must also make it plain that I expect to be permitted to remain undisturbed during the last weeks of my office here, up to and including 31 March 1982, and that I would not welcome visits from, or the presence of, any of these persons during this period.[15]

For his part Philipson was very upset that John refused to get together with him before his starting date of April 1982. But he was particularly annoyed that John, as one of his final acts as DG, gave tenure to five staff members—Hajo Delius, Jacques Dubochet, Arthur Jones, Kevin Leonard, and Nick Strausfeld. Philipson had requested that John not grant tenure to any additional staff members without consultation with himself or the chairman of the council. Shortly after taking over as DG Philipson spoke with his attorney to assess whether the five tenure contracts could be nullified, but no action was ever taken.

Summing Up

At a council meeting in 1982 it was made clear that all of European molecular biology owed a great debt of gratitude to John without whom there would have been no EMBL. They felt that only John with his scientific vision and coherency, diplomatic and organizational skills, charm and stubbornness could have brought about the creation of EMBL.

In John's biography for the Royal Society's Biographical Memoirs, Ken Holmes commented that

> The actual EMBL turned out to be rather different from the plan. It was a great place for young scientists to get a toehold, particularly if they had been in the USA as postdoctoral workers . . . The scientists were young and vigorous, and as soon as the new building was opened the laboratory took off with amazing rapidity. Much of the strength turned out to be in establishing networks, rather than monolithic "big science." It was a truly pan-European laboratory that quickly became so indispensable that it was difficult to imagine how one could have done without it.[16]

When John left EMBL in March 1982 the laboratory had a staff of about 300 people and an annual budget of 40 million DM (about $18 million). During the 45 years since John's appointment as the first DG of EMBL, Lennart Philipson (1982–1993), Fotis Kafatos (1993–2005), Iain Mattaj (2005–2019), and Edith Heard (2019–) followed him in the role. EMBL currently has more than 1,600 employees with backgrounds in biology, chemistry, physics, and computer science. It consists of 24 member states, with more than 60 percent of its total support, more than 100 million Euros (about $125 million), provided by France, Germany, Italy, and the UK. EMBL now operates from six different cities in Europe : Heidelberg (flagship laboratory), Hamburg

(DESY, structural biology), Hinxton (EBI, bioinformatics), Grenoble (ILL, structural biology), Rome (epigenetics and neurobiology), and Barcelona (tissue biology and disease modeling).

In 2014, 40 years after the founding of EMBL, its Alumni Relations Office produced a brochure to commemorate the occasion and wrote the following tribute to John:

> EMBL was created to provide Europe with a critical international resource for its scientists and to cultivate both cooperative action and intellectual excellence. It still does so. The Heidelberg campus and EMBL's expanded set of Outstations continue to offer priceless resources to investigators across Europe and throughout the world. It continues to train some of Europe's brightest young investigators in cutting-edge scientific techniques and leadership skills—and those who pass through its revolving doors still export those important skills to academic and industrial research settings throughout Europe.
>
> Each of these scientists carry with them a deeply embedded belief in the value of independent research, yet couple this with a highly collaborative philosophy. Almost every EMBL-trained scientist will testify to his or her rich network of professional relationships in laboratories across Europe.
>
> EMBL still works—more effectively now than ever. This was a gift to Europe—a gift from Sir John Kendrew, protected and extended by three Directors General who have followed in his footsteps, as well as the scientists and all the support that they could rely on, who have worked so hard to realise Kendrew's original lofty expectations.[17]

In spring 1992, while retired and living in The Old Guildhall, John ended a letter to Konrad Müller, a former head of personnel and administration at EMBL, by saying that he thought that "his time at EMBL was the most exciting, interesting, and enjoyable period of his whole life." A decade had passed since John had left EMBL and apparently his anger had lessened considerably about the process by which he was replaced as DG. However, John did not return to visit EMBL until April 1997, shortly before his death.

14

St. John's, Retirement, and Death
(1982–1997)

[T]he college enjoyed something of a golden period under his presidency.

Paul Harris

St. John's College

Thomas White (1492–1567), master of the Merchant Taylors' Company in London, founded St. John's in Oxford in 1555, the first Oxford college to be founded by a merchant not a member of the clergy. The company began in 1327 as the Company of Tailors and Linen Armouras, a guild of tailors known as the Fraternity of St. John the Baptist. The site chosen for St. John's was St. Bernard's College, a monastery and house founded in 1437 for Cistercian monks studying at Oxford. The property was purchased from Christ Church, and the first members took up residence in 1557. Early on St. John's was primarily a college for educating Anglican clergymen, but over time it gained a reputation for history, philosophy, economics, politics, law, and medicine.

St. John's is one of 38 colleges at Oxford and today is the wealthiest of the colleges due largely to its land holdings acquired during suburban development in the 19th century. The college occupies about 14 acres on St. Giles', with most of the buildings organized around seven quadrangles (fig. 14.1). Margaret J. Snowling (b. 1955), a highly respected British psychologist, was elected the 36th president of St. John's in 2012 and is the first female president of the college in its history, stretching more than 450 years.

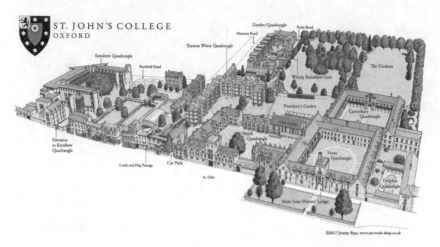

Fig. 14.1 Black-and-white reproduction of St. John's College pictorial map by Jeremy Bays at the Art-Work-Shop, Oxford (www.art-work-shop.co.uk). The Kendrew Quadrangle is depicted on the extreme left of the pictorial map

New President

Richard Southern (1912–2001), a preeminent English medieval historian and author, was elected the 32nd president of St. John's in 1969 and retained the position for 12 years. In 1981 Southern approached the end of his term and was encumbered by progressive deafness. A search committee headed by Mark Freedland, at the time vice president of St. John's and professor of employment law, was formed to select a successor to Southern. The vacant position was not advertised.

Two members of the search committee who were fellows of St. John's, one a chemist and the other a biologist, proposed John for the position. It was an unusual proposal since a scientist had never been president of St. John's and, although born in Oxford, John had never been affiliated with the university during his career. Only one other man not previously affiliated with Oxford had been elected college president. However, this was considered an exciting suggestion and appointing John as president would be considered a revolutionary move by Oxford. It was pointed out by members of the committee that while there were very good scientists at Oxford, John was a preeminent scientist and truly in a different category. He was a Nobel laureate who had the breadth of cultivation necessary for the job and could readily overcome the hurdles of electing a scientist as president of St. John's.

In due course Freedland paid a visit to John at EMBL in Heidelberg, broached the possibility of him becoming a candidate for president of St. John's, and described some of the conditions for the appointment. Of course John knew at the time that his appointment as DG of EMBL was soon coming to an end, but this was not revealed to Freedland during his visit. John expressed some interest in the position and subsequently paid a visit to St. John's where he greatly impressed everyone. John was selected by the 50 or so members of St. John's Governing Body, was offered the position, and after a few weeks delay became the 33rd president of the college for a five-year term beginning in 1982 (fig. 14.2). It was anticipated that John would retire from St. John's in 1987 at the age of 70.

The college president's major role is to serve as chair of the Governing Body which meets four times each term and as chair of College Committees which meet three times each term. The president also meets with undergraduates

John Cowdery Kendrew, 1917-1997

Fig. 14.2 Black-and-white reproduction of Presidential portrait of John Kendrew, 33rd President of St. John's College, Oxford (1982–1987). The portrait by Ruskin Spear is reproduced with permission of the President and Fellows of St. John's College, Oxford

individually at the end of every term, three times annually, and with graduate students individually once a year. The president may also become involved in the administration of the university, but apparently John did not. According to Ross McKibbin, a fellow at St. John's, "[John] concluded on arrival [at St. John's] that the College was very well-run, as it was, and not even Nobel prize-winners should interfere with things that are very well-run . . . ."[1] On the other hand, consistent with his penchant for architecture and good taste, John did arrange and oversee a significant renovation of the president's quarters at St. John's. A Portuguese woman, Olivia Faria, functioned as John's housekeeper and chef during his presidency.

Some colleagues found John socially distant and reluctant to do what is generally expected of a college president, such as leading the fellowship into chapel on Sundays. John was present at chapel on very formal occasions, but was habitually absent on Sundays, spending most of his weekends at The Old Guildhall in Linton rather than in Oxford. It was common knowledge among members of the college that John's automobile, a high-powered Saab, routinely was pointed in the direction of Linton on Friday afternoons. John boasted that he had devised the shortest and quickest route between Oxford and Linton. Overall, John led a quiet, contained existence at St. John's and did not socialize very much.

John was most concerned with academic prowess at the college and his presidency was marked by significant academic achievement. In both 1983 and 1984 St. John's regained its place at the top of the academic rankings of Oxford colleges (Norrington Table); it had ranked fourth from the top in 1982. Paul Harris, a fellow at St. John's, recalled that, whereas Richard Southern had been an active president of the college, "John Kendrew's presidency was different in style. His very presence constituted both a good model and an inspiration for intellectual achievement. Indeed, the college enjoyed something of a golden period under his presidency, particularly in terms of undergraduate achievement. If and when problems emerged, he stepped in effectively and sorted it out. In retrospect, his presidency seems almost uneventful, but that was part of his success."[2]

Other Activities

In addition to his duties as president, John continued his many roles outside the college, including as editor-in-chief of *JMB* and president of the International Council of Scientific Unions (ICSU).

In August 1984 John attended an international conference on science and technology education and future needs in Bangalore, India, as president of ICSU. Described in the Indian press as elderly but sprightly, John addressed several significant issues during the conference, including the teaching of science, social responsibility of scientists, world climate research, problems of the environment, and dangers of a nuclear arms race. About the latter he commented that nuclear proliferation continued in the world and advocates for slowing it down should become much more forceful. He characterized the situation as very depressing. John acknowledged to the press that he had moved away from research to formulating science policy and teaching and was actively involved with several international organizations. High praise for John's contributions to such organizations came from a colleague who found his "skills at inducing convergence from the most unpromising boundary conditions, have been honed to razor sharpness by years of experience in international scientific organizations. He possesses a most enviable flair for choosing the right path at the right moment when a shift in debate, imperceptible to the other participants, appears to him as a highway to consensus."[3]

In 1984 John was appointed chairman of a high-energy particle physics (HEPP) committee reviewing UK participation at CERN, the largest particle physics laboratory in the world. CERN's annual budget came from subscriptions based on each member country's gross national product. Nearly 10 percent of the UK science budget, £570 million (about $800 million), was allocated for nuclear physics, of which more than 60 percent went to CERN, making the UK the third largest subscriber. The review was commissioned as part of Prime Minister Margaret Thatcher's (1925-2013) overall government inquiry and was charged with evaluating whether the country was spending too much on HEPP in view of the serious shortage of funds for support of basic research. The government agreed that if it was decided to decrease its support for CERN, the funds saved would be redeployed within the UK science budget, then in crisis.

The committee chaired by John included Douglas Hague (1926-2015), a member of the Economic and Social Research Council; Jack Lewis (1928-2014), a physical chemist; Kenneth Pounds (b. 1934) an X-ray astronomer; Francis Tombs (b. 1924), a member of the UK Atomic Energy Authority; and Christopher Llewellyn-Smith (b. 1942), a high-energy physics theoretician who served as a specialist advisor. The committee obtained written and oral evidence from many individuals and organizations; and in June 1985 John informed the Parliamentary and Scientific Committee in London that "Britain is spending too much on high-energy particle physics as a proportion of the science budget. We should try to stay in the subject . . . But we should reduce our support for the subject substantially for the sake of other areas of research . . . ."[4] John's committee recommended cutting the UK subscription to CERN and support for domestic HEPP research by 25 percent by 1990 which would result in savings of £56 million (about $78 million) through 1992. The recommendation caused an immediate stir in the UK press with headlines such as "Kendrew dismays physicists" in the *London Times* and "Kendrew takes knife to high-energy physics" in *Nature*. John became the target of widespread criticism from many notable scientists and politicians, scientific journals, and the news media. Exacerbating the response was a dissenting opinion from Llewellyn-Smith who did not believe that a reduction could be attained without significant detriment to CERN's position in HEPP.

The recommendation to decrease the UK's domestic particle physics expenditure was implemented, however, the proposed cuts to the UK's CERN subscription were not since it was fixed by international treaty. Another committee chaired by Anatole Abragam (1914–2011) was set up by the CERN Council in 1986 to look at the implications of a budget cut on CERN. They recommended cost saving measures that would result in a 7 percent reduction in the UK's contribution to CERN and emphasized that a larger reduction would definitely jeopardize their preeminence in particle physics.

John was especially active with ICSU, founded in 1931 to promote international scientific activity in different branches of science and its application for the benefit of humanity. It is a non-governmental organization whose headquarters are in Paris and consists of 122 members representing national scientific bodies of 142 countries and 30 members representing international scientific unions.

John was secretary general of ICSU from 1974 to 1980, vice president from 1982 to 1983, president from 1983 to 1988, and past president from 1988 to 1990. As president he attended biennial ICSU General Assemblies held in Ottowa, Canada (1984), Berne, Switzerland (1986), and Beijing, China (1988). In addition, John took part in regular meetings of the Executive Board and Standing Committees and undertook assignments between 1974 and 1988 that took him to Istanbul, Paris, Washington, Budapest, Athens, Brussels, Amsterdam, and other destinations.

After John's death in 1997, two people very familiar with his activities at ICSU, Richard Keynes (1919–2010) and Julia Marton-Lefèvre (b. 1946), provided some insight into John's accomplishments at ICSU. At a memorial for John in Cambridge, Richard Keynes recalled that

In 1974 John was appointed as Secretary General of ICSU, at a time when the organization was in need of a new injection of life, and after six years in this post, and then another as President-elect and five as President, he left ICSU in a more flourishing condition. One of his most significant contributions was undoubtedly to appoint Julia Marton-Lefèvre as Assistant Executive Secretary in 1978, for she and he later worked together with great success ...

Having served on ICSU's Executive Committee myself for much of this period, I greatly admired John's invariable competence and patience as President, especially during the interminable negotiations leading to the readmission of the representatives of the China Association of Science and Technology in Beijing as a National Member of ICSU. The difficulty was not that ICSU didn't warmly welcome the adherence of the People's Republic of China, but that their condition for joining was that Taiwan must first be expelled. And since Taiwan had some years earlier succeeded to the old-established membership of the Academia Sinica, there was nothing in ICSU's Statutes to justify their expulsion. The battle waged to and fro endlessly for many months, but thanks to John's superb diplomacy, a successful compromise was eventually reached with the People's Republic and Taiwan perhaps not sitting side by side at meetings, but at least in the same room.

It was under John's presidency that ICSU launched what has now become one of its most successful projects, to conduct a multi-disciplinary scientific examination of the earth system known as IGBP, the International Geosphere-Biosphere Programme: a Study of Global Change. It was also under his leadership that ICSU began to undertake studies in genetics and biotechnology, and that the committees entitled COGENE [Committee on Genetic Experimentation] and COBIOTECH [Committee for Biotechnology] grew up in the forest of acronyms. He also backed strongly ICSU's initiatives in partnership with UNESCO [United Nations Educational, Scientific, and Cultural Organization] to help the scientists of the developing countries to help themselves through formation of the International Biosciences Networks, founded in 1981 well before "networking" had become the buzzword that it is today.

On a less grand scale it was John who encouraged the ICSU Secretariat to join the informatics age, and as Maureen and Tish, who we are glad to see here today as ICSU's representatives [at the Memorial in Cambridge for John], will attest, he was their mentor and guide when they coped with their first computers and joined INTERNET.

John was the longest serving Officer that ICSU has ever had, and he will long be remembered with gratitude and affection for his invaluable contribution to the organization of international science.[5]

In an obituary that appeared in *Science International* in December 1997, Julia Marton-Lefèvre, an assistant executive secretary at ICSU, also commented on John's role at ICSU:

Those of us who were part of John's international world know how much this also meant to him and know also how very much he contributed to our efforts. Sir John Kendrew held the position of ICSU Secretary General from 1974 to 1980 and then President-elect and President from 1982 to 1988. No other person has served as an Officer for so long since ICSU's creation in 1931. John Kendrew's tenure in these important ICSU positions was by no means a symbolic one: he truly cared about international cooperation and about ICSU, and was there at the time when momentous decisions were made which have survived and which will be an important part of ICSU's mission well into the next century. It was under his presidency that ICSU convened a meeting at the Schloss Ringberg in 1985 which began to look at a New Agenda for International Science and which is being followed up by the ambitious assessment ICSU is undergoing presently. It was also under his presidency that ICSU took the step of undertaking what has turned out to be the most ambitious and comprehensive study of the earth system. In fact John Kendrew felt so strongly about these issues that he even chaired the planning group which led in 1986 to the official launching of the International Geosphere-Biosphere Programme: A Study of Global Change (IGBP). It was similarly under his leadership that ICSU began to undertake activities in genetics and subsequently in biotechnology, that the ICSU's activities in developing countries grew in importance. John Kendrew again felt so strongly about the latter that he was even Chairman of COSTED [Committee on Science and Technology for Developing Countries] for a few years. He was an outspoken supporter of UNESCO even when his own country decided to leave that organization, and I am sure that it warmed his heart to know, before his death, that the UK had rejoined that important international forum. On a less grand scale it was John Kendrew who encouraged the ICSU Secretariat to join the informatics age and he was our mentor and our guide for many years when we coped with our first computers. He continued to follow ICSU's progress in that domain through Internet, of which he was an accomplished user.

An entire history book on ICSU could be constructed around the contributions of this single individual who managed in one lifetime to be so deeply involved in so many facets of this organization. In the ICSU world, John's private passions for music and art found many sympathizers, and for a while when he was in office we even had an informal opera and art club, managing to add a little time to our official meetings to visit the world's art treasures and to take in an opera or two while we traveled the world on ICSU business.[6]

Retirement

The final decade of John's life was spent in retirement, living in The Old Guildhall in Linton where he was able to enjoy his garden and surroundings, though he travelled abroad extensively during this period with trips to Belgium, China, France, Germany, India, Israel, Italy, Japan, the Netherlands, Spain, Switzerland, and the United States. He also spent considerable time in Cambridge, London, and Oxford, visiting friends, dining in college, and attending college functions and concerts. John gave occasional talks, as for example at the 1988 American Crystallographic Association annual meeting in Philadelphia in 1988 where he spoke about "Protein Crystallography in the Cavendish Laboratory in the 1950s." He also took on the editorship of the *Encyclopedia of Molecular Biology* which appeared in 1994, included more than 150 contributors, and consisted of nearly 1,200 pages. But beginning in about 1990, after being treated for prostate cancer, John had to deal with escalating health problems that lingered for the remainder of his life. In this context, a female companion pointed out that "John liked an exciting emotional life and when he was ill life was not exciting anymore."[7]

Near the end of April 1997, about one month after John's 80th birthday, a three-day celebration was held in his honor in Heidelberg. The celebration was organized by Ken and Mary Holmes and hosted by Fotis Kafatos (1940–2017), the third DG of EMBL. John had not been back to EMBL

since 1982 and he was terribly ill and quite frail. However, he was determined to attend the festivities and managed to travel to Heidelberg, but only
after obtaining some help from his physician in Cambridge. In conjunc-
tion with the Heidelberg celebration a one-day scientific symposium, titled
"Proteins as Engines," was organized by Ken Holmes with Sydney Brenner,
Robert Huber, Hugh Huxley, Anthony Hyman, Joel Sussmann, and others
as speakers. Several others in attendance, including Don Caspar, Bernard
Dobberstein, Jacques Dubochet, Freddie Gutfreund, Ken Holmes, Karl Korz,
Heiner Schirmer, and Hans Weiss, paid tribute to John for founding EMBL
and for serving as its first DG. John gave an after-dinner speech in which he
reminisced about events and people that led to the establishment of EMBL
and about the early days there. Ken Holmes recalled that one of the things
that John valued highly at the celebration was the chance to have dinner with
the workshop, technical, and service staff that he had known at EMBL. The
latter was in keeping with John's excellent rapport with workshop staff at the
Cavendish and LMB many years earlier. For John this celebration proved to
be a final get-together with friends and colleagues, some of whom he had
known for several decades, as well as a final opportunity to recall and share
old memories.

In summer 1997 John received an Honorary Doctor of Law degree from
Cambridge University and a party was held at Peterhouse to celebrate the
occasion (fig. 14.3). He was clearly delighted to receive both the honorary degree and the attention, however, he continued to remain in very poor health
throughout the summer months and in August was admitted into hospital.

Death and Bequest

On Saturday, August 23, 1997, John passed away at the age of 80 from
complications of metastatic prostate cancer at the Evelyn Hospital in
Cambridge. He was cremated in a private ceremony without any religious
observances as stipulated in John's final will, and his ashes were deposited in
Bagley Wood in Oxford by three friends from St. John's College.

John's obituary appeared soon after in newspapers around the world, including the *Independent, New York Times*, and *Washington Post*, as well as
in several scientific journals (Appendix A.2). Thus ended the remarkable
life of an extraordinary scientist, scholar, administrator, and internationalist. A Nobel laureate in chemistry who together with a few other gifted

Fig. 14.3 Photo of John Kendrew (ca. 80 yoa), right, and Max Perutz (ca. 83 yoa), left, in 1997 at Peterhouse, Cambridge, on the occasion of John receiving an Honorary Doctor of Law degree from Cambridge University. Reproduced from an item in the Kendrew Archives, Weston Library, Oxford

individuals in Europe and America founded the discipline of molecular biology and thereby brought about the biological revolution of the second half of the 20th century.

In the 1970s the executors of John's will were his long-time companion Ruth Harris and close friend John Graham. However, the executors of John's last will, dated December 19, 1991, were John Montgomery, emeritus fellow of St. John's, Oxford, and Noel Cannon, certified accountant, Cambridge. Among those receiving legacies in the will were the Musicians' Benevolent Fund, National Art Collections Fund, and the Glynn Research Foundation Ltd.; a Kendrew Fund was established at St. John's for promotion of music and the visual arts. In addition, relatively small sums were provided for a few individuals, including John's housekeeper and gardener in Linton, and lifetime annuities were provided for two longtime female companions.

For more than 60 years John was truly a Cambridge man, with very strong ties to two colleges, Trinity where he was a major scholar as an undergraduate and Peterhouse where he was a postgraduate research fellow and very active member for several decades. He was elected an honorary fellow of Trinity in 1972 and Peterhouse in 1975 and received an Honorary Doctor of Law degree from Cambridge University in 1997, shortly before his death.

Despite very strong ties to Cambridge, unexpectedly John left the bulk of his considerable estate, amounting to more than £1.3 million (about $2 million), to St. John's where he had served as president for five years. The bequest was welcomed by St. John's, but came as a complete surprise to officers of Peterhouse who anticipated receiving at least a portion of John's assets following his death. It is possible that certain unpleasant experiences at Peterhouse, particularly at a time when John was going through a painful divorce, influenced his decision not to leave any of his estate to Peterhouse. On the other hand, in assessing John's bequest to St. John's, Ross McKibbin noted that "His bequest to the College [St. John's] sums him up: the affection for St. John's, the absolute exactness with which he specified how the bequest was to be spent, and the generosity of mind and spirit which underlay it."[8] It seems likely that John's strong ties to Oxford during his early childhood and late adulthood contributed significantly to his final decision to leave his considerable estate to St. John's.

Quadrangle and Memorials

About 13 years after John's death, in October 2010, St. John's opened the Kendrew Quadrangle on St. Giles' to the north of the main body of the College in John's honor. The quadrangle is only steps away from the Inland Revenue Office at 15 St. Giles' where his grandfather worked at the turn of the 20th century. According to Michael Scholar (b. 1942), who at the time was the 35th president of St. John's, the Kendrew Quadrangle may be the last new quadrangle in the center of Oxford. It is a large complex, about 6,000 square meters, that centers on a garden and large beech tree, providing substantial student and fellow housing; a gym and café; teaching, music, and event rooms; a law library; and other facilities. It commemorates John's service as the 33rd president of St. John's and as one of the college's greatest recent benefactors.

About two months after John's death, in November 1997, a Memorial Meeting in his honor was held in the University Music School, West Road, Cambridge. The meeting was presided over by John Meurig Thomas, master of Peterhouse where John had been a fellow for several decades. Following introductory remarks by Meurig Thomas, the ceremonies included addresses by several distinguished guests, including Hermann Bondi, Paul Harris, Ken Holmes, Ephraim Katzir-Katchalsky, Richard Keynes, Aaron Klug, Max Perutz, Kai Simons, and Jim Watson. The addresses were interspersed with musical items by Haydn, Beethoven, and Schubert, known to be among John's favorites.

As the Memorial Meeting concluded, Meurig Thomas recalled that "A few days before John Kendrew passed away a friend of his read to him a short poem on *Eternity* by William Blake:

> *He who bends to himself a joy*
> *Does the winged life destroy,*
> *But he who kisses the joy as its flies*
> *Lives in eternity's sunrise*

John remarked that this poem should never be forgotten."[9]

A similar Memorial Meeting for John was held in October 1997 in St. John's Chapel, Oxford. The gathering was presided over by Bill Hayes who followed John as the 34th president of St. John's. The commemoration included reminiscences by Max Perutz, an address by Ross McKibbin, John's colleague and friend at St. John's, and three musical interludes, including John's favorite suite in C major for unaccompanied cello by J.S. Bach.

Several accounts of aspects of John's life were presented by friends at the Oxford and Cambridge Memorials and two of these accounts, by Ross McKibbin and Paul Harris, are reproduced here, with their permission.

Ross McKibbin's Account

John Kendrew was the first scientist to be elected President of St. John's. He was a man of outstanding intellect who succeeded another outstanding scholar, Richard Southern, and consecutively they presided over a remarkable twenty years in the history of the College. Indeed, in the range of his intellectual interests Kendrew perhaps is equalled only by Laud himself;

but he also had common sense, which Laud, to his cost, frequently did not. Kendrew was born in 1917. His father W.G. Kendrew, was a geographer who became university reader in climatology. His mother, Evelyn Sandberg-Vavalà, was a distinguished historian of early renaissance Italian art. Although his parents separated when he was young he had affectionate, though complicated, relations with both of them and inherited many of their own intellectual concerns: from his father his interest in architecture, in nature and photography; from his mother his interest in art and music—Florence, where his mother lived for most of her adult life, he called his "second home." He was educated at Dragon School and Clifton, whose science teachers he admired. In 1936 he went to Trinity College Cambridge and graduated in 1939 with first-class honors in Chemistry. At the outbreak of war he went to the Air Ministry as a Junior Scientific Officer working on radar. In 1940 he joined Operations Research with Honorary RAF rank. Thereafter, he worked primarily on anti-surface vessel warfare in the Middle East and South East Asia. He was exceptionally successful in this role—he was one of a handful to have access to the ultra-enigma codes—and finished the war with the rank of wing commander.

By the end of the war he had decided to return to Cambridge to work on protein structure and thus began the celebrated partnership with Max Perutz. The determination of myoglobin's structure was immensely difficult and required tenacity, insight and a mastery of different techniques and technologies. His success in doing so was one of the great achievements of post-war British science and in 1962 he was awarded the Nobel Prize, jointly with Perutz. In 1947 he became a fellow of Peterhouse, continuing as director of studies and an active supervisor even after he received his Nobel Prize. That he took his teaching seriously is to be seen from the meticulous notes on scholarship and entrance examinations which survive. We see here also the intellectual breadth: he was librarian and curator of pictures at Peterhouse—appropriately subjecting the pictures to X-ray analysis—and he taught the history and philosophy of science papers when they were established within the tripos.

Although he continued research into the 1960s and produced progressively refined models of myoglobin, he became increasingly committed to the development of molecular biology as a discipline. He was first editor of the Journal of Molecular Biology and was instrumental in the foundation of what eventually became the European Molecular Biology Laboratory, of which he was Project Leader and first Director General. The establishment of the Laboratory was an exhausting and political business and Kendrew was not always happy with what happened, but it was unquestionably another great achievement. His energies, however, were not confined to the Laboratory. He was a member of many British and international organizations—mostly scientific, but not all. His trusteeship of the British Museum was one of the positions which gave him most satisfaction. In 1981 he became President of St. John's and retired as President in 1987.

It would be presumptuous, indeed dangerous, for me to talk in any detail of his work as a scientist—in any case, that is public knowledge. But historians are supposed to delve in the past and it is possible to delve into John Kendrew's past. In doing so, the historian is surprised at the extent to which we can see the future President of St. John's in the confident young man of the 1940s. As President he was a meticulous record-keeper. But so he was all his life. He kept not only all correspondence to and from him, but, where he could lay his hands on it, correspondence about him. The Kendrew Archive, for example, contains, quite apart from his exercises and essays, 9,000 pages of notes he took as a schoolboy and undergraduate. All of these also contain detailed comments on his teachers, on their lectures and tutorials, as, I have no doubt, he made detailed comments on all the fellows of St. John's. The style is recognizable to those who knew him as President: austere, elegant, often witty. And the letters display the kind of self-confidence which undoubtedly was a part of his later life. In November 1943, of a proposed transfer to South East Asia from the Middle East, he wrote: "I think I am prepared to go if they make me W/Cdr, O.i.C., and Scientific Advisor to the Air C-in-C. A big order, but them's my terms." And they were the terms he got. In contemplating this compulsive preservation and categorization, utterly beyond what most of us would ever consider, the historian is struck by two things. The first is Kendrew's desire to be in control of his life and future, and the second, it seems to me, is a sense, even as an operations research officer in the war, that he had a place in history and that others would wish to know and understand this place. The archive suggests, in embryo, the rather grand and authoritative figure we knew as President. And he was grand. On the first occasion I spoke to him the conversation rather improbably turned to King Leopold III of Belgium. When I commented that his marriage to the Princess de Rethy had made him unpopular, Kendrew said that was very unfair. He knew the Princess de Rethy: in fact he had been driving through Brussels with the Princess de Rethy just the previous week. This reply, however, was definitely not name-dropping; just something he thought the well-informed person should know.

As President I think most of us regarded him, inter alia, as a problem-solver. It is true, he thought that there were real problems and other problems, and perhaps failed to realise that most of the problems a head of house is obliged to handle are "other" rather than "real." But when he was convinced of a problem's reality he approached it with both decision and care. This was so of the young Kendrew. Although his work during the war was primarily with anti-surface vessel operations, he actually worked on a wide variety of projects. To all of these he brought an immensely practical and methodical intelligence, best represented by what he called the "bible," the astonishing handbook he devised for bomber pilots which eventually became a standard text for all the Allied airforces. Kendrew was aware of his "practical" leanings and was, in a slightly defensive manner, suspicious of too much theory. Although, for instance, he freely acknowledged J.D. Bernal's influence on his later career, he was not always wholly an admirer of Bernal. In February 1943 he wrote of him that " . . . strange as it seems he has a number of good ideas," and the following month: "I saw a good deal of [Zuckerman] while he was here, and formed the impression that he is a good thing, with all the right ideas—once separated from his somewhat theoretically minded companion [Bernal]."

All this was no doubt related to his love for new technologies. Those who saw his music-reproduction systems can be in no doubt of this: he must surely have been the only President about whose noise undergraduates complained. It was not just computers or X-ray diffraction or hi-fi, it was also photography: it is doubtful if the country's most distinguished architectural historian would ever have been able to use his new photographic lens had not John Kendrew shown him how. Again, this fascination with technology was always so. There is in 1944 an almost self-parodying letter of a lost Parker pen which he wanted a friend to replace, he wrote: "As you know I like a moderately fine, moderately (but not too) hard, straight (i.e. not oblique) nib, and I enclose a specimen of my writing done with the old pen. I also like a pen with the largest possible ink capacity, discreet in color (preferably black). I should prefer to try one of these new-fangled devices which are sported by Menzies [a colleague] and yourself (Parker 51??), though Menzies' nib is a bit hard for my liking; I can also bear the grey and silver color of this pattern. I hear however that these may be difficult to come by these days, and if so my second choice would be a Parker of the standard pattern, plunger filling, large capacity, name I believe vacumatic, black . . . " And he ended: "Incidentally I should also be grateful if you can tell me of any other nice gadgetry now available in Washington D.C." But it would be misleading to quote only this, for the same letter includes an equally characteristic deflation: if his correspondent cannot get the Parker, then he should get any damn thing.

The love of gadgetry available in Washington D.C. might account for the noticeable Americanisation of his speech: for those who wrote to the young Kendrew, officers in the RAF or civil servants, men were almost invariably "chaps" or even "chappies"; but to Kendrew they were "good guys" or even "phonies," just as they were when he was President of St. John's.

Kendrew was a political man, and not just big science politics. He was a founder-signatory of the SDP [Social Democratic Party], the party created by the gang of four when they seceded from the Labour Party in 1982. It is easy to see why he did this. As a young man in the services, like so many young men of his age and education, he was strongly affected by wartime collectivism. In Jan. 1945 he wrote that he was "much . . . impressed by [Laski's] "Faith, Hope and Civilisation." I have also read miscellany including [C.H.] Waddington's "Science and Ethics" (leading to many arguments with the latter when he was here): Huxley, "Evolution—the Modern Synthesis": Bodmer, "The Loom of Language": Raven, "Science, Religion and the Future": Cole, "Fabian Socialism" and some others." He would certainly have voted Labour in 1945 had he not been accidentally disenfranchised by that army of "third-rate, neurotic and moronic clerks' against whom the young Kendrew was constantly battling, and he celebrated the Labour victory by getting "mildly drunk with one or two kindred spirits." He saw that victory presaging a New World of which he wanted to be a part. His New World was a particular kind of educated class-technological-governmental progressivism which was a powerful strand of 1930s and 1940s thought, particularly in his kind of science. Kendrew was himself attracted to government and he only reluctantly gave up the idea of the civil service in 1945. Throughout his career he was closely associated with the Ministry of Defense and government science policy—and it was for this as much as for his Nobel Prize that he was knighted. To a large extent it was this educated-class, pan-European progressivism that the SDP was trying to preserve, and his support for it should not surprise us.

Kendrew's world-view, as we might expect from a man of his politics, was remorselessly secular and rationalist. Furthermore, he retained his secularity to the last hour with a determination few of us possess. When life is over, life is over, and that is it. In fact, this was a position he reached not without effort. As a young man he was clearly interested in religion and human spirituality, even in something as comparatively recondite as Tibetan Buddhism, and in the relationship between science and religion. Nor, it seems likely, was he interested in them merely as intellectual problems. I suspect that his rejection of religion and his adherence to a rather positivist rationality was not instinctive but came after something of a mental struggle. And we might conjecture that "spirituality" was never banished from his personality: his intense preoccupation with art and music, especially music, was surely not entirely secular, and we know that he continued to be engaged with the relations between religion and science.

Finally, Kendrew the man. Mrs. Alton, who has catalogued much of Kendrew's archive for the Bodleian, wrote this of him: "on the one hand [there is] his methodical and analytical power, his meticulous not to say obsessive insistence on accuracy and comprehensive documentation . . . on the other hand, an aloofness or elusiveness of temperament which sets certain limits to personal relations. There are steadiness and control, a detachment combined with seemingly tireless application which constitute a formidable intellectual armoury and which are present from the earliest records." Anyone at St. John's during his Presidency, particularly anyone of his Vice-Presidents, was aware of his social unease; the reluctance to get involved. Yet the picture was certainly more complicated. During the war he was extremely sensitive to the needs of those under him and to their families and he was, after all, Director General of a very large organization. Furthermore, though he was rather awkward with undergraduates en masse, he was excellent with them as individuals. Sometimes too good: as a result of Presidential warmth and confidence-boosting we once lost a man who almost certainly would have got a first in history to psychology—where he did get a first. The aloofness was, in fact, a side of one of his virtues: his balance and his slight suspicion of excitability. Jim Watson, one of his post-docs, in *The Double Helix*—I quote from memory—gives an amusing example. He describes the scene in the Kendrew kitchen when, after a night on the town, he revealed to John and Elizabeth Kendrew that DNA was indeed double helical—Elizabeth was excited at the news; John took it more calmly. It is a scene most of us can imagine. A reluctance to overstate things was also part of Kendrew's modesty. He was asked once why he thought

Cambridge biochemistry was so brilliant in those years after the war: "Oh," he said, "it was Bragg and the money. I went for years getting wrong results and still the money kept coming in." There was something of the same in his attitude to the College: he had concluded on arrival that the College was very well-run, as it was, and not even Nobel prize-winners should interfere with things that are very well-run. And those who knew him only in retirement would probably be surprised to hear of the reputed coolness. As an ex-President he was loyal (a regular attender at domus dinners and progresses and surprisingly well-informed about those who were not), relaxed and charming; even within a few weeks of his death. His bequest to the College sums him up: the affection for St. John's, the absolute exactness with which he specified how the bequest was to be spent, and the generosity of mind and spirit which underlay it.

Ross McKibbin (b. 1942) is a renowned historian of modern Britain and was a lecturer in history at the University of Sydney, research lecturer at Christ Church, and tutor in modern history at St. John's, Oxford; he is currently an emeritus research fellow at St. John's. He has published many highly regarded works on modern British history and is a fellow of the British Academy and the Australian Academy of the Humanities. McKibbin was a colleague and friend of John during his presidency at St. John's and thereafter.

Paul Harris's Account

I want to tell you something of John Kendrew's later years—from 1981 to 1987—when he served as President of St. John's College in Oxford. After the description of scientific achievement that we have just heard, "Notes from College Life"—even when they concern the life of a college President— might seem anti-climactic. Still, the office is a varied one, with its own pitfalls and possibilities. It necessarily reveals a good deal about the holder.

Traditionally, St. John's had found it perfectly satisfactory to elect its presidents from within its own fellowship. The election of Sir Richard Southern—John Kendrew's predecessor—broke with that tradition, although he had admittedly been educated next door in Balliol. In that sense, the election of John Kendrew to the presidency was much more innovative, even exotic. Not only had he been educated in Cambridge, he was also the first scientist to hold the position, for more than 3 centuries.

Richard Southern had vigorously sought to raise the academic aspirations and indeed, the standing of the college. He was—as the expression goes—an "active" president. John Kendrew's presidency was different in style. His very presence constituted both a good model and an inspiration for intellectual achievement. Indeed, the college enjoyed something of a golden period under his presidency, particularly in terms of undergraduate achievement. If and when problems emerged, he stepped in effectively and sorted it out. In retrospect, his presidency seems almost uneventful but that was part of his success. Colleges are prone to dithering and dissent but both of those characteristics were kept well in check during his presidency.

He was particularly effective as chairman—dispassionate and authoritative. Even the chemists, not usually regarded in the college as an acquiescent or deferential group, were led along the path of accommodation. I recall myself being gently brought to a halt. We were interviewing a potential research fellow who was to work on language. I, as a psychologist, and another fellow, a social anthropologist—convinced that we knew something about the topic—and perhaps even convinced that we each knew more than the other—proceeded to argue about just how the topic should be pursued. John Kendrew intervened by suggesting that at least it sounded as if we agreed on the merits of the candidate. She was elected without further ado.

If the more intense side of his personality was rarely visible in his public persona, there were clues to his private passions, particularly his involvement in music, the visual arts, and technology. The undergraduates came to explain at one stage that the president's highly sophisticated CD system, put to powerful use in the lodgings, was keeping them awake at night. One of his first actions, on arrival at the college, was to refurbish the SCR guest rooms. None of us, he quietly pointed out, had ever stayed in one whereas we had accommodated him there on his initial visit to the college in connection with the election procedures. The point was taken and the refurbishment was a great success. Howard Colvin, an architectural historian in the college, was nonplussed by the operation of a new camera that he had bought in order to take photographs of buildings. The only person in the college who could explain its workings was John Kendrew.

His capacity for emotional engagement was easiest to detect when it concerned the students. Although he was somewhat wary of boisterous undergraduate celebrations, he was invariably supportive to those individual undergraduates who sought his help or advice. One of my own students had started off reading history but was tempted by the sirens of psychology. Too diffident to speak to me and anxious about offending the tutors in history, he eventually went to see John Kendrew. After listening to the dilemma, Kendrew's advice was simple. He told the student to stop worrying about his tutors' reactions and to pursue his own intellectual goals. The student saw me the following day, and the change of course went ahead.

After his retirement, and following his death, we at St. John's have increasingly realized the strength of feeling that he had for the college. He visited us regularly during his retirement, and his mood on these visits, even on the last, shortly before his death, was invariably relaxed and benign. The great generosity of his bequest to the college will enable us to nurture three of the attachments that I have mentioned: his love of music, his love of the visual arts, and his concern for the aspirations of students, especially those from the Third World.

In reading the obituaries of John Kendrew, one phrase caught my attention. He had suggested it as a title for a talk to be given by Francis Crick: "What mad pursuit." You'll recall that it comes from a description by Keats of a scene on a Grecian Urn—Keats' exquisite description conveys the unended nature of that pursuit, frozen as it is on the urn. At the time, it may have been an all too apt description of the apparently interminable task that Kendrew had set himself. Happily, however mad, the pursuit was brought to an end in real life. Few except John Kendrew could have managed it.

Paul Harris (b. 1946) is a renowned psychologist who has taught at the University of Lancaster, Free University of Amsterdam, and London School of Economics. He was lecturer in psychology, fellow, and tutor at St. John's, and professor of developmental psychology, Oxford. Harris is currently professor of education at Harvard University and emeritus fellow at St. John's.

Final Word

By any measure Sir John Cowdery Kendrew led an extraordinary life, enriched by his intellect and acumen, his love of science and technology, and his love of the arts. John was the epitome of a well-rounded or complete man. In background and demeanor he was quintessentially British, but John's words and deeds suggest he was a true European.

John and Max Perutz founded what became the MRC LMB in 1962. This highly acclaimed research institution has been home to a dozen or more Nobel Prize winners and more than 50 fellows of the Royal Society. What began as a two-man "Unit for Research on the Molecular Structure of Biological Systems" at the Cavendish Laboratory in 1947 has morphed in 55 years into a state-of-the-art facility in Cambridge employing more than 400 scientists.

As a staff member at the Cavendish Laboratory in the 1950s, John accomplished what was thought to be impossible at the time—determination of the atomic structure of a protein. John was the first to use X-ray diffraction, multiple isomorphous replacement, electronic computing, and model building to solve the three-dimensional structure of the protein myoglobin. With his *Nature* publication in 1958, the low-resolution structure of myoglobin, and in 1960, the high-resolution structure of myoglobin, he changed a "mad pursuit" into a monumental success. John's remarkable success in research earned him tremendous admiration and respect among scientists, as well as a share of the Nobel Prize in Chemistry in 1962 and many other awards and honors that were to follow. In 1958 John founded *JMB* and served as its editor-in-chief for 30 years. *JMB* was the first scientific journal dedicated exclusively to advances in the emerging field of molecular biology. As a result of his research and editorial achievements John became a key figure responsible for the rise of molecular biology in the late 1950s and its rapid expansion thereafter.

However, John was always his own man. In the early 1960s, while still relatively young and at the apex of his research career, John chose to leave the laboratory and take on what he considered to be far more important pursuits. He believed that science is an international endeavor that can be used to promote cooperation and peace. John worked diligently and resolutely over many years in different venues as a staunch advocate for international cooperation in science, especially among European nations. His considerable insight, efforts, diplomacy, and dedication led eventually to the founding of EMBO in 1964 and EMBL in 1973. John served as the first DG of EMBL and in doing so set the highest scientific standards by which EMBL continues to operate today.

The structure of myoglobin, MRC LMB, *JMB*, EMBO, and EMBL provide ample testimony to John's enormous contributions during his lifetime. Few scientists, even among Nobel laureates, have had such a significant, long-term impact on the practice of science. Today John is remembered by the annual Kendrew Memorial Lectures at the MRC LMB, Cambridge, and the Weizmann Institute of Science, Rehovot; a Kendrew Young Scientist

Award at EMBL, Heidelberg; a Kendrew Lecture Theater at the European Bioinformatics Institute (EBI), Hinxton; and the Kendrew Quadrangle, Kendrew Scholarship, and annual Kendrew musical events at St. John's, Oxford. In the history of modern science, John Kendrew shall be remembered as one of the most gifted, accomplished, and influential pioneers among 20th century molecular biologists.

Description of John Kendrew's Archives

Catalogues I, II, and Supplement

There are three catalogues that describe the Sir John Cowdery Kendrew Archives held by the Department of Western Manuscripts, Special Collections, Bodleian Library, Oxford University. Catalogues I and II describe material deposited by Kendrew from April 1987 to April 1989, compiled by the late Jeannine Alton. The third, the Supplementary Catalogue, describes material deposited by Kendrew's executor, John A. Montgomery, in 1998, compiled by Adrian Nardone and Peter Harper. Shown here are sections A–R, subjects and pages covered, with the National Cataloguing Unit for Archives of Contemporary Scientists (NCUACS) designations in parentheses.

Contents of Catalogue I

Contents of Catalogue II

Contents of Supplementary Catalogue

Jeannine Alton's Description of Archives

The following is a description of the Sir John Cowdery Kendrew Archives (NCUACS 11.4.89) by the late Jeannine Alton who catalogued the collection that is held in the Department of Western Manuscripts, Special Collections, Bodleian Library, Oxford University:

The material is very extensive and provides information not only on virtually all aspects of Kendrew's own career but on many of the individuals and organizations connected with it. The papers are presented as shown in the List of Contents. Additional explanatory notes or information accompany the separate sections and many of the sub-sections and individual entries in the body of the catalogue. The following paragraphs are intended only to draw attention to material of particular interest.

It should be said at the outset that the entire collection reflects on almost every page what are probably the best-known features of Kendrew's personality: on the one hand his methodical and analytical power, his meticulous not to say obsessive insistence on accuracy and comprehensive documentation, shown in his lifelong interest in record-keeping and the devising of recondite systems for information storage and retrieval; on the other hand, an aloofness or elusiveness of temperament which sets certain limits to personal relations. There are a steadiness and control, a detachment combined with seemingly tireless application which constitute a formidable intellectual armoury and which are present from the earliest records.

Thus, Section A (Notebooks, Notes and Essays) though mainly covering early school and undergraduate work 1930–39 is remarkable for its maturity and for the quantity and quality alike of the content. It is classified under a system of Kendrew's own devising which, though intrinsically clear and flexible, involved careful pagination with at least two referents, and an appreciable measure of cross-referencing in order to amalgamate school and university work; as there are some 9000 manuscript pages of notes, not counting notebooks and essays, the degree of labour required would not have been contemplated, let alone undertaken and carried through, by many. A modern historian of intellectual or educational development will be grateful for the scrupulous indexing of topics, the very full notes of lecture courses and the careful identification of lecturers. The latter included most of the leading figures in Cambridge science immediately before and after the Second World War and some visiting lecturers (J.E. Lennard-Jones, F.W. Aston, R.G.W. Norrish, E.K. Rideal, F.P. Bowden, J.A. Ratcliffe, W. Cochran, F.G. Hopkins, M. Dixon, F.G. Mann, W.J. Pope, D.D. Woods, J. Needham, A. Neuberger, D. Keilin, K. Bailey, I. Langmuir may be cited among very many others). The total sequence therefore provides an exceptionally comprehensive picture of the education available at that time at a well-run school science department and a major "science" university.

Section B (Second World War) chronicles Kendrew's overseas service and contributions to operational research, and contains original reports from various commands by him and others. It also includes several letters or private reports on his analysis of the war situation at various dates, of the current and future state of operational research and of his own career plans, his suggestions and proposals for the postwar organization of government science, and a punched card, designed by him, to be used by aircraft patrol crews to record incidents.

Section C (Research) is one of the major components of the collection and furnishes a very full record of Kendrew's research, including some of his collaborators' work. It is concerned almost entirely with protein structure analysis, beginning in 1946, and includes the devising of computer programs to handle and process data on machines of progressive sophistication (EDSAC-I, EDSAC-II, Mercury, IBM 7090 and others), the many years of experimentation with types of myoglobin until in December 1952 sperm-whale emerged as the most promising crystal source, and the ensuing protracted sequence of diffraction pictures, phase determination, scaling, manual and computer calculations leading up to the establishment of contour maps and the final 3-dimensional picture at successive Angström resolutions. There is also material on the concurrent and subsequent work on amino-acid sequencing by chemical methods in correlation with the crystallographic analysis. In addition to the main sequence of notebooks, notes and data, there is material relating to specimens, apparatus and models, and correspondence with collaborators and colleagues extending over more than twenty years. The progress of the work can thus be followed in great technical detail. On a more immediately accessible level, there are reports, project diaries, summaries of experimental findings, charts of progress, work-allocations and the like written from the bench or in correspondence throughout the period, and covering key points such as the identification of the best crystalline protein, the excitement of the eventual syntheses and their publication. Attention is drawn to these and other items of interest in the introduction to Section C and at various points in the body of the text.

Section D (The MRC Laboratory Cambridge) is a relatively short section, but contains correspondence on the new building and its extension conducted by the Director M.F. Perutz, and some material on equipment and staff. There are also agendas, minutes and research proposals for Laboratory Board meetings, and a full record of committee meetings of Kendrew's own Structural Studies Division from 1969 (no.1) until his secondment to Heidelberg in 1975.

Section E (Cambridge: University and College) is another short section and does not fully reflect Kendrew's work as lecturer and teacher, or his committee service at Cambridge. His own extremely detailed notes on entrance and scholarship examinations for Peterhouse and the "King's Group" of colleges in the 1950s are again of interest for the history of education.

Sections F, G and H are all devoted to the history of European molecular biology and constitute another major component of the collection, covering twenty years 1962–82. The extent of the material made it expedient to present it in separate sections for the Organization EMBO (Section F), the Conference EMBC (Section G) and the Laboratory EMBL (Section H). A general introduction to the topic and to the material has been prepared as well as the specific introductions preceding each section. As has already been stated, Kendrew was closely involved in the movement from its earliest inception; he held high office in its key committees and secretariat and was also active in several channels of the science policy establishment in Britain such as the Council for Scientific Policy and the International Relations Committee of The Royal Society. His unique place at or near the centre of events makes his record of special value in several regards. It is remarkably complete and contains the early history or "founding papers" of all three European bodies, many of them in the form of manuscript or informal letters exchanged with distinguished colleagues throughout Europe as well as in Britain where the Cambridge MRC Laboratory itself provided the first Chairman of EMBO, M.F. Perutz, and many founder members such as S. Brenner, H.E. Huxley, F.H.C. Crick and A. Klug. In addition, the first Secretary-General of EMBO, J. Wyman, has at Kendrew's request made over his own papers and correspondence on the subject for incorporation in the present collection. All aspects of the European molecular biology movement are thus covered: negotiations, discussions and agreements at personal, official, national and international level; research projects, membership and elections; buildings and staffing; budgets, costings and funding—made dangerously unpredictable for the laboratory project by the global inflation and fluctuating European exchange rates in the 1970s; and as time advances a proliferation of committees and working parties. The element of doubt, unfavourable comment and hostility is not neglected; the efforts of founder members to keep the project alive by personal contacts is evident in many contexts. Kendrew's own notes of discussions, planning and problems at all stages are of the greatest interest, especially for the laboratory project; Section H also includes documentation of his own career at EMBL.

Sections J and K are concerned respectively with UK and with international societies and organizations. Each has an alphabetical list of contents and most of the items have a descriptive entry or introductory note. Of special interest in Section J are the "founding papers" of the British Biophysical Society, papers of the Council for Scientific Policy including those of the Working Groups on molecular biology, on the proposal for EMBL and on the Dainton and Rothschild reports on the organization of research, and the Council's Standing Committee on International Relations. There is also a full record of the High Energy Particle Physics Review Group chaired by Kendrew, and material relating to various committees and sub-committees of the Royal Society. Correspondence with W.L. Bragg and others at the Royal Institution provides useful links with Kendrew's research career during the myoglobin project; material for the Design Research Unit shows him in an unexpected role in the 1940s submitting ideas for industrial and design applications of scientific advances.

On the international scene (Section K) there are records of Kendrew's extensive service on the councils or advisory boards of institutions and laboratories and of their research programmes—examples among others are the Basel Institute for Immunology, the Laboratory of Molecular Embryology Naples, the Max Planck Gesellschaft Munich, the Molecular Biology Department Free University of Brussels, the United Nations University Tokyo—and good documentation for his long and continuing association, begun in 1963, with the Weizmann Institute Israel. The international scientific unions are represented by Biochemistry (IUB) and Pure and Applied Biophysics (IUPAB), but chiefly by the International Council of Scientific Unions (ICSU) which remains one of Kendrew's most important commitments and in which he held high office from 1974.

Section L (Lectures, Publications, Reviews) was considerably expanded at a late stage in the compilation of the catalogue by Kendrew's decision to include his folders of lectures and talks 1946–1987. These appear as addenda and have been extensively cross-referenced to related material, conferences, invitations and other events elsewhere in the collection. They contain many more lectures than previously available and include substantial courses given in America and Japan, and various special invitation lectures.

Section M (Journal of Molecular Biology) has been treated separately from other publications because of Kendrew's long involvement with the journal as

Editor-in-Chief from its inception and as a Director of Academic Press. Once again there are "founding papers," careful notes and analyses by Kendrew, and miscellaneous material on the fortunes and vicissitudes of the journal and its publishers over a time-span of thirty years.

Section N (Visits and Conferences) is not extensive and is far from reflecting the extent of Kendrew's travels. His notoriously peripatetic existence becomes apparent by the accumulation throughout the collection of references to journeys in connection for example with European molecular biology, or on behalf of ICSU and its constituent unions, or the official visits for the Council for Scientific Policy, or the regular visits and meetings at other laboratories and institutions.

Sections O, P and R are all short sections. Section O (Correspondence) is not extensive since Kendrew kept almost all his correspondence in the files or notebooks to which it related. One touches here on the reticent element in his temperament; the correspondence is open and friendly yet rarely develops into long-term exchanges. Section P (References and Recommendations) covers a long period and is international in range; some is subject to restricted access. Section R (Biographical) contains some interesting material on Kendrew's career and appointments, including many offers of posts in Britain and abroad. Unsurprisingly, it contains little of a personal nature.

It will be seen that the collection comprises material of very different nature and of potential interest in several fields of enquiry. There is the full history of a major scientific discovery as such (the structure of myoglobin) and, in the many research proposals put forward by individuals and institutions including EMBL, examples of how leading scientists saw the key developments in their subject at a particular time. There is material on intellectual training and the diffusion of changing scientific concepts by filtering to a wider public awareness through the educational process at various levels. The international aspects of science can be studied in the organization, assemblies, membership and evolving preoccupations of the scientific unions and especially of their central body ICSU; while there is exceptional coverage of one field of European co-operation in the history of EMBO, EMBC and EMBL. The material relating to the Journal of Molecular Biology reflects an important aspect of the scientific process: the evaluation and communication of research findings and the criterion of peer-judgment. Of perhaps more specialised interest is the evidence of the scientific contribution to the war effort during the Second World War, seen here in the correspondence and reports on operational research in several fields of hostilities and in the responsibility exercised by people still in their early twenties. It should be recalled that the documents on all these topics include both official papers in the shape of minutes, reports and the like, and also Kendrew's own meticulous notes, drafts and comments which greatly supplement the official material, and background correspondence and discussions which rarely form any part of them.

Collection Level Description of Archives

Correspondence and Papers of Sir John Cowdery Kendrew

The following is a Collection Level Description of the Correspondence and Papers of Sir John Cowdery Kendrew compiled by the Bodleian Library in 2017. The Shelf Marks/Call Numbers are provided and designated as MS-Manuscript, MSS-Manuscripts, Eng- English, and b, c, d, e-Size of File. The Shelf Marks/Call Numbers were assigned by the Bodleian Library upon closure in 2009 of the NCUACS that had been located at the University of Bath.

Reference: MSS. Eng. b. 2010–2018, c. 2385–2611, d. 2105–2223, e. 2317–2358

Title: Correspondence and papers of Sir John Cowdery Kendrew

Dates of Creation: 1927–1988

Extent: 397 boxes

Language of Material: English

Administrative/Biographical History

John Cowdery Kendrew was born in Oxford on 24 March 1917. He was educated at the Dragon School, Oxford, (1923–1930), and Clifton College, Bristol, (1930–1936). In 1936 he entered Trinity College, Cambridge as a Major Scholar, reading chemistry, physics, biochemistry, and advanced mathematics (the latter two as half subjects) for the first part of the natural sciences tripos, and chemistry for the second. In June 1939 he graduated with First-Class Honours in chemistry and immediately began research in reaction kinetics in the Department of Physical Chemistry, under E.A. Moelwyn-Hughes.

At the outbreak of the Second World War, Kendrew was encouraged to continue his research, but he was eager to contribute to the war effort. In early 1940 he was appointed Junior Scientific Officer at the Air Ministry Research Establishment, and worked on the development of airborne radar. In September 1940 he was attached to the staff of Sir Robert Watson Watt for operational research duties, with special reference to anti-submarine warfare, bombing accuracy, and radio aids. Most of his war service was spent abroad: in Cairo with the Middle East Command and then in India and Ceylon with the South-East Asia Command, where he was officer in charge of operational research and Scientific Adviser to the Allied Air Commander-in-Chief.

Wartime travels and encounters were to have major effects on his future career. J.D. Bernal, whom he met in Cairo, India, and Ceylon, persuaded him of the importance of research into protein, and this was reinforced by a meeting with L.C. Pauling in the course of a roundabout journey home via Australia and America in the spring of 1945. Though he hesitated for some time and explored the possibility of remaining in Government service to continue operational research and planning for peacetime policies, he decided to return to Cambridge.

In September 1945, having been awarded an ICI research fellowship, Kendrew began a collaboration with M.F. Perutz at the Cavendish Laboratory. They started investigating the structure of hemoglobin, embarking on a comparison of fetal and adult hemoglobin. This work gained him his PhD in 1949. From the beginning, however, they also attempted the crystal analysis of myoglobin, the protein responsible for oxygen storage in muscle. This project was hampered by the difficulty of growing crystals of a size suitable for X–ray analysis. Protein crystallography in the Cavendish, under the guidance of W. Lawrence Bragg, was put on a more secure footing by the creation, in 1947, of the Medical Research Council unit for the molecular structure of biological systems. After exploring many possible problems and materials Kendrew chose myoglobin and in particular sperm-whale myoglobin as the most suitable for analysis by X-ray crystallography; he and his collaborators eventually succeeded in producing a 3-dimensional model at a resolution of 6-Å in 1957 and 2-Å in 1959. The crystallographic calculation for both models relied decisively on the use of the first electronic digital computers built at Cambridge, EDSAC-I and -II, of which Kendrew made pioneering use. This work gained Kendrew, jointly with Perutz, the 1962 Nobel prize for chemistry, the same year that F.H.C. Crick and J.D. Watson (both of the MRC Unit) shared the Prize in Physiology or Medicine with M.H.F. Wilkins for the determination of the structure of DNA.

Alongside the laboratory work, Kendrew had maintained his links with university life principally through Peterhouse, which had welcomed him during the early postwar years as a Research Fellow 1947–1953 and later as a Supernumerary Fellow. He was Director of Studies in Natural Sciences for many years, with responsibility for the selection and tuition of undergraduate members, as well as holding several College offices. He later became an Honorary Fellow of Peterhouse, as also of his undergraduate college, Trinity.

From the late 1950s Kendrew became more involved in scientific matters in the wider world. He was a founding member and first Honorary Secretary of the British Biophysical Society; in 1959 he undertook the Editorship of the new *Journal of Molecular Biology*, which he retained until 1987; he was Deputy Chief Scientific Adviser in the Ministry of Defense 1960–1963; and he served on committees and advisory boards of the Royal Society where he had been elected to the Fellowship in 1960. With the award of the Nobel Prize this involvement gained momentum and an altogether new dimension in international terms with the development of the European Molecular Biology Organization (EMBO) and its Laboratory (EMBL).

Many other commitments to national and international science policy also belong to these years. In Britain they include service on the Council for Scientific Policy and chairmanship of some of its committees and working parties 1964–1972, service on the Defense Scientific Advisory Council 1969–1974, continuing service on the Council and other committees of the Royal Society and on other learned societies, in particular the British Biophysical Society and the Institute of Biology. Examples of increasing involvement in international science and scientific policy can be seen in his appointment as Governor of the Weizmann Institute, Israel, in 1964 and the Vice-Presidency and Presidency of the International Union of Pure and Applied Biophysics 1964–1972.

During the 1960s Kendrew continued his research on myoglobin, refining the resolution to 1.4-Å and determining the co-ordinates of virtually all the 2500 atoms in the molecule. In the later 1960s, however, his other commitments increasingly absorbed his time and energy and his official move to Heidelberg as Director-General of EMBL in 1975 marked the end of his active research. The creation of the EMBL as a physical entity, and more importantly as an international centre of excellence where several teams and research projects could co-exist and collaborate was a lasting achievement. In addition, or in consequence, Kendrew's diplomatic skills, mastery of detail and experience in chairmanship made him constantly in demand on a wider stage. He served, often as chairman, on the scientific councils or advisory boards of laboratories or research institutions in Naples, Basel, Brussels, Stockholm, Heidelberg and others, on various UNESCO committees, and on many electoral boards for honours and appointments in Britain and abroad. His formal association with science at the international level may be said to have culminated in his service with the International Council of Scientific Unions (ICSU) as Secretary-General 1974–1980 and President 1983–1988.

Kendrew's original contract of secondment from the Medical Research Council and appointment as Director-General of the EMBL was renewed twice, until 1982, when he retired on reaching the age of 65. His last appointment brought him back to Oxford, where he served as President of St. John's College until 1987. He died of cancer in Cambridge on 23 August 1997.

Scope and Content

The collection contains:

- Notebooks, notes, and essays
  - Notebooks
    - School [MSS. Eng. e. 2317–2341]
    - War service [MS. Eng. e. 2342]
    - University [MS. Eng. e. 2343]
  - Notes and essays
    - School [MS. Eng. c. 2385]
    - University [MS. Eng. c. 2385; MSS. Eng. d. 2105–2128]
    - War service [MS. Eng. d. 2128]
    - Later miscellaneous notes [MS. Eng. d. 2128]
- Second World War
  - Career [MS. Eng. c. 2386]
  - Correspondence and papers, 1941–1946 [MSS. Eng. c. 2386–2387]
  - Notes [MS. Eng. e. 2344]
  - Reports
    - Coastal Command [MS. Eng. c. 2388]
    - Bomber Command [MS. Eng. c. 2388]
    - Anti-submarine warfare [MS. Eng. c. 2388]
    - Combined operations [MS. Eng. c. 2388]
    - Methodology [MS. Eng. c. 2388]
    - Middle East Command [MSS. Eng. c. 2389–2392]
    - South East Asia Command [MS. Eng. c. 2393]
    - Postwar papers [MS. Eng. c. 2394]
  - History of Operational Research (OR) [MS. Eng. c. 2394]
- Research
  - Early research
    - Reaction kinetics [MS. Eng. c. 2395]
  - Information retrieval/Data processing
    - Information retrieval [MS. Eng. c. 2395]
    - Computation on EDSAC–I [MS. Eng. c. 2396]
  - Protein analysis projects
    - Adult and fetal sheep hemoglobin [MS. Eng. c. 2397]
    - Muscle [MS. Eng. c. 2397]
    - Protein solubility [MS. Eng. c. 2397]
    - Procollagen [MS. Eng. c. 2397]
    - Controlled shrinkage of protein crystals [MS. Eng. c. 2397]
    - Polypeptide configuration [MS. Eng. c. 2397]

- X-ray experiments [MS. Eng. c. 2397]
- Chymotrypsinogen [MS. Eng. c. 2397]
- Correspondence "New proteins" [MS. Eng. c. 2397]
- Early research reports
  - Reports, 1946–1953 [MS. Eng. c. 2397]
- Myoglobin notebooks (Kendrew and collaborators)
  - Preliminary work [MSS. Eng. d. 2129–2134; MSS. Eng. e. 2345–2351]
  - Main myoglobin programme [MSS. Eng. d. 2135–2194]
  - Collaborators' notebooks [MS. Eng. c. 2398; MSS. Eng. d. 2195–2203; MSS. Eng. e. 2352–2358]
  - Atomic co-ordinates/amino-acid sequencing [MSS. Eng. d. 2204–2212]
  - Miscellaneous [MSS. Eng. d. 2213–2216]
- Myoglobin notes and data (Kendrew and collaborators)
  - Preliminary work [MSS. Eng. b. 2010–2014; MS. Eng. c. 2399]
  - Main myoglobin programme [MS. Eng. b. 2015; MS. Eng. c. 2400]
  - Collaborators' notes and data [MS. Eng. b. 2016]
  - Miscellaneous [MS. Eng. c. 2401]
- Myoglobin materials and apparatus (Kendrew and collaborators)
  - Supplies and specimens [MS. Eng. b. 2017]
  - Optical diffractometer [MS. Eng. b. 2017]
  - Microcamera [MS. Eng. b. 2017]
  - Densitometer [MS. Eng. b. 2017]
  - Computer time [MS. Eng. b. 2017]
- Myoglobin collaborators and staff
  - Individual files [MSS. Eng. c. 2402–2403]
  - Chronological files [MS. Eng. c. 2404]
- Myoglobin correspondence
  - Aspects of myoglobin [MSS. Eng. c. 2405–2406]
  - Atomic co-ordinates/amino-acid sequencing [MS. Eng. c. 2407]
  - Publications [MS. Eng. c. 2408]
- Myoglobin models
  - Skeletal model [MS. Eng. c. 2409]
  - Ball-and-spoke model [MS. Eng. c. 2409]
  - Science Museum, London [MS. Eng. c. 2409]
  - Correspondence [MS. Eng. c. 2410]
- Myoglobin miscellaneous
  - Pantographs [MS. Eng. d. 2217]
- MRC Laboratory of Molecular Biology, Cambridge
  - Buildings [MS. Eng. c. 2411]
  - Apparatus and equipment [MS. Eng. c. 2412]
  - Staff [MS. Eng. c. 2413]
  - Research and administration [MSS. Eng. c. 2414–2415]
  - Historical [MS. Eng. c. 2415]
- European Molecular Biology Organization (EMBO)
  - Early history
    - Preliminary meetings and correspondence, 1963 [MS. Eng. c. 2418]
    - Formal constitution and statutes [MS. Eng. c. 2418]
    - Relations with other organizations [MSS. Eng. c. 2419–2420]

- Policy document [MS. Eng. c. 2421]
- Funding [MSS. Eng. c. 2422–2423]
- Miscellaneous correspondence [MS. Eng. c. 2423]
- Membership
  - Nominations and elections, 1963–1982 [MS. Eng. c. 2424]
  - Circulars and lists, 1964–1975 [MS. Eng. c. 2425]
- Council
  - Correspondence and meetings, 1964–1980 [MSS. Eng. c. 2426–2428]
  - Membership and elections, 1963–1974 [MS. Eng. c. 2429]
  - Minutes and circulars, 1963–1981 [MS. Eng. c. 2430]
- Fund Committee
  - Membership, 1965–1974 [MS. Eng. c. 2431]
  - Correspondence and papers, 1964–1974 [MS. Eng. c. 2431]
  - Fellowship applications, 1964–1974 [MS. Eng. c. 2432]
  - Minutes, 1965–1973 [MS. Eng. c. 2433]
- Course Committee
  - Membership, 1965–1974 [MS. Eng. c. 2433]
  - Correspondence and papers, 1965–1974 [MS. Eng. c. 2433]
  - Minutes and circulars, 1966–1969 [MS. Eng. c. 2433]
- Laboratory Committee
  - Meetings, correspondence and papers, 1963–1973 [MSS. Eng. c. 2434–2436]
- Administration
  - Appointments [MS. Eng. c. 2437]
  - Finance and accounts [MS. Eng. c. 2437]
  - General administrative correspondence [MS. Eng. c. 2437]
- Symposium Committee
  - Correspondence and papers, 1974–1981 [MS. Eng. c. 2438]
  - Minutes, 1974–1980 [MS. Eng. c. 2438]
- Miscellaneous
  - Press releases, articles, comments, 1963–1971 [MS. Eng. c. 2438]
  - Annual reports, 1966–1981 [MS. Eng. c. 2439]
- European Molecular Biology Conference (EMBC)
  - Early history
    - "Swiss Initiative," 1964–1967 [MS. Eng. c. 2440]
    - Intergovernmental meetings and negotiations, 1967–1969 [MS. Eng. c. 2441]
    - Signing and ratification of Agreement, 1969–1970 [MS. Eng. c. 2442]
  - General correspondence and papers, 1969–1979 [MS. Eng. c. 2443]
  - Membership, 1968–1976 [MS. Eng. c. 2444]
  - Subcommittees and working groups
    - Laboratory Working Group I—role of the laboratory [MS. Eng. c. 2445]
    - Laboratory Working Group II—organization, structure, administration [MS. Eng. c. 2445]
    - Laboratory Working Group III—site of the laboratory [MS. Eng. c. 2446]
    - Laboratory Working Group IV—financial aspects [MS. Eng. c. 2447]
    - Steering Group of Laboratory Working Groups [MS. Eng. c. 2448]
    - Enlarged legal sub-group [MS. Eng. c. 2448]
    - "Andres" Working Group on the future of Conference [MSS. Eng. c. 2448–2449]

- Conference and committee papers [MSS. Eng. c. 2450–2458]
- European Molecular Biology Laboratory (EMBL)
  - Agreements
    - Laboratory agreement [MS. Eng. c. 2459]
    - Headquarters agreement [MS. Eng. c. 2459]
  - The building of the laboratory
    - Temporary accommodation in Heidelberg [MS. Eng. c. 2460]
    - Early planning and costing [MS. Eng. c. 2461]
    - Building Committee [MSS. Eng. c. 2462–2463]
    - Architects, tenders, plans [MSS. Eng. c. 2464–2466]
    - Furnishing and interior design [MS. Eng. c. 2467]
    - Inauguration [MS. Eng. c. 2468]
  - Research committees and working groups
    - Provisional Scientific Advisory Committee (PSAC) [MS. Eng. c. 2469]
    - Scientific Advisory Committee (SAC) [MSS. Eng. c. 2470–2479]
    - Nuclear Magnetic Resonance (NMR) Working Group [MS. Eng. c. 2480]
    - Computer Policy Working Group [MS. Eng. c. 2480]
    - Recombinant DNA (rDNA) Committee [MS. Eng. c. 2481]
    - Scientific Purchases Committee [MSS. Eng. c. 2482–2486]
    - Workshops [MS. Eng. c. 2487]
  - Research divisions—Heidelberg
    - Cell Biology [MSS. Eng. c. 2488–2491]
    - Biological Structures [MS. Eng. c. 2492]
    - Instrumentation [MSS. Eng. c. 2493–2494]
  - Outstation at Deutsches Elektronen Synchrotron (DESY)—Hamburg
    - Early history [MS. Eng. c. 2496]
    - Relations with DESY [MS. Eng. c. 2496]
    - DESY committees [MS. Eng. c. 2497]
    - Staff [MS. Eng. c. 2497]
    - Equipment [MS. Eng. c. 2498]
    - Research [MSS. Eng. c. 2499–2500]
  - Outstation at Institut Max von Laue—Paul Langevin (ILL)—Grenoble
    - Early history [MS. Eng. c. 2501]
    - ILL—EMBL building [MS. Eng. c. 2502]
    - ILL Scientific Council [MS. Eng. c. 2503]
    - Staff [MS. Eng. c. 2503]
    - Equipment [MS. Eng. c. 2503]
    - Research [MS. Eng. c. 2504]
  - Research programmes [MS. Eng. c. 2504]
  - Seminars, lectures, courses [MS. Eng. c. 2505]
  - Visitors and staff [MS. Eng. c. 2506]
  - Membership [MS. Eng. c. 2507]
  - Administration
    - Organization and planning [MS. Eng. c. 2508]
    - Committees [MS. Eng. c. 2509]
    - Finance [MSS. Eng. c. 2510–2511]
    - Staff Association [MS. Eng. c. 2511]
    - Miscellaneous [MSS. Eng. c. 2512–2513]

- Appointments [MS. Eng. c 2514]
- Director-General's notes and correspondence
  - Notes [MS. Eng. c. 2515]
  - Kendrew's appointments and career at EMBL [MS. Eng. c. 2516]
  - Personal correspondence [MS. Eng. c. 2516]
  - Director-General's correspondence [MSS. Eng. c. 2517–2522]
  - Finance Committee and Council minutes [MSS. c. 2523–2532]
- UK societies, organizations, and consultancies
  - Bristol University [MS. Eng. c. 2533]
- British Association for the Advancement of Science (BAAS) [MS. Eng. c. 2535]
- British Biophysical Society [MS. Eng. c. 2535]
- British Broadcasting Corporation Science Consultative Group [MS. Eng. c. 2535]
- British Council [MS. Eng. c. 2535]
- British Museum [MS. Eng. c. 2535]
- Campaign for Better Broadcasting [MS. Eng. c. 2535]
- CIBA Foundation [MS. Eng. c. 2535]
- Commonwealth Science Council (CSC) [MS. Eng. c. 2536]
- Council for Scientific Policy (CSP) [MSS. Eng. c. 2537–2547]
- Design Research Unit [MS. Eng. c. 2548]
- High Energy Particle Physics Review Group (HEPP) [MSS. Eng. c. 2549–2554]
- Imperial Chemical Industries LTD. (ICI) [MS. Eng. c. 2555]
- Medical Research Council (MRC) [MS. Eng. c. 2555]
- Ministry of Defense [MS. Eng. c. 2555]
- Ministry of Overseas Development [MS. Eng. c. 2556]
- Queen Elizabeth House, Oxford [MS. Eng. c. 2557]
- Royal Institution [MS. Eng. c. 2558]
  - Royal Society [MSS. Eng. c. 2559–2564]
- International societies, organizations, and consultancies
  - Asian Molecular Biology Organization (AMBO) [MS. Eng. c. 2565]
- Basel Institute for Immunology [MS. Eng. c. 2565]
- Eidgenossische Technische Hochschule, Zurich [MS. Eng. c. 2566]
- European Communities Commission [MS. Eng. c. 2566]
- European Research and Development Committee (CERD) [MS. Eng. c. 2566]
- French Academy of Sciences [MS. Eng. c. 2566]
- International Council of Scientific Unions (ICSU) [MSS. Eng. c. 2567–2568]
- International Foundation for Science (IFS) [MS. Eng. c. 2569]
- International Institute of Cellular and Molecular Pathology (ICP) [MS. Eng. c. 2569]
- International Union of Biochemistry (IUB) [MS. Eng. c. 2570]
- International Union for Pure and Applied Biophysics (IUPAB) [MS. Eng. c. 2571]
- Laboratorio di Embriologia Molecolare, Naples [MS. Eng. c. 2572]
- Max-Planck-Gesellschaft [MS. Eng. c. 2573]
- National Academy of Sciences (NAS) [MS. Eng. c. 2574]
- National Scientific Research Center, Iran [MS. Eng. b. 2018]
- North Atlantic Treaty Organization (NATO) [MS. Eng. b. 2018]
- Stazione Zoologica, Naples [MS. Eng. b. 2018]
- UNESCO [MS. Eng. b. 2018]
- Union Internationale Contre le Cancer (UICC) [MS. Eng. b. 2018]

- United Nations University (UNU) [MSS. Eng. c. 2575–2576]
- Universite Libre de Bruxelles [MS. Eng. c. 2577]
- Weizmann Institute of Science and Weizmann Institute Foundation [MSS. Eng. c. 2578–2579]
- World Health Organization [MS. Eng. c. 2580]
- Lectures, publications, and broadcasts
  - Lectures, publications, and reviews [MSS. Eng. c. 2581–2587; MS. Eng. d. 2218]
  - Radio, television, and films [MS. Eng. c. 2588]
  - Correspondence with publishers and editors [MSS. Eng. c. 2589–2592]
  - Addenda: lectures, 1946–1987 [MSS. Eng. d. 2219–2223]
- *Journal of Molecular Biology*
  - Founding papers, 1957–1959 [MS. Eng. c. 2593]
  - Correspondence with editors and authors [MSS. Eng. c. 2593–2594]
  - Correspondence with Academic Press [MSS. Eng. c. 2595–2596]
- Visits and conferences [MSS. Eng. c. 2597–2600]
- Correspondence [MSS. Eng. c. 2601–2602]
- References and recommendations
  - Theses and higher degrees [MS. Eng. c. 2603]
  - Appointments and staffing [MSS. Eng. c. 2603–2604]
  - Grant applications/research funding [MSS. Eng. c. 2604–2605]
  - Miscellaneous [MS. Eng. c. 2605]
- Biographical [MSS. Eng. c. 2606–2611]

John Kendrew's Obituaries (1997)

New York Times, August 30, 1997, "John C. Kendrew Dies at 80; Biochemist Won Nobel in '62," by Wolfgang Saxon

John Cowdery Kendrew, a British molecular biochemist and 1962 Nobel Laureate who decoded part of the mystery of life by revealing the detailed structure of a key protein, died last Saturday in Cambridge, England. He was 80.

Dr. Kendrew started his quest at the end of World War II when he and a cohort of determined biochemists at Cambridge University set out to decipher the structure of proteins.

The equipment on hand was barely adequate, but persistence prevailed when he and a colleague, Max Perutz, identified the properties of globular protein molecules.

Dr. Kendrew unraveled the vast complexities of myoglobin, the red protein in muscles, while Dr. Perutz did the same for hemoglobin, the substance in red blood cells that absorbs oxygen and releases it in the tissues. Their work took science a giant step toward an understanding of how proteins do their vital work.

It helped establish the new field of molecular biology at Cambridge University's Cavendish Laboratory. For their feats, Dr. Kendrew and Dr. Perutz shared the Nobel Prize in Chemistry in 1962.

They had used X-ray diffraction to outline the spiral structure of the globular proteins that enable humans and animals to breathe. In introducing atoms of heavy metals, like gold or mercury, into a molecule, they were able to trace its structure by means of irradiation.

Chemical means were used to determine the parts of the molecule to which the metal had attached itself. Electronic computers quickened the drudgery of translating the results of their studies into precise measurements for building models of the proteins.

Protein molecules in their myriad varieties form the basic constituent of all living matter, from bacteria to plants to higher animals. Dr. Perutz and Dr. Kendrew, who started out as his assistant, were honored for being the first to work out the structure of two of the most important.

John Kendrew, a mild-mannered bachelor, was born in Oxford, the son of a noted climatologist, Wilfrid George Kendrew. He was educated at Oxford and Cambridge, where he received a PhD in physics in 1947. During World War II he worked for the Ministry of Aircraft Production as a junior scientific officer in radar technology and advised the Allied air command in Southeast Asia.

In Asia he met the crystallographer J. D. Bernal, who suggested to him the imminent convergence of physics and biology. On a side trip to California, he encountered a fellow future Nobel Laureate, the physical chemist Linus Pauling, whose protein work inspired his own.

He then returned as a doctoral candidate to Cambridge and the Cavendish Laboratory, where Dr. Perutz, a former student of Dr. Bernal's, directed the medical research council in the fledgling molecular biology section.

Dr. Perutz, who had researched hemoglobin for years, used a procedure suggested by Dr. Bernal by which a heavy-metal atom bonded with a protein to serve as a marker.

Dr. Kendrew was promoted from the assistant's job when Dr. Perutz assigned him to puzzle out the structure of the simpler of the two most interesting proteins, myoglobin, which stores oxygen in muscle tissue and gives red meat its color.

Little was known about the substance except that it was a protein, and his task was to identify the position of each atom in the myoglobin molecule and to plot the molecule's shape.

He did well enough to come in first. Dr. Kendrew formulated its 3-dimensional structure before Dr. Perutz cornered his prey, hemoglobin.

"In a real sense proteins are the 'works' of living cells," Dr. Kendrew wrote in Scientific American in 1961. "Almost all chemical reactions that take place in cells are catalyzed by enzymes, and all known enzymes are proteins."

Dr. Kendrew continued to work on the structure of myoglobin after receiving the Nobel Prize but became drawn into administrative and government advisory functions. The department that he and Dr. Perutz created is now the Laboratory for Molecular Biology, and he acted as its chairman until 1974.

The next year he established the European Molecular Biology Laboratory at Heidelberg and served as its director until 1982. He also founded The Journal of Molecular Biology and edited it until 1987, when he retired as president of St. John's College, Oxford.

He was the author of "The Thread of Life: An Introduction to Molecular Biology" (Harvard University Press, 1966). The recipient of many honors, he was knighted and awarded the Order of the British Empire in 1963.

Independent, September 12, 1997, "Sir John Kendrew," by John Meurig Thomas*

John Kendrew will be remembered for many contributions to science, but three stand out: his determination, in atomic detail, of the structure of myoglobin, a protein found in muscle, the function of which is to take oxygen from hemoglobin; his role as the founder editor (1959) of the Journal of Molecular Biology; and the setting up of the European Molecular Biology Laboratory (EMBL) in Heidelberg, which will forever be regarded as his monument.

From 1947, for 27 years, Kendrew was College Lecturer, Official Fellow and Director of Studies in Natural Science at Peterhouse, Cambridge; in the latter position he was responsible for selection and tuition of undergraduate members of Peterhouse. As is typical of Oxbridge Fellows, he also undertook other college duties, serving successively as Librarian, Proelector, Steward, Wine Steward and Custodian of the college's paintings and portraits. And when Cambridge University introduced its new Tripos on the History and Philosophy of Science, he also supervised undergraduates reading those subjects.

But Kendrew, while serving as a teaching Fellow, won the Nobel Prize in Chemistry in 1962 (which he shared with his fellow Petrean Max Perutz), and also served as President of the British Association for the Advancement of Science in 1974. And when he handed on his responsibilities to his successor, Dr. Klug (now Sir Aaron Klug OM, President of the Royal Society), whom he was instrumental in recruiting to Peterhouse, he too won the Nobel Prize in Chemistry (in 1982) while a teaching fellow.

He was born in Oxford in 1917, the son of a climatologist father and an art historian mother who carried out distinguished work in Italy on Veronese and Florentine painting.

He attended the Dragon School, Oxford, from 1923 to 1930 and Clinton College from 1930 to 1936. Entering Trinity College, Cambridge, as a Scholar in 1936, he became, in due course, Senior Scholar in Natural Sciences when he took the Tripos Part I in Chemistry, Physics, Mathematics and Biochemistry with first class Honors and another First in Part II Chemistry.

After graduating in 1939, he spent the first few months of the Second World War doing research on reaction kinetics under the supervision of Dr. E.A. Moelwyn-Hughes. He then became a member of the Air Ministry Research Establishment (later Telecommunications Research Establishment) and worked on radar. In 1940 he joined the staff of Sir Robert Watson Watt, and for the rest of the war was engaged in operational research at Royal Air Force Headquarters, successively in coastal command, Middle East and South East Asia; he held the Honorary rank of Wing Commander, RAF.

During the war years, his scientific interest became more biological, largely because of the influence of two great scientists. First, the brilliant polymathic physicist J.D. Bernal, with whom he rubbed shoulders in Ceylon/Sri Lanka. When the Japanese surrendered he returned from the Far East via Pasadena where he spoke to Linus Pauling at the California Institute of Technology. Among the varied provinces of Pauling's protean genius, his penetrating insights into the structural elucidation of (small) biological molecules was particularly exciting. Pauling's stimulus greatly influenced Kendrew, so that, when he returned to Cambridge in 1946, he had already decided to commence work on the structure of proteins.

At the Cavendish Laboratory he began his collaboration with Max Perutz (who had earlier identified hemoglobin as his target) under the direction of Sir Lawrence Bragg. What stimulated Perutz and what also sustained him and Kendrew for over a decade was the certain knowledge, traceable to a seminal paper on the structures of pepsin in the mid-1930s by Bernal and Dorothy Crowfoot (later Hodgkin), that crystalline macromolecules (like the blood protein hemoglobin and myoglobin consisting respectively of 12,000 and 2,600 atoms) had each of the constituent atoms situated in a precise site, and that the rabbinically complicated task of determining these sites could, in principle, be retrieved from the tens of thousands of diffraction spots that crystals of these proteins yielded when exposed to X-rays.

The scientific methods harnassed and brilliantly extended by Kendrew and Perutz has some of their origins in the universities of Glasgow, where J.M. Robertson invented the heavy atom substitution method, and of Utrecht, where the Dutch crystallographer Bijvoet showed how the 3-dimensional structure of complicated molecules could be retrieved from the X-ray diffraction patterns of two different heavy-atom variants. Kendrew, assisted by two visiting American scientists who came to the Cavendish, Howard Dintzis and Richard Dickerson, succeeded in obtaining crystals of gold- and palladium-substituted myoglobin. These proved crucial. Kendrew with his formidable mathematical skills, could also take advantage of the emergence in Cambridge of the EDSAC-I and EDSAC-II digital computers, which he exploited for the Fourier analysis of his diffraction data.

When, in 1957, Kendrew solved the structure of myoglobin, Perutz confessed to being envious. But shortly thereafter, he too, using mercury-substituted hemoglobin, reached the promised land. What was particularly exhilarating was the realisation that the twisted and folded helical chains that Kendrew found in myoglobin, were also present in Perutz's hemoglobin. This galvanised activity in molecular biology world-wide.

Adolf Butenandt, the eminent German Nobel Laureate, set his colleagues in Munich the task of using chemical methods (such as those pioneered by Frederick Sanger) to trace the sequence of amino acids in the proteins studied by Kendrew and Perutz. The chemical

APPENDIX A.2

results harmonised beautifully with those of crystallography. (It was Butenandt who nominated Kendrew and Perutz for the Nobel Prize.)

Under Lawrence Bragg's aegis, in 1949, Perutz and Kendrew formed the newly constituted Medical Research Council Unit for Molecular Biology in the Cavendish, the forerunner of the MRC's Laboratory of Molecular Biology; and after Bragg moved to become Director of the Royal Institution (RI) in London, they were both appointed Honorary Readers of the Davy Faraday Research Laboratory there, posts that they held from 1954 to 1968. In 1963, at a famous Friday Evening Discourse at the RI, they unveiled to a dazzled lay audience the secrets of their discovery.

Kendrew's own interest in fundamental research began to wane in the mid 1960s as he gradually turned his brilliant mind to matters of policy. He had already served as Deputy Chief Scientific Adviser to the Ministry of Defense (from 1960 to 1963); and then he became Chairman and Secretary General of the International Council of Scientific Unions from 1974 to 1980, a body of which he became President in 1988 to 1990.

Fluent in Italian, German and French and immersed in Renaissance culture and music, Kendrew was a committed European. When Vicki Weisskopf and Leó Szilárd called him and Jim Watson (immediately after the Stockholm Nobel Ceremony in 1962), to discuss the prospects of establishing a European Molecular Biology Organization (Embo) like the Nuclear Science Centre (Cern) in Geneva, he responded enthusiastically. Supported by other European molecular biologists, notably Perutz, Jacob (France), Friedrich-Freksa (Germany), Ole Maaløe (Denmark), Jeffries Wyman (Italy) and, crucially, Ephraim Katzir-Katchalsky of Israel (who pursuaded Golda Meir to give $20,000 towards the nascent Embo), Kendrew led the way.

Amongst other things it culminated in the creation of one of the finest biological research centres in the world, the European Molecular Biology Laboratory (EMBL) at Heidelberg of which he was the founding Director (for 20 years).

Kendrew served as Trustee of the British Museum, Chairman of the National Science Advisory Board of the UK National Commission for Unesco, member of the BBC Scientific Advisory Group, and member of the Board of Governors of the Weizmann Institute in Israel from 1964 to the time of his death.

In 1981, he took on the Presidency of St. John's College, Oxford, where his diplomatic skills, artistic and cultural tastes and formidable intellect were appreciated by the Fellows and students. Upon his retirement, he returned to live in Cambridge, where he was Honorary Fellow of both Trinity and Peterhouse.

He was appointed CBE in 1963 and knighted in 1974. He was a Fellow of the Royal Society and an Honorary Foreign Member of many national academies. He received numerous Honorary doctorates, one of the first from Complutense University, Madrid and the last, just a few weeks before he died, from his Alma Mater.

John Kendrew was a shy, private person who exuded dignified charm. It was always a pleasure meeting him. Even his answerphone message: "Please be patient and I shall get back to you as soon as possible," captured one's mind and heart. He will be mourned by many throughout the world.

*John Meurig Thomas is a renowned chemist who has been Head of the Department of Physical Chemistry and Professorial Fellow at King's College, Cambridge, Director and Fullerian Professor of Chemistry at the Royal Institution of Great Britain and Davy Faraday Research Laboratory, London, Deputy Pro-Chancellor of the Federal University of Wales, Master of Peterhouse College, Cambridge, and is currently Honorary Professor of Materials Science, Cambridge. He was knighted in 1991 for "Services to chemistry and popularization of science."

Nature, September 25, 1997, "Pioneer in Structural Biology and Collaborative Biological Research in Europe," by Kenneth C. Holmes*

". . . and in a jungle in Ceylon, J. D. Bernal talked with John Kendrew about the way it should be possible to use X-rays to solve the structure of proteins and so to understand their function in every living organism." Thus wrote Dorothy Hodgkin about how wartime operational research, concerned with elephants and bombs, inspired Kendrew's ensuing research career.

Sir John Kendrew died on 23 August, aged 80. He was educated at schools in Oxford and Bristol, and took his undergraduate degree (a first in chemistry) in 1939, at Trinity College, Cambridge. Like so many other scientists, during the Second World War he worked on radar, before being posted to southeast Asia as scientific adviser to the Allied Air Command.

At the outbreak of peace he sought out Max Perutz, an erstwhile student of Bernal's. Impressed by Kendrew's uniform as a wing commander, Perutz took him on as his second in command to work in the fledgling Medical Research Council unit for molecular biology at the Cavendish Laboratory, Cambridge, under Sir Lawrence Bragg (Kendrew may also have benefited from his entitlement to the residue of a senior scholarship from Trinity, which meant he didn't cost too much). This extraordinarily fruitful collaboration led to the joint award of the Nobel prize for chemistry just over a decade later. Moreover, they recruited Francis Crick, Hugh Huxley, Jim Watson and Sydney Brenner to the unit, all of whom became prime movers in modern biology, and a dazzling array of postdoctoral students passed through the lab.

Perutz set out to solve the structure of hemoglobin, the oxygen carrier in blood. Kendrew chose myoglobin, hemoglobin's small brother. The first attempt was with horse myoglobin but this would not crystallize properly. Try another species? Kendrew realized that myoglobin, an oxygen storage protein found in muscle, should be abundant in aquatic mammals. For reasons connected with the war, the nearby "low-temperature laboratory" had cold rooms stuffed with whale meat, which duly yielded crystals of the quality necessary for X-ray diffraction.

How does one work out the structure of a molecule containing thousands of atoms? In principle it's easy: you collect the X-ray diffraction data from the crystal in all orientations and compute the Fourier transform to give the electron density. Unfortunately, the diffracted intensities yield only the amplitudes; the phases are lost, but it was thought that they might be recovered by adding a heavy metal at a specific site in each molecule. To everyone's surprise this method worked and was used by Perutz to calculate a projection map of hemoglobin in 1953. Kendrew applied this method to myoglobin using 3-dimensional data and many different heavy atoms, and in 1957 produced the first low-resolution map of a protein. It showed myoglobin to be almost entirely alpha-helical, the helices being coiled round each other and round the heme group.

Four years later Kendrew had the coordinates of the 1,250 non-hydrogen atoms in myoglobin, an epoch-making achievement. The enormous impact that these discoveries were to have on biochemistry was only slowly appreciated by classical biochemists, which in 1958 motivated Kendrew to found the *Journal of Molecular Biology*.

Kendrew ran his lab as chief of staff. He himself developed numerical techniques for finding the relative positions of the heavy atoms. Automatic methods for measuring the intensities of some 100,000 Bragg reflections were devised. Numerous computing ladies were entrusted with the collection and collation of all the data. Bigger and better X-ray sources were built. Early experiences had convinced Hugh Huxley that computing Fourier summations by hand was not a task befitting the human condition. As a result he had introduced Kendrew to a friend, John Bennett from the Cambridge computer lab (EDSAC). In 1952, Bennett and Kendrew wrote a 3-dimensional Fourier program, probably the first.

In 1962 Perutz and Kendrew were awarded the Nobel prize. That same year they moved into the splendid new MRC Laboratory of Molecular Biology on Hills Road, Cambridge, with Perutz as director, and Kendrew as deputy director and head of the division of structural studies.

However, he was restless, in search of new organizational peaks to conquer. In parallel to his Cambridge appointment he was deputy scientific adviser to the Ministry of Defense and later chairman of the Defense Scientific Advisory Committee, which earned him his knighthood. At the same time plans for a European Molecular Biology Laboratory, to be modelled on the particle-physics research facility CERN, were taking shape. Kendrew was appointed project leader and carried out all the intricate negotiations leading to the successful founding of this laboratory in Heidelberg in 1974 (he would have preferred Nice, but France had just got the new CERN ring so Germany must have EMBL).

In Heidelberg, he made the theme technology for molecular biology. He set up the EMBL outstations at DESY (Hamburg) and ILL (Grenoble) for synchrotron light as an X-ray source and for neutron scattering; the DESY outstation was the first in the world to use synchrotron radiation as a source for X-ray diffraction. In the Heidelberg laboratory he promoted many technologies, particularly cryo-electron and optical microscopy. Furthermore, he hired bright young project leaders and gave them freedom to develop. By the time he retired as Director General in 1982 he had built up a world-class laboratory.

Between 1982 and 1988 John Kendrew was vice-president and then governor of the Weizmann Institute, Israel, where he had many friends. On leaving EMBL he was appointed president of St John's College, Oxford, and was very happy to return to the city of his childhood and once more to work with students. In his Cambridge days he was a Fellow of Peterhouse, and shortly before his death was awarded an Honorary doctorate by the university, which occasioned him great pride.

Kenneth Charles Holmes is a renowned structural biologist who was a member of Staff of the MRC LMB, Cambridge, Director of the Department of Biophysics, Max Planck Institute for Medical Research, Heidelberg, Professor of Biophysics at the University of Heidelberg, and is currently Emeritus Professor at the Max Planck Institute for Medical Research, Heidelberg. Ken and his wife, Mary, knew John Kendrew for more than 40 years in Cambridge and Heidelberg.

MRC News, Autumn 1997, "Sir John Kendrew," by Max F. Perutz*

In October 1945 a young man in a smart Wing Commander's uniform walked into my room at the Cavendish Laboratory in Cambridge and said that he wanted to become my research student. He was John Kendrew. I was flattered because I had never had a research student, let alone one hardly my junior who had distinguished himself in the war, but I also felt embarrassed because my work on the structure of hemoglobin promised no quick route to a PhD. Walking across to the Molteno Institute, I ran into Joseph Barcroft, the great respiratory physiologist, who suggested that Kendrew might make a comparative study of adult and fetal sheep hemoglobin for which Barcroft would supply the blood. Much relieved, I suggested this to Kendrew who keenly took up the project.

Today's readers can hardly imagine how courageous Kendrew's decision was then to take up protein crystallography. I had no research students because responsible dons advised graduates against joining such a forlorn undertaking, but Kendrew's spirit of adventure won. He was born in Oxford in 1917, educated at Dragon School and then Clifton College, where an outstanding chemistry professor inspired him. In 1936 he entered Trinity College, Cambridge, with a scholarship. Graduating in chemistry in 1939, he began working for a PhD in physical chemistry. The war diverted him to radar and later to operational research. He ended the war as advisor to Lord Mountbatten's South Eastern Command in Ceylon.

In 1947, soon after Kendrew joined me, the MRC agreed to make us into a research unit which later became the Unit for Molecular Biology. When the present Laboratory of Molecular Biology opened in 1962, Kendrew became its deputy director, a post he held until 1974.

I found in Kendrew an outstandingly able, resourceful, meticulous, brilliantly organized, knowledgeable hard worker and stimulating companion with wide interests in science, literature, music and the arts. Having carried sheep hemoglobin as far as was then possible, he moved on to the structure of myoglobin. At only a quarter of the molecular weight of hemoglobin it seemed a more hopeful candidate for X-ray study. After a long struggle with myoglobin from horse heart which refused to yield crystals large enough for X-ray analysis, Kendrew realised that diving mammals offered a better prospect, because nearly one-tenth of the dry weight of their muscles consists of myoglobin which they use as an oxygen store. A chance encounter enabled me to get him a large chunk of sperm-whale meat from Peru, and to our delight its myoglobin yielded large sapphire-like crystals which gave beautiful X-ray diffraction diagrams. However, there was still a seemingly insuperable obstacle.

The X-ray diffraction pattern from a crystal contains only half the information needed to solve its structure: the amplitude of the diffracted rays, but not their phases. There seemed to be no way of determining these. Fortunately, I discovered in 1953 that the phase problem could be solved by comparing the diffraction patterns of the protein with, and without, heavy atoms attached to it. Kendrew, together with several colleagues found ways of attaching heavy atoms to myoglobin. By 1957 they had obtained an electron density map at a resolution which allowed Kendrew to build a rough molecular model. Two years later they extended the resolution to enable him to build an atomic model, the first of any protein. For this work, for the introduction of the heavy atom method and for the solution of the hemoglobin structure we were awarded the Nobel Prize for Chemistry in 1962.

That year also marked the end of Kendrew's scientific research. He became part-time deputy to Solly Zuckerman, then Chief Scientific Advisor to the Ministry of Defense. Later, Kendrew's prime interest shifted to international scientific collaboration. He spoke fluent German, French and Italian and was a devoted European. Kendrew was deeply concerned that European universities and research institutes were slow in grasping the promise of molecular biology, and that Europe was falling behind the United States in training young people in the subject.

In 1963, he became one of the founders of the European Molecular Biology Organization which started the programme of travelling fellowships and summer schools that has been such an outstanding success. He also founded and remained for many years editor-in-chief of the Journal of Molecular Biology. He took great pride in it, and it became the journal where nearly all important papers in the subject appeared. In 1974, after four years of skillful diplomacy, he persuaded governments to build a European Laboratory of Molecular Biology in Heidelberg and became its firs director. This great laboratory stands as his monument.

Kendrew died in Cambridge on 28th August aged 80.

The late Max Ferdinand Perutz was a renowned structural biologist who founded and directed the MRC Unit for Research on the Molecular Structure of Biological Systems at the Cavendish Laboratory, Cambridge, and became Chairman of the MRC LMB, Cambridge. He was made Commander of the British Empire in 1962, the same year that he shared the Nobel Prize in chemistry with his student and colleague of more than 25 years, John Kendrew.

Structure, December 15, 1997, "Sir John Kendrew, 1917–1997," by Bror Strandberg*

Sir John Kendrew, who died in Cambridge on August 23, was one of a small group of scientists who laid down the foundations of molecular biology and structural biology. In 1962, John Kendrew and Max Perutz shared the Nobel Prize in Chemistry, awarded for "their studies of the structures of globular proteins."

Kendrew was born in Oxford, where his father was Reader in Climatology and his mother was an art historian, her main interest being Italian Primitives. John Kendrew studied at Clifton College, Bristol, and then in 1936 went on to Trinity College, Cambridge, where he graduated in Chemistry in 1939. He then started research in the field of reaction kinetics, but after a few months he was called up for military service and became a member of the Air Ministry Research Establishment, working on radar. Kendrew had a distinguished war career, during which time he was engaged in operational research with the Coastal Command in the Middle East and in South-East Asia.

A short description of Kendrew's connections with a few other key scientific persons will help understand how structural biology was started and John's part in this process. While in military service in South-East Asia, John met J.D. Bernal, who explained to him the great and challenging possibility of determining protein structures by means of X-ray analysis. During the 1930s, Bernal was the leader of an active research group at the Cavendish Laboratory in Cambridge. Bernal's group had been successful in starting structural studies of biologically important molecules and macromolecules. Two enthusiastic and talented students of Bernal were Max Perutz, who started work on hemoglobin in 1937, and Dorothy Hodgkin (then Dorothy Crowfoot), who in 1934, together with Bernal, took the first X-ray photograph of a protein crystal—that of pepsin. Bernal's description of his research no doubt had a big influence on Kendrew. This interest in protein crystallography was further amplified during a military visit to California, where Kendrew happened to meet Linus Pauling, the well-known chemist, and found that he too was interested in the structures of amino acids, peptides, and proteins. This made a great impression on John. Finally, Sir Lawrence Bragg's vast knowledge in X-ray structure analysis as well as his enthusiasm and strong support of Max and John was of utmost importance during their first difficult years of structural studies of biological macromolecules. In 1938, when Max showed Bragg, the new Professor in Physics at the Cavendish Laboratory, an X-ray picture of hemoglobin, Bragg immediately saw the challenging possibilities and the great importance of being able to determine the structure of hemoglobin.

Soon after the war (in 1947) Bragg, the physicist, succeeded in obtaining a grant from the Medical Research Council, to support the work of Max Perutz, John Kendrew (who started to work with Max in 1946) and two research assistants. This marked the creation of the MRC Unit for Molecular Biology (later the MRC Laboratory of Molecular Biology) at Cambridge. It was in this new unit, housed in a small "Hut" in the courtyard of Cavendish Laboratory, that Max and John started their difficult and pioneering work of solving the structures of hemoglobin and myoglobin by means of X-ray analysis. It was also in this humble laboratory that Francis Crick and Jim Watson a few years later determined the structure of DNA.

The projects of John and Max were immensely difficult, but there was no doubt that hemoglobin and myoglobin were two important protein molecules. As became evident later on the choice was also excellent from a structure determination point of view. The high content of alpha helices (around 70 percent) simplified the interpretation, especially in the early stages, and the close structural similarities between the two molecules enabled rewarding comparisons to be made and also strongly increased the confidence in the correctness of both structures. John and Max, however, had some very tough years full of hard work and worries about the outcome of their projects. So much was new, scientifically and technically. There were needs for things like a microdensitometer, methods for phase angle determination, a computer with reasonable power, computer programs, etc. It certainly did not make things easier when many colleagues like Astbury (who worked on keratin) wondered if the problems were too difficult and if John and Max were just wasting their time. At this stage, the encouragement from two scientists with deep insight in structure determination was of great value. Both Sir Lawrence Bragg and Dorothy Hodgkin argued as follows: the diffraction data were obviously possible to collect, even at high-resolution, although it would imply very hard work. This meant that sooner or later, when the techniques improved, new methods would be developed and make the whole process possible. No doubt they were right. In 1953, a very important breakthrough came. Max and his collaborators were able to show that the isomorphous replacement method could be used for the calculation of phase angles in structure determination of biological macromolecules. John and Max started to "see the light at the end of the tunnel."

It is by no means easy to distinguish between the many important contributions made by Max and John. In the following, however, I will concentrate on John's way of thinking and working. After the discovery of the possibility to use the isomorphous replacement method for phase angle determination, John was in an ideal situation. He was calm, had sane judgement, was very accurate, and had strict order and method in his planning and work. Furthermore, he was an excellent leader of a research team. All these qualifications were of course of utmost importance during the gigantic work which followed. The structure of myoglobin was determined in steps: firstly to a resolution of 6 Å and then to 2 Å, using film data, and four and three heavy-atom derivatives, respectively. Some years later, the resolution was pushed to 1.5 Å, using data collected on a diffractometer. The 6 Å map and model, which appeared in 1957, were most encouraging. Several stretches of high electron density had the dimensions of alpha helices (in accordance with Linus Pauling's model, described a few years earlier), and one area with the highest density indicated the position of the heme group. The 2 Å resolution step, which followed, involved much more work, new methods and, of course, a lot of new worries. Nobody knew if, at that resolution, the phase angles could be determined with an accuracy good enough to allow interpretation of the mainchain and the sidechains. Again, John was keen to apply the highest possible accuracy, in order to obtain the best possible map. It was very lucky for the project that the new computer in Cambridge, EDSAC-II (probably the fastest computer in the world at that time), was just ready to be used for all the extensive calculations.

Richard Dickerson has in his article "A little ancient history" [1] described many of the activities during the 2 Å myoglobin work, which culminated on a sunny morning in the beginning of August 1959. After a whole night of calculating the electron-density map, a tired group of about 10–15 people could, even as the sections were being printed, see that the heme group had the correct density distribution and that a couple of the helices, running obliquely through the map sections, were hollow. The result was no doubt excellent, clearly better than we could have anticipated.

With these results at hand, it was most useful that John was a master in writing papers, as well as an excellent lecturer with a special knowledge to speak to mixed audiences of people with different backgrounds. From the two papers describing the 2 Å structure of myoglobin [2,3], it is very easy to follow how the model was constructed: brass skeleton models, at a scale of 5 cm = 1 Å (known as the "Caltech models"), were fitted into a forest of steel rods and colored clips were attached to indicate the levels of electron density. The heme group and all the eight helices were easily interpretable and most of the sidechains could be positioned. Many internal checks (including the heme group, the helices, comparisons of the structure with the amino acid sequence, etc.) as well as the very interesting comparisons with the 5.5 Å structure of hemoglobin, determined by Max Perutz and his colleagues also in August 1959, gave a clear assurance that both the structures were essentially correct. This fact probably explains why the Nobel Prize was awarded to Max and John only three years after these discoveries.

Two incidents during the 2 Å myoglobin work are good illustrations of John Kendrew's calm and sound judgement: firstly, when the vial containing the crystals of the gold derivative (which needed months of soaking time) dried out in the middle of the data collection because the cap was not properly screwed on, John's only comment was: "We just have to see where we are with the gold derivative"; secondly, during the refinement of the heavy atom parameters, when the scale factors behaved completely abnormally (the native data had been punched on absolute scale and the derivatives on a very different relative scale) and the atmosphere in the myoglobin team was somewhat chaotic, John only remarked: "Let us be systematic." The unexpressed piece of advice he gave us was obvious: "Turn frustration into creative thinking and action."

During the intensive work on the myoglobin structure, John was already involved in the planning and administration of various scientific activities. During the late the 1950s, he shouldered the heaviest work when the Journal of Molecular Biology was started. He then served as its Editor-in-chief until 1987. During the early 1960s, he became even more engaged in committee work and organization of scientific matters. He was one of the founders of EMBO, the European Molecular Biology Organization, for which he was Secretary-General for many years. He was also extensively involved in the creation of EMBL, the European Molecular Biology Laboratory, located in Heidelberg, where he worked for many years as its first Director General. Furthermore, John was a member of many other committees and boards, both public and private. He chaired, for example, the

Defense Scientific Advisory Council (1971–1974) and he served as a trustee of the British Museum (1974–1979).

Many people might think that it was somewhat surprising that John only a few years after the 2 Å work on myoglobin almost lost touch with active scientific research and became so deeply involved in committee work and the organization of scientific matters. There may be many reasons for this development. I would just like to point out a few things. John was extremely capable of the organizing and planning of difficult activities and he apparently liked very much to work on such things. He probably felt challenged in solving these often difficult problems. Furthermore, he would only accept to do things very well.

John Kendrew will be remembered for his outstanding contribution to the birth and building-up of molecular biology; in part, for his enormous work in organizing important scientific matters and thus serving both his scientific colleagues, as well as the society outside science; but foremost, he will be remembered for his brilliant work in solving the structure of myoglobin and thereby, together with a few other pioneers, laying down the foundations of structural biology.

References

1. Dickerson, R.E. (1992). A little ancient history. *Protein Sci.* 1, 182–186.
2. Kendrew, J.C., Dickerson, R.E., Strandberg, B.E., Hart, R.G., Davies, D.R., Phillips, D.C. & Shore, V.C. (1960). Structure of myoglobin: A three dimensional Fourier synthesis at 2 Å resolution. *Nature* 185, 422–427.
3. Kendrew, J.C., Watson, H.C., Strandberg, B.E., Dickerson, R.E., Phillips, D.C. & Shore, V.C. (1961). The amino-acid sequence of sperm-whale myoglobin: Partial determination by X-ray methods, and its correlation with chemical data. *Nature* 190, 666–670.

The late Bror Erik Strandberg was a renowned structural biologist who was Professor in the Department of Molecular Biology at the University of Uppsala, Biomedical Center, Uppsala, Sweden. In the late 1950s he collaborated with John Kendrew at the Cavendish Laboratory, Cambridge, on the high-resolution 2 Å structure of myoglobin.

Preface to John Kendrew's PhD Dissertation (1949)

In May 1949 John submitted his PhD dissertation, "X-ray Studies of Certain Crystalline Proteins: The Crystal Structure of Fetal and Adult Sheep Hemoglobins and of Horse Myoglobin." The preface to his thesis follows.

This Dissertation gives an account of two studies of crystalline proteins, both using X-ray techniques. The first was an examination of the crystal structures of adult and fetal sheep hemoglobins, the main object of which was to obtain new evidence regarding the differences between the adult and fetal hemoglobins of a single species; a secondary aim—which in the event could not be pursued very far—was to discover something of the structural basis of these differences. The second was a study of the structure of the heme-containing protein myoglobin; in this case the principal object was to obtain information about the molecular structure of this protein which, being one of the simplest known, was expected to be a particularly favorable subject for analysis. As a subsidiary project a new method was developed for studying the shrinkage of protein crystals on drying.

A paper dealing with fetal and adult sheep hemoglobins has already been published with Dr. M.F. Perutz (Kendrew and Perutz, 1948); a copy is appended to the Dissertation. In this research the practical work was entirely my own, and the presentation of the results and the conclusions drawn are also my own, with an exception mentioned in the text. I am glad to acknowledge my indebtedness to Dr. Perutz for suggesting the problem to me, for help in the early stages of the work, and for much useful discussion throughout its course. The rest of the work in the Dissertation was entirely carried out by myself, with one or two exceptions noted in the text. Short accounts of the research on myoglobin have already been published (Kendrew, 1948a, 1949).

A further paper is submitted as an additional memorandum. This deals with the kinetics of mutarotation, and was published jointly with Dr. E.A. Moelwyn-Hughes in 1940 (Kendrew and Moelwyn-Hughes, 1940). The work described in it was carried out by myself under the general supervision of Dr. Moelwyn-Hughes; the programme of which it was to form the first part was interrupted by the War, and its subject is too far removed from the main topics of this Dissertation for it to be included as an integral part thereof.

I wish to record my special gratitude to Dr. E.A. Moelwyn-Hughes, who gave me my first training in the methods of research, and to Dr. M.F. Perutz, who taught me the rudiments of crystallography and who has given me invaluable advice and encouragement throughout. I am also heavily in debt to the late Sir Joseph Barcroft, F.R.S., who first suggested the fetal hemoglobin problem and maintained an active interest in its progress up to the time of his death; and to Professor Sir Lawrence Bragg, F.R.S., Professor D. Keilin, F.R.S., and Dr. W.H. Taylor, all of whom have given me very much active help and encouragement during the past three years. Finally I wish to thanks Dr. J. Keilin, who collaborated with me in attempts to crystallize various myoglobins; Mr. M.W. Rees, who provided the crystals of horse myoglobin used in the analysis; and the Department of Scientific and Industrial Research and the Medical Research Council for financial support.

John Kendrew's Scientific Publications (1940–1994)

Primary Research Publications (1940–1968)

Kendrew, J.C. (1940) The kinetics of muta-rotation in solution. *Proceedings Royal Society London A* **176**, 352.

Kendrew, J.C., and Perutz, M.F. (1948) A comparative X-ray study of fetal and adult sheep hemoglobins. *Proceedings Royal Society London A* **194**, 375–398.

Kendrew, J.C. (1948) Preliminary X-ray data for horse and whale myoglobin. *Acta Crystallographica* **1**, 336.

Kendrew, J.C. (1949) X-ray studies of certain crystalline proteins: The crystal structure of fetal and adult sheep hemoglobins and of horse myoglobin. PhD Thesis, Trinity College, Cambridge.

Kendrew, J.C. (1950) The crystal structure of horse met-myoglobin. I. General features— arrangement of polypeptide chains. *Proceedings Royal Society London A* **201**, 62.

Bragg, W.L., Kendrew, J.C., and Perutz, M.F. (1950) Polypeptide chain configurations in crystalline proteins. *Proceedings Royal Society London A* **203**, 321.

Bennett, J.M., and Kendrew, J.C. (1952) Computation of Fourier syntheses with a digital electronic calculation machine. *Acta Crystallographica* **5**, 109.

Huxley, H.E., and Kendrew, J.C. (1952) Extractability of the Lotmar-Picker material from dried muscle. *Nature* **170**, 882.

Huxley, H.E., and Kendrew, J.C. (1953) Discontinuous lattice changes in hemoglobin crystals. *Acta Crystallographica* **6**, 76.

Kendrew, J.C., and Trotter, I.F. (1954) A pseudo-orthorhombic form of horse myoglobin. *Acta Crystallographica* **7**, 347.

Kendrew, J.C., Parrish, R.G., Marrack, J.R., and Orlans, E.S. (1954) The species specificity of myoglobin. *Nature* **174**, 946–949.

Kendrew, J.C., and Parrish, R.G. (1955) Imidazole complexes of myoglobin and the position of the heme group. *Nature* **175**, 206–207.

Ingram, D.J., and Kendrew, J.C. (1956) Electronic spin resonance in myoglobin and hemoglobin. *Nature* **178**, 905–906.

Bluhm, M.M., and Kendrew, J.C. (1956) The crystal forms and molecular weight of alpha-chymotrypsinogen: An X-ray study. *Biochemica Biophysica Acta* **20**, 562–563.

Kendrew, J.C., Pauling P.J., and Bragg, W.L. (1956) The crystal structure of myoglobin II. Finback whale myoglobin. *Proceedings Royal Society London A* **237**, 255.

Kendrew, J.C., and Parrish, R.G. (1957) The crystal structure of myoglobin. III. Sperm whale myoglobin. *Proceedings Royal Society London A* **238**, 305.

Kendrew, J.C., Bodo, G., Dintzis, H.M., Parrish, R.G., Wyckoff, H., and Phillips, D.C. (1958) A three-dimensional model of the myoglobin molecule obtained by X-ray analysis. *Nature* **181**, 662–666.

Bluhm, M.M., Bodo, G., Dintzis, H.M., and Kendrew, J.C. (1958) The crystal structure of myoglobin. IV. A Fourier projection of sperm whale myoglobin by the method of isomorphous replacement. *Proceedings Royal Society London A* 246, 369.

Bodo, G., Dintzis, H.M., Kendrew, J.C., and Wyckoff, H.W. (1959) The crystal structure of myoglobin. V. Low-resolution 3-dimensional Fourier synthesis of sperm whale myoglobin crystals. *Proceedings Royal Society London A* 253, 70.

Kendrew, J.C., Dickerson, R.E., Strandberg, B.E., Hart, R.G., Davies, D.R., Phillips, D.C., and Shore, V.C. (1960) The structure of myoglobin: A three-dimensional Fourier synthesis at 2 Å resolution. *Nature* 185, 422–427.

Watson, H.C., and Kendrew, J.C. (1961) The amino-acid sequence of sperm whale myoglobin. Comparison between the amino-acid sequences of sperm whale myoglobin and human hemoglobin. *Nature* 190, 670–672.

Kendrew, J.C., Watson, H.C., Strandberg, B.E., Dickerson, R.E., Phillips, D.C., and Shore, V.C. (1961) The amino-acid sequence, X-ray methods, and its correlation with chemical data. *Nature* 190, 666–670.

Dickerson, R.E., Kendrew, J.C., and Strandberg, B.E. (1961) The crystal structure of myoglobin: Phase determination to a resolution of 2 Å by the method of isomorphous replacement. *Acta Crystallographica* 14, 1188.

Watson, H.C., Kendrew, J.C., and Stryer, L. (1964) The binding of p-chloro-mercuribenzene sulphonate to crystals of sperm whale myoglobin. *Journal Molecular Biology* 8, 166–169.

Stryer, L., Kendrew, J.C., and Watson, H.C. (1964) The mode of attachment of the azide ion to sperm whale met-myoglobin. *Journal Molecular Biology* 8, 96–104.

Schoenborn, B.P., Watson, H.C., and Kendew, J.C. (1965) Binding of xenon to sperm whale myoglobin. *Nature* 207, 28–30.

Banaszak, L.J., Watson, H.C., and Kendrew, J.C. (1965) The binding of cupric and zinc ions to crystalline sperm whale myoglobin. *Journal Molecular Biology* 12, 130–137.

Perutz, M.F., Kendrew, J.C., and Watson, H.C. (1965) The structure and function of hemoglobin: II. Some relations between polypeptide chain configuration and amino acid sequence. *Journal Molecular Biology* 13, 669–678.

Nobbs, C.L., Watson, H.C., and Kendrew, J.C. (1966) Structure of deoxy-myoglobin: A crystallographic study. *Nature* 209, 339–341.

Kretsinger, R.H., Watson, H.C., and Kendrew, J.C. (1968) Binding of mercuri-iodide and related ions to crystals of sperm whale met-myoglobin. *Journal Molecular Biology* 31, 305–314.

Primary Reviews (1954–1966)

Kendrew, J.C. (1949). Fetal hemoglobin. *Endeavour* 8, 80.

Kendrew, J.C. (1954) The crystalline proteins: Recent X-ray studies and structural hypotheses. *Progress in Biophysics* 4, 244.

Kendrew, J.C. (1954) Structure of proteins. *Nature* 173, 57.

Crick, F.H.C., and Kendrew, J.C. (1957) X-ray analysis and protein structure. *Advances in Protein Chemistry* 12, 134.

Kendrew, J.C., and Perutz, M.F. (1957) X-ray studies of compounds of biological interest. *Annual Review of Biochemistry* 26, 327.

Kendrew, J.C. (1958). Architecture of a protein molecule. *Nature* 182, 764.

Kendrew, J.C. (1959) 3-Dimensional structure of globular proteins. *Reviews of Modern Physics* 31, 94.

Kendrew, J.C. (1959). Structure and function in myoglobin and other proteins. *Federation Proceedings* 18, 740.

Kendrew, J.C. (1961) The three-dimensional structure of a protein molecule. *Scientific American* 205, 96.

Kendrew, J.C. (1962) Side-chain interactions in myoglobin. *Brookhaven Symposium on Biology* 15, 216.

Kendrew, J.C. (1963) Myoglobin and the structure of proteins (Nobel Lecture). *Science* 139, 1259.

Kendrew, J.C. (1966) Stabilizing interactions in globular proteins. *Ciba Symposium on Principles of Biomolecular Organization* 86.

Kendrew, J.C. (1970). Some remarks on the history of molecular biology. *Biochemical Society Symposia* 30, 5.

Books

Kendrew, J.C. (1966) *The Thread of Life: An Introduction to Molecular Biology,* Harvard University Press, Cambridge, MA.

Kendrew, J.C., ed. (1994) *The Encyclopedia of Molecular Biology,* Blackwell Science Ltd., Oxford, UK.

John Kendrew's *Science* Article (1963)

29 March 1963, Volume 139, Number 3561

SCIENCE

Myoglobin and the Structure of Proteins

Crystallographic analysis and data-processing
techniques reveal the molecular architecture.

John C. Kendrew

When I first became interested in the question of solving the structure of proteins, during the latter part of World War II, I had no doubt that this problem above all others deserved the attention of anyone concerned with fundamental aspects of biology. Had my interests been awakened a few years later I would, no doubt, have recognized that there were in fact two such basic unanswered questions—the structure of proteins and the structure of nucleic acids. As events turned out, the second question was posed later and answered sooner. For me in the early 1940's, however, there seemed to be only one question uniquely qualified to engage the interest of anyone wishing to apply the disciplines of physics and chemistry to the problems of biology. It also seemed that the only technique offering any chance of success in determining the structures of molecules as large and complex as proteins was that of x-ray crystallography. As I look back on that time it occurs to me that my own almost total ignorance of this

The author is affiliated with the British Medical Research Council's Laboratory of Molecular Biology, Cambridge, England. This article is adapted from the lecture which he delivered in Stockholm, Sweden, 11 December 1962, on receiving the Nobel prize in chemistry, a prize which he shared with Max F. Perutz. It is published with the permission of the Nobel Foundation. It will also be included in the complete volume of Nobel lectures in English which is published yearly by the Elsevier Publishing Company, Amsterdam and New York.

method was fortunate, in that it concealed from me the extent to which contemporary x-ray crystallographic techniques fell short of what was needed to solve the structures of molecules containing thousands of atoms; it was indeed a case of ignorance being bliss. For a number of years this situation persisted; many roads were explored, but none of them seemed to offer real hope of a definitive solution until my colleague Max Perutz showed that the method of isomorphous replacement, until then applied rather rarely in crystallography generally and never in the field under discussion, was in fact ideally suited to the protein problem. His first successful application of this method to the hemoglobin structure provided the basis of all subsequent work in the field, my own included. In this discussion (1) I shall refer to questions of methodology only insofar as they have special relevance to my own work.

As I have indicated, my choice of problem and of method seemed straightforward. The choice of material was not so simple. One looked for a protein of low molecular weight, easily prepared in quantity, readily crystallized, and not already being studied by x-ray methods elsewhere. Myoglobin seemed to satisfy these criteria, and it had the additional advantage of being closely related to hemoglobin, which had been

the object of Perutz's attention for many years, and of sharing with hemoglobin a most important and interesting biological function, that of reversible combination with oxygen. As was revealed more clearly later, myoglobin consists of a single polypeptide chain of about 150 amino acid residues, associated with a single heme group. Its one-to-four relationship with hemoglobin, already suggested in early days by a comparison of molecular weights, turned out to be not coincidental but a fundamental structural relationship, as has now been shown by comparing the molecular models of the two proteins. At the beginning, however, one was more concerned with practical problems, which took a number of years to solve, than with hypothetical structural relationships.

First of all it was necessary to find some species whose myoglobin formed crystals suitable, both morphologically and structurally, to the purpose at hand; the search for this took us far and wide, through the world and through the animal kingdom, and eventually led us to the choice of the sperm whale, *Physeter catodon*, our material coming from Peru or from the Antarctic. There were some close runners-up, including the myoglobin of the common seal; its structure is now being studied by Helen Scouloudi at the Royal Institution in London. Once the method of isomorphous replacement had been shown to be capable, in principle, of solving the structure, one was faced with the task of attaching a small number of very heavy atoms at well-defined sites to each protein molecule in the crystal. Myoglobin lacks the sulfhydryl groups whose presence in hemoglobin was so successfully exploited by Perutz and Ingram for the attachment of mercurial reagents; we had to look for other ways, and our attempts to use the unique heme group for the attachment of ligands which contained heavy atoms having proved for the most part unsuccessful (our ligands were always rapidly ejected by even very small traces of oxygen which were almost impossible to exclude), we were thrown back to a more empirical approach. This consisted in crystallizing myoglobin in the presence of metallic ions and then seeing whether any changes in the x-ray pattern could be detected; further analysis was required to determine whether, as we desired, substitution had taken place at a single site. In the absence of any sound foundation of theory, it was necessary to examine a very large number of possible ligands—several hundreds—before two or three were found which satisfied all the rather rigid criteria. Such laboriously empirical procedures are still forced upon all workers in this field and very drastically limit the exploitation of the isomorphous replacement method. A rational and generally applicable solution to this problem is yet to be discovered; it would do more than any other single factor to open up the field.

General Strategy of the Analysis

Turning now to the strategy actually adopted for the solution of the structure, we may remember that Perutz's first application of the isomorphous replacement method in hemoglobin, as well as our own in myoglobin, had been to produce a two-dimensional projection of the structure. For such a projection the number of x-ray reflections required is fairly small, and the solution of the phase problem is simple; even with a single isomorphous replacement the results are unambiguous. But the amount of structural information which could be derived from a projection was almost nil, owing to the high degree of overlapping of the elements of so complex a structure. It was immediately clear that to exploit the method we would have to apply it in three dimensions, to produce a spatial representation of the electron density throughout the crystal. This involved the study of a much larger number of reflections and the calculation of general phases, and required, for an unambiguous solution, comparison of several heavy-atom derivatives substituted in different parts of the molecule.

The whole diffraction pattern of a myoglobin crystal consists of at least 25,000 reflections. In 1955, when the three-dimensional work was begun, there were no computers fast enough to calculate Fourier syntheses containing so many terms; besides, the method was unproved, and it seemed advisable to test it first on a smaller sample of data.

We may regard a typical x-ray photograph of a myoglobin crystal (Fig. 1) as a two-dimensional section through a three-dimensional array of reflections; each reflection corresponds to a single Fourier component, and the whole structure can be reconstructed by using all the components as terms of a Fourier synthesis. As Perutz has

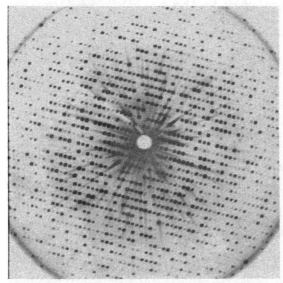

Fig. 1. X-ray precession photograph of a myoglobin crystal.

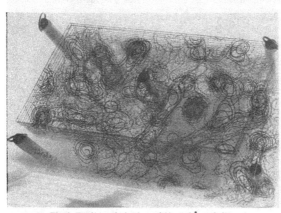

Fig. 2. Fourier synthesis of myoglobin at 6-Å resolution.

indicated, the components of higher frequency (higher harmonics), which are responsible for filling in the fine details of the structure, lie toward the outside of the pattern.

Thus, one can obtain a rendering of the molecule at low resolution by using simply the reflections within a spherical surface at the center of the pattern. By doubling the radius of the sphere (which now encloses eight times as many reflections) we double the resolution of the density distribution. We decided to undertake the solution of the structure in three stages. The first, completed in 1957, involved 400 reflections and gave a resolution of 6 angstroms; the second (in 1959) included nearly 10,000 reflections and gave a resolution of 2 angstroms; the third (not yet complete) includes all the observable reflections—about 25,000 —and gives a resolution of 1.4 angstroms. It may be recalled that polypeptide chains pack together at center-to-center distances of 5 to 10 angstroms; that atoms (other than hydrogen) of neighboring groups in van der Waals contact, or brought together by hydrogen bonds or charge interactions, lie 2.8 to 4 angstroms apart; and that the separation between covalently bonded atoms is 1 to 1.5 angstroms. It follows that the three stages chosen would be expected to separate polypeptide chains, groups of atoms, and individual atoms, respectively. In the third stage, with its resolving power of 1.4 angstroms, neighboring covalently bonded atoms should be just distinguishable, but this is as far as the analysis can go because, beyond this point, the diffraction pattern fades away. This limit represents a lower degree of order than is usual in crystals of molecules of low or moderate complexity; in fact, myoglobin crystals possess a higher degree of order than do crystals of almost any other protein, and this was an additional reason for my choice of this protein for analysis.

Before considering the results of the three stages of the analysis, let us revert to the question of computers. As will be evident from what follows, the amount of useful structural information obtainable increases rapidly with the resolving power. Indeed, it seems probable that for most proteins the dividend obtained from high resolution would be even greater, for it has been found that the helix content of myoglobin (75 percent) is a good deal higher

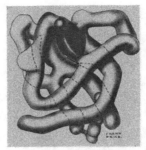

Fig. 3. Model of the myoglobin molecule, derived from the 6-Å Fourier synthesis. The heme group is a dark grey disk, center top.

than that of most other proteins, and identification of structural features in myoglobin at less than atomic resolution is greatly dependent on the presence of many helical segments of polypeptide chain, readily identifiable even at 6-angstrom resolution, and at 2-angstrom resolution already showing well-defined takeoff points for side chains, often making it possible to identify these even though the individual atoms cannot be distinguished. But, as already indicated, the amount of computation required increases very rapidly with the resolving power. Even at the first stage of the analysis we made use of an electronic computer, EDSAC I, which, though small and slow by modern standards, was at the time one of the very few such instruments in operation in the world. These early Fourier syntheses of the myoglobin data were, to the best of my belief, the first crystallographic computations ever carried out on an electronic computer and initiated a practice which later (and, incidentally, after a time lag of several years) became universal among crystallographers. At each stage of the myoglobin analysis the computers employed were among the most rapid available at the time; we are now using very fast and large computers such as EDSAC II and IBM 7090; most protein molecules are larger than the myoglobin molecule, and their analysis will require even bigger computers. There is also the problem of data collection and data handling. In the myoglobin analysis the data for the 6-angstrom and 2-angstrom stages were mainly collected by conventional photographic

methods, but at the 2-angstrom stage the solution of the phase problem for 9600 reflections involved densitometric measurements of some quarter of a million spots, from different heavy-atom derivatives and exposures of different lengths. Such a task approaches the limit of the practicable, especially as we were aiming for, and achieved, a mean error of 2 to 4 percent in the determination of amplitude. I would not care to have to undertake such a task a second time. In any case, serious effects of radiation damage to the crystals make photographic techniques increasingly difficult if not impossible at the higher resolutions. Fortunately, the automatic diffractometer designed by my colleagues U. W. Arndt and D. C. Phillips became available just in time for the final stage of the work. With this apparatus the intensities of successive reflections, measured with a proportional counter, are recorded on punched tape which can be fed directly into a high-speed computer. There is no doubt that automatic data-collecting equipment and very fast, large computers are highly desirable for all, and essential for most, x-ray studies of proteins.

Myoglobin at 6-Angstrom Resolution

The three-dimensional electron density distribution in a crystal is most conveniently represented as a series of contour maps plotted on parallel transparent sheets. Such a map for myoglobin at 6-angstrom resolution is shown in Fig. 2. A cursory inspection of the map showed it to consist of a large number of rodlike segments, joined at the ends and irregularly wandering through the structure; a single dense flattened disk in each molecule; and sundry connected regions of uniform density. These could be identified, respectively, with polypeptide chains, with the iron atom and its associated porphyrin ring, and with the liquid that fills the interstices between neighboring molecules, its boundaries being demarcated by the adjoining liquid. A scale model is shown in Fig. 3. For the most part the course of the single polypeptide chain could be followed as a continuous region of high density, but some ambiguities remained, especially at the irregular regions between two straight rods. The most striking features of the molecule were its irregularity and its total lack of symmetry; this made all

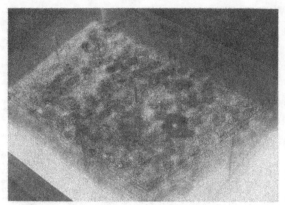

Fig. 4. Fourier synthesis of myoglobin at 2-Å resolution, showing a helical segment of polypeptide chain end-on.

the more remarkable the later finding by Perutz that each of the four subunits of hemoglobin closely resembled the myoglobin molecule, in spite of wide differences in species and in amino acid composition. As had been expected, it was not possible at 6-angstrom resolution to draw any conclusions regarding the nature of the folding of the polypeptide chain, or to see, let alone identify, side chains.

Myoglobin at 2-Angstrom Resolution

To achieve a resolution of 2 angstroms it was necessary to determine the phases of nearly 10,000 reflections and then to compute a Fourier synthesis with the same number of terms. As already indicated, this task represented about the extreme limit of what is practicable with photographic techniques, and the Fourier synthesis itself (exclusive of preparatory computations of considerable bulk and complexity) required about 12 hours of continuous computation on a very fast machine (EDSAC II). The electron density function was calculated at about 100,000 points in the molecule and was represented on the three-dimensional contour map shown in Fig. 4. In Fig. 4 we are looking at the density distributions directly along the axis of one of the straight rodlike sections of polypeptide chain identified at low resolution; it may be seen that the rod has now

developed into a straight hollow cylinder. Study of the density distribution on the surface of this (and other) cylinders showed that it fits the arrangement of atoms in the α-helix postulated by Pauling and Corey in 1951 as the chain configuration in the so-called α-family of fibrous proteins. Careful analysis of the density distribution, carried out on the computer, shows that the helical segments are nearly all precisely straight, and that their co-ordinates correspond to those given by Pauling and Corey within the limits of error of the analysis. Furthermore, it is possible to see directly the orientation of each side chain relative to the atoms within the helix, and hence, from a knowledge of the absolute configuration of an L-amino acid, to show that all the helices are right-handed.

Another view (Fig. 5) of the contour map shows the heme group edge on, now appearing as a flat disk with the iron atom at its center. To our surprise we found that the iron atom lay more than ¼ angstrom out of the plane of the group; it was only later that we heard from Koenig at Johns Hopkins University that he had observed the same phenomenon in his structure analysis of hemin. We were also able to see that the iron atom was attached to one of the helical segments of polypeptide chain by a group which we were later able to identify as histidine—a striking confirmation of suggestions which had been made as many as 30 years earlier

to the effect that histidine was the heme-linked group in hemoglobin and myoglobin.

In our preliminary publication about this Fourier synthesis, in 1960, we pointed out that at a resolution of 2 angstroms neighboring covalently bonded atoms are not resolved, and we gave it as our opinion that systematic identification of side chains would not be possible at this resolution. Events proved that we had been too pessimistic; by studying carefully the shapes of the lumps of density projecting at the proper intervals from the polypeptide chain we were often able to identify them unambiguously with one or another of the 17 different types of side chain known, from the overall composition, to be present in the myoglobin molecule. We were able to seek confirmation and extension of our results from a quite different source. At the time when the myoglobin program was getting seriously under way I discussed with W. H. Stein and Stanford Moore at the Rockefeller Institute in New York the possibility that some member of their laboratory might undertake a determination of the complete amino acid sequence of myoglobin, using the methods originally employed by Sanger in his studies of insulin and later developed and extended at the Rockefeller Institute for the analysis of ribonuclease. They were kind enough to arrange for Allen Edmundson, at that time a graduate student working in their laboratory under the supervision of C. H. W. Hirs, to undertake this task. By the time our 2-angstrom synthesis was available, Edmundson had studied most of the peptides obtained by tryptic digestion of myoglobin and had determined their composition and in a few cases the sequence of residues within them. We found that, by laying his peptides along the partial and tentative sequence derived from the x-ray analysis, we were able in many cases to observe correspondences which confirmed both our identifications and his analysis, and to clear up ambiguities and confusions in each (Fig. 6). All in all, it was possible to identify about two-thirds of all the residues in the molecule with some assurance, though certain pairs of residues of similar shape were difficult to distinguish. We were able to summarize the results of the analysis up to this stage in the form of a model (Fig. 7) which showed the positions in space of the helical polypeptide chain segments, of the heme group, and of most of the side chains; it included, less precisely, the positions of the atoms in most of the nonhelical regions and in many of the remaining side chains.

Myoglobin at 1.4-Angstrom Resolution

During the past 2 years we have been concerned with improving the resolution of the electron density map by including virtually all the observable reflections in the pattern, about 25,000, and extending the analysis to spacings of 1.4 angstroms; we now plot the electron density at half a million points in the molecule. As I have already pointed out, this extension of the analysis was made possible by the availability of automatic data-collecting equipment with proportional counters, and of still larger computers, such as the IBM 7090. Even so the task would have been a formidable one if we had continued to use the method of isomorphous replacement, involving the collection of data from a number of different isomorphous derivatives. Instead, we reverted to a more conven-

tional method, that of successive refinement, and abandoned the use of heavy-atom derivatives.

From a study of the 2-angstrom Fourier synthesis we were able to assign spatial coordinates to about three-quarters of the atoms in the molecule. Owing to the limited resolving power of this synthesis, the accuracy with which atoms could be located was a good deal less than is desirable, but this imprecision was compensated for by their number, a good deal higher in proportion to the size of the structure than is generally necessary for the success of the refinement method. This method consists in calculating the phases of all the reflections from the coordinates of the atoms which have already been located; a Fourier synthesis is then computed from *observed* amplitudes and *calculated* phases. This synthesis necessarily shows all the atoms which have been used for calculating phases but should reveal "ghosts" of additional ones, with reduced density; it also indicates any minor errors in the positions of the atoms previously located, if their positions are found not to coincide exactly with those assumed.

One is now in a position to embark

on the next cycle of refinement, using the previous set of atoms with corrected coordinates, together with additional atoms located after the first cycle. After a few such cycles the successive Fourier syntheses should converge to give a precise representation of the whole structure.

We have so far carried out two cycles of refinement, including 825 atoms in the first and 925 atoms in the second (the myglobin molecule contains, in all, 1260 atoms exclusive of hydrogen: in addition, there are about 400 atoms of liquid and salt solution, of which some are bound to fixed sites on the surface of the molecule). One or two further cycles of refinement will probably be necessary, but in the meantime the 1.4-angstrom Fourier synthesis based on the second cycle gives very much better resolution than the 2-angstrom synthesis. In many cases neighboring covalently bonded atoms are just resolved, the background between groups of atoms is much cleaner than in the earlier syntheses, and many of the disturbances found in the region of the heavy-atom sites in the 2-angstrom synthesis have disappeared. Figures 8 to 10 and the

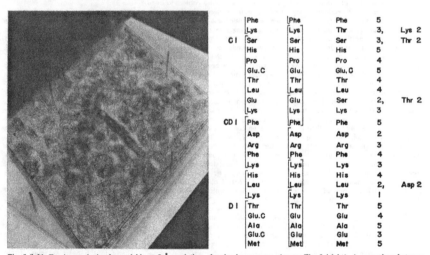

Fig. 5 (left). Fourier synthesis of myoglobin at 2-Å resolution, showing heme group edge on. Fig. 6 (right). A comparison between chemical and x-ray evidence for part of the amino-acid sequence of myoglobin. First column, tryptic peptides; second column, chymotryptic peptides; third column, x-ray evidence. Peptides are enclosed by brackets. The figures give the degree of confidence in the identification (5, complete confidence; 1, a guess).

29 MARCH 1963

cover photograph will give some impression of this synthesis.

Meanwhile Edmundson has greatly advanced his study of the amino acid sequence of myoglobin. In particular he has characterized a large number of chymotryptic peptides, in addition to the tryptic peptides previously mentioned. From the results of the x-ray and chemical studies, today some 120 amino acid residues are known with almost complete certainty and many of the remaining 30 are known with a fair degree of probability. There is little doubt that the residual ambiguities will shortly be resolved and that the positions of all the atoms in the structure will be known with reasonable accuracy, with the exception of a few long side chains (such as lysine), which are apparently flexible and do not occupy defined positions in the crystal.

General Nature of the Structure

What is the nature of the molecule which has emerged with progressively increasing clarity from successive Fourier syntheses? Some 118 out of the total of 151 amino acid residues make up eight segments of right-handed α-helix, seven to 24 residues long. These segments are joined by two sharp corners (containing no nonhelical residues) and five nonhelical segments (of one to eight residues); there is also a nonhelical tail of five residues at the carboxyl end of the chain. The whole is folded in a complex and unsymmetrical manner to form a flattened, roughly triangular prism with dimensions about 45 by 35 by 25 angstroms. The whole structure is extremely compact; there is no water inside the molecule, except for a very small number (less than five) of single water molecules presumably trapped at the time the molecule was folded up; there are no channels through it, and the volume of internal empty space is small. The heme group is disposed almost normally to the surface of the molecule, one of its edges (that containing the polar propionic acid groups) being at the surface and the rest buried deeply within.

Turning now to the side chains, we find that almost all those containing polar groups are on the surface. Thus, with very few exceptions, all the lysine, arginine, glutamic acid, aspartic acid, histidine, serine, threonine, tyrosine, and tryptophan residues have their polar groups on the outside (the rare

exceptions appear to have some special function within the molecule—for example, the heme-linked histidine). The interior of the molecule, on the other hand, is almost entirely made up of nonpolar residues, generally close packed and in van der Waals contact with their neighbors.

We may ask what forces are responsible for maintaining the integrity of the whole structure. The number of contacts between neighboring groups in the molecule is very large, and to analyze these it has been necessary to use a large computer to calculate all the interatomic distances and to determine which of these lie within the limits corresponding to each type of bonding. These results have not yet been studied in detail, but it is clear that by far the most important contribution comes from the van der Waals forces between nonpolar residues which make up the bulk of the interior of the molecule. It is true that there are a number of charge interactions and hydrogen bonds between neighboring polar residues on the surface of the molecule, but one gains the strong impression that many, or even most, of these are "incidental"; a polar group on the surface is quite content to bond with a water molecule or ion in the ambient solution and only links up with a neighboring side chain if it can do so without departing too far from its normal extended configuration. I should perhaps qualify this statement by remarking that one observes a number of polar interactions of side chains such as glutamic acid, aspartic acid, serine, and threonine with free imino groups on the last turn of helical segments, and it may be that these have some significance in determining the point at which a helix is broken and gives way to an irregular segment of chain. If so, these special interactions will be important in a wider context as determinants of the three-dimensional structure of proteins, and they might be of service in predicting the nature of the structure from a knowledge of the amino acid sequence.

The interactions of the heme group deserve special consideration. It is these which are responsible for the characteristic function of myoglobin, since an isolated heme group does not exhibit the phenomenon of reversible oxygenation. At present we can merely enumerate the heme group interactions; to explain reversible oxygenation in terms of them is a task for the future. As

already mentioned, the fifth coordination position of the iron atom is occupied by a ring nitrogen atom of a histidine residue, the so-called heme-linked histidine. On the other (distal) side of the iron atom, occupying its sixth coordination position, is a water molecule, as would be expected in ferrimyoglobin, the form of myoglobin used for x-ray analysis; beyond the water molecule, in a position suitable for hydrogen-bond formation, is a second histidine residue. It is noteworthy that the same arrangement of two histidines also exists in hemoglobin. For the rest, the environment of the heme group is almost entirely nonpolar; the group is held in place by a large number of van der Waals interactions. In hemoglobin it seems that the environment of the heme group is closely analogous. For both proteins, it is clear, a rich field of knowledge awaits exploration, and we may hope that the very extensive studies of the oxygenation reaction made during the past half century may now be interpreted in precise structural terms.

Some Implications

The oxygenation reaction of myoglobin and hemoglobin may be held to be interesting and important enough in its own right to justify the choice of these two proteins for study. In fact, as was indicated at the beginning of this article, the choice was originally made on different grounds, such as availability, ease of crystallization, and molecular weight. There are very many proteins which have specific functions as important, or more important. Every enzyme—and many hundreds of these have been characterized—has its own specific function vital to some particular process in cell function. A number of enzymes are being studied by x-ray methods in laboratories all over the world, and in several cases the analysis is on the brink of success. A knowledge of the detailed structure of each of these enzymes will give insight into some essential biological process, by resolving the molecular architecture of the active site and permitting the same kind of interpretation of function, in molecular terms, that we may soon achieve in the heme proteins. From this point of view there is no foreseeable limit to the number of proteins whose structure is worth analyzing, since each will have its unique function

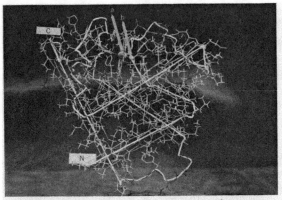

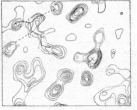

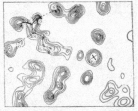

Fig. 7 (above). Model of the myoglobin molecule, derived from the 2-Å Fourier synthesis. The white cord follows the course of the polypeptide chain; the iron atom is indicated by a grey sphere (a) and its associated water molecule by a white sphere (b). C and N are the C and N terminals. Fig. 8 (right). Comparison between the same section through the sperm whale myoglobin molecule at 6-Å resolution, isomorphous phases (top), at 2-Å resolution, isomorphous phases (center), and at 1.4-Å resolution, calculated phases (bottom). The top left portion of each shows the longitudinal section through a helix; the right center of each shows the heme group edge on. The atoms marked are part of the distal histidine (see text). Note that several neighboring atoms are resolved

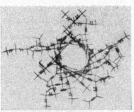

at 1.4 Å. Fig. 9 (left), End-on view of a helix at 2-Å and 1.4-Å resolutions. A model helix (top) allows for comparison. Fig. 10 (below). Part of the 1.4-Å Fourier synthesis. Center, the heme group (edge on) shows heme-linked and distal histidines and the water molecule attached to the iron atom. Top right, a helix end on. Bottom, a helix seen longitudinally, together with several side chains.

2Å

1.4Å

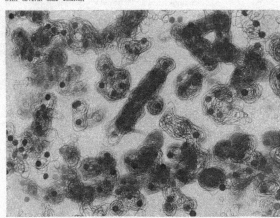

which demands explanation in structural terms.

From another angle we may, rather, ask what features are common to all proteins and study the structure of myoglobin as a typical member of this vast class of substances. Probably more experimental work has been done on proteins than on any other kind of compound, and a huge corpus of knowledge has been built up by the organic chemist and the physical chemist. Many generalizations have been made, but always they have been limited in scope by the fact that they could not be based on a precise molecular model. The emergence of such a model, even for a single protein such as myoglobin, makes it possible to test and to add precision to the chemist's generalizations. Already sperm whale myoglobin is being studied by biochemists in a number of laboratories with this end in view. To give only a few examples, it is being examined from the standpoint of optical rotatory power and helix content, of titration behavior, of metal binding, of chemical modification of side chains, and of hydrodynamic characteristics. Such studies, and others like them, will serve to deepen our understanding of the ways in which proteins behave and of the reasons why they are uniquely capable of occupying so central a position in living organisms.

The geneticists now believe, though the point is not yet rigorously proved, that the hereditary material determines only the amino acid sequence of a protein, not its three-dimensional structure. That is to say, the polypeptide chain, once synthesized, should be capable of folding itself up without being provided with additional information. This capacity has, in fact, recently been demonstrated by Anfinsen in vitro for one protein—ribonuclease. If the postulate is true it follows that one should be able to predict the three-dimensional structure of a protein from a knowledge of its amino acid sequence alone. Indeed, in the very long run it should only be necessary to determine the amino acid sequence of a protein in order to predict its three-dimensional structure. In my view this day will not come soon, but when it does come the

x-ray crystallographers can go out of business, perhaps with a certain sense of relief; it will also be possible to discuss the structures of many important proteins which cannot be crystallized and which therefore lie outside the crystallographer's purview.

We have taken a preliminary look at the structure of myoglobin from this point of view and have to confess that the difficulties are formidable. The structure is highly irregular; the seven "corner" regions between helical segments are all different, so that generalization is impossible; the interactions between side chains are numerous and of many different types, and one cannot easily see which are crucial in determining the structure. The complexity of myoglobin is very great, yet myoglobin is probably simpler than most proteins, not only by virtue of its low molecular weight but also in respect to its high helix content, probably much higher than that of most other proteins. As things stand, we cannot even hazard a guess as to why the helix content of myoglobin is so high, let alone see how to predict its structure in detail.

Much help with these problems may come from a comparison of myoglobin with the subunits of hemoglobin, which Perutz has shown to resemble it very closely in spite of notable differences in amino acid sequence. By laying side by side the sequences of myoglobin and of the α and β chains of hemoglobin, and by making certain plausible assumptions to explain the (fairly small) differences between the lengths of the chains, it is possible to observe homologies—points at which the same amino acid appears in a corresponding position in all three chains. The number of these homologies is surprisingly small, but presumably it is these which are responsible for the crucial interactions which determine that all three chains have the same three-dimensional arrangement (though some of the homologies may be accidents of evolutionary development). Study of homology will soon be extended by examination of myoglobin and hemoglobin of other species—human myoglobin and human, horse, rabbit, and human fetal hemoglobin—and of the aberrant hemoglobins whose "mistakes" in amino

acid sequence have been shown in recent years to be associated with so many hereditary diseases of the blood.

Perutz and I, with our collaborators, have already spent some time looking at these homologies, and a number of interesting facts have come to light. Yet, even in this narrow field our studies are in their infancy, and in any case I suspect that only generalizations of limited scope can be made from myoglobin and hemoglobin alone. The detailed structures of a few other proteins should soon become known, but it will be clear from many of the topics I have touched upon that we have pressing need to know the structures of very many others, for proteins are unique in combining great diversity of function and complexity of structure with a relative simplicity and uniformity of chemical composition. In determining the structures of only two proteins we have reached, not an end, but a beginning; we have merely sighted the shore of a vast continent waiting to be explored.

Bibliography

M. M. Bluhm, G. Bodo, H. M. Dintzis, J. C. Kendrew, *Proc. Roy. Soc. London* **A246**, 369 (1958).
G. Bodo, H. M. Dintzis, J. C. Kendrew, H. W. Wyckoff, *ibid.* **A253**, 70 (1959).
J. C. Kendrew, "Brookhaven Symp. Biol. No. 15," 216 (1962).
——, G. Bodo, H. M. Dintzis, R. G. Parrish, H. W. Wyckoff, D. C. Phillips, *Nature* **181**, 666 (1958).
J. C. Kendrew, R. E. Dickerson, B. E. Strandberg, R. G. Hart, D. R. Davies, D. C. Phillips, V. C. Shore, *Nature* **185**, 422 (1960).
J. C. Kendrew and R. G. Parrish, *Proc. Roy. Soc. London* **A238**, 305 (1956).
J. C. Kendrew, H. C. Watson, B. E. Strandberg, R. E. Dickerson, D. C. Phillips, V. C. Shore, *Nature* **190**, 666 (1961).
H. C. Watson and J. C. Kendrew, *ibid.* **190**, 670 (1961).

Note

1. The work described in this article was done by many hands, and a list of those who have contributed to it would be long. They come from many countries and many disciplines, and any such list must be incomplete. I nevertheless wish to record the following names of colleagues whose ideas and whose collaboration have been particularly important and sometimes essential. J. M. Bennett, C. Blake, Joan Blows, M. M. Bluhm, G. Bodo, Sir Lawrence Bragg, C. I. Branden, D. A. G. Broad, C. L. Coulter, Ann Cullis, D. R. Davies, R. E. Dickerson, H. M. Dintzis, A. B. Edmundson, R. G. Hart, Ann Hartley, W. Hoppe, V. M. Ingram, L. H. Jensen, J. Kraut, R. G. Parrish, P. Pauling, M. F. Perutz, D. C. Phillips, Mary Pinkerton, Eva Rowlands, Helen Scouloudi, Violet Shore, B. E. Strandberg, I. F. Trotter, H. C. Watson, Joyce Wheeler, Ann Woodbridge, and H. W. Wyckoff. I am indebted also to the staff of the Mathematical Laboratory, Cambridge.

Science

Myoglobin and the Structure of Proteins: Crystallographic analysis and data-processing techniques reveal the molecular architecture

John C. Kendrew

Science **139** (3561), 1259-1266.
DOI: 10.1126/science.139.3561.1259

ARTICLE TOOLS	http://science.sciencemag.org/content/139/3561/1259.citation
PERMISSIONS	http://www.sciencemag.org/help/reprints-and-permissions

Science (print ISSN 0036-8075; online ISSN 1095-9203) is published by the American Association for the Advancement of Science, 1200 New York Avenue NW, Washington, DC 20005. 2017 © The Authors, some rights reserved; exclusive licensee American Association for the Advancement of Science. No claim to original U.S. Government Works. The title *Science* is a registered trademark of AAAS.

Gunnar Hägg's Nobel Presentation Speech (1962)

Your Majesties, Your Royal Highnesses, Ladies and Gentlemen.

In the year 1869 the Swedish chemist Christian Wilhelm Blomstrand wrote, in his at that time remarkable book *Die Chemie der Jetztzeit* (*Chemistry of Today*): "It is the important task of the chemist to reproduce faithfully in his own way the elaborate constructions which we call chemical compounds, in the erection of which the atoms serve as building stones, and to determine the number and relative positions of the points of attack at which any atom attaches itself to any other; in short, to determine the distribution of the atoms in space."

In other words, Blomstrand gives here as his goal the knowledge of how compounds are built up from atoms, i.e. knowledge of what is nowadays often called their "structure." Moreover, structure determination has been one of the biggest tasks of chemical research, and has been approached using many different techniques. For several reasons, the structure determination of carbon compounds, the so-called organic compounds, experienced an initial rapid development. At this stage the techniques were generally those of pure chemistry. One drew conclusions from the reactions of a compound, one studied its degradation products, and tried to synthesize it by combining simpler compounds. The structure thus arrived at, however, was in general rather schematic in character; it showed which atoms were bonded to a given atom, but gave no precise values for interatomic distances or interbond angles. However, for an up-to-date treatment of the chemical bond and in order to derive a correlation between structure and properties, these values are needed, and they can only be obtained using the techniques of physics.

The physical method which, more than any other, has contributed to our present-day knowledge of these mutual dispositions of the atoms is founded on the phenomenon which occurs when an X-ray beam meets a crystal. This phenomenon, called diffraction, results in the crystal sending out beams of X-rays in certain directions. These beams are described as reflections. The directions and intensities of such reflections depend on the type and distribution of the atoms within the crystal, and can therefore be used for structure determination. It is 50 years ago this year since Max von Laue discovered the diffraction of X-rays by crystals, a discovery for which he was awarded the 1914 Nobel Prize for Physics. This work opened up a whole new range of possibilities for studying both the nature of X-rays and the structure of compounds in the solid state. The initial application of structure determination was developed first and foremost by the two English scientists, Bragg father and son, and as early as 1915 they were rewarded with the Nobel Prize for Physics. The techniques have since been considerably refined, and it has been possible to solve more and more complicated structures. However, considerable difficulties were encountered as soon as any other than very simple structures were considered. There is no simple general way of progressing from experimental data to the structure of the compound under investigation. Moreover, the mathematical calculations are exceedingly time consuming. However, by about the middle of the 1940s a point had been reached where it was becoming possible to carry out X-ray determinations of the structures of organic compounds which were so complicated that they defied all attempts using classical chemical methods.

In 1937 Max Perutz performed some experiments in Cambridge to find out whether it might be possible to determine the structure of hemoglobin by X-ray diffraction, since no other method could be imagined for this purpose. Sir Lawrence Bragg, who tirelessly continued the work begun jointly with his father, in 1938 became the head of the Cavendish Laboratory in Cambridge. When he saw the results obtained by Perutz, he encouraged him to continue and has ever since lent a very efficient support. Hemoglobin belongs to the proteins which play such an enormous part in life processes, and which are a basic material in living organisms. Hemoglobin is a component of the red blood corpuscles. It contains iron which can take up oxygen in the lungs and later give it up to the body's other tissues. Hemoglobin is counted among the globular proteins, whose molecules are nearly spherical. It was chosen for the initial attempt, partly because it could develop good crystals, and partly because the hemoglobin molecule is quite small for a protein molecule. About ten years later, John Kendrew joined Perutz' research group, and the task allotted to him was to try to determine the structure of myoglobin. Myoglobin is another globular protein, closely related to hemoglobin, but with a molecule only a quarter as large. It is found in the muscles, and enables oxygen to be stored there. Particularly large amounts of myoglobin are found in the muscular tissues of whales and seals, which need to be able to store large quantities of oxygen when diving.

However, Perutz and Kendrew encountered considerable difficulties. In spite of exceptionally comprehensive work, the result was not forthcoming until 1953, when Perutz succeeded in incorporating heavy atoms, namely those of mercury, into definite positions in the hemoglobin molecule. By this means the diffraction pattern is altered to some extent, and the changes can be utilized in a more direct structure determination. The method was already known in principle, but Perutz applied it in a new way, and with great skill. Kendrew also succeeded, by an alternative method, in incorporating heavy atoms, generally mercury or gold, into the myoglobin molecule, and could subsequently proceed in an analogous manner.

A necessary condition for this technique is that the addition of the heavy atoms should not alter the positions of the other atoms of the molecule within the crystal. In this connection it is simply because of its enormous dimensions that the molecule remains practically unaltered. Bragg has rather aptly said that "the molecule takes no more notice of such an insignificant attachment than a maharaja's elephant would of the gold star painted on its forehead."

But even if the path was now open for a direct structure determination of hemoglobin and myoglobin, there was still an enormous amount of data to be processed. Myoglobin, the smaller of the two molecules, contains about 2,600 atoms, and the positions of most of these are now known. But for this purpose, Kendrew had to examine 110 crystals and measure the intensities of about 250,000 X-ray reflections. The calculations would not have been practicable if he had not had access to a very large electronic computer. The hemoglobin molecule is four times as large, and its structure is known less thoroughly. In both cases, however, Kendrew and Perutz are currently collecting and processing an even greater number of reflections in order to obtain a more detailed picture.

As a result of Kendrew's and Perutz' contributions it is thus becoming possible to see the principles behind the construction of globular proteins. The goal has been reached after twenty-five years' labour, and initially with only modest results. We therefore admire the two scientists not only for the ingenuity and skill with which they have carried out their work, but also for their patience and perseverance, which have overcome the difficulties which initially seemed insuperable. We now know that the structure of proteins can be determined, and it is certain that a number of new determinations will soon be carried out, perhaps chiefly following the lines which Perutz and Kendrew have indicated. It is fairly certain that the knowledge which will thus be gained of these substances which are so essential to living organisms will mean a big step forward in the understanding of life processes. It is thus abundantly clear that this year's prize-winners in chemistry have fulfilled the condition which Alfred Nobel laid down in his will, they have conferred the greatest benefit on mankind.

Doctor Kendrew and Doctor Perutz. One of you recently said that today's students of the living organism do indeed stand on the threshold of a new world. You have both contributed very efficiently to the opening of the door to this new world and you have been among the first to obtain a glimpse of it. Through your combined efforts there is now in view, as it has been stated by yourself, a firm basis for an understanding of the enormous complexities of structure, of biogenesis and of the functions of living organisms both in health and disease.

It is with great satisfaction, therefore, that the Royal Swedish Academy of Sciences has decided to award you this year's Nobel Prize for Chemistry for your brilliant achievement.

On behalf of the Academy I wish to extend to you our heartiest congratulations, and now ask you to receive from the hands of His Majesty the King the Nobel Prize for Chemistry for the year 1962.

Gunnar Hägg was a professor of general and inorganic chemistry at Uppsala University, Uppsala, Sweden and a member of the Nobel Committee for Chemistry of the Royal Swedish Academy of Sciences. He chaired the Nobel Committee in 1976.

John Kendrew's EMBL Signature of Agreement Speech (1973)

Mr. President, Your Excellencies, Ladies and Gentlemen: Today's ceremony is the culmination of a ten-year endeavor. A ten-year endeavor on the part of many individuals, and of many governments, and with help from various international organizations, for example CERN, where we meet today, which has been our friendly host and counselor ever since the beginning, the European Communities who have lent us staff at no cost to ourselves, and the World Health Organization which funded some of our early meetings. As far as I was concerned it all began ten years and five months ago when I met in this building with the then Director General of CERN, Professor Viki Weisskopf, with the distinguished ex-Hungarian American biologist, Leó Szilárd, and with my own ex-pupil Dr. Jim Watson, and there we discussed the idea of founding a European laboratory in molecular biology. From that first discussion emerged an informal meeting of European biologists, with some Americans too, at Ravello near Naples in August of 1963, and from that meeting came the foundation of EMBO as a private body composed of individual biologists elected on the basis of their own personal qualifications, not on their national allegiances. The motives underlying the decision to create EMBO centered on a general concern about the state of molecular biological research in Europe. Many of the fundamental discoveries leading to the previous decade of extraordinarily rapid advance in the field had been made in European laboratories. But increasingly it was the Americans who were making the running, and European biologists were emigrating to the USA in numbers sufficient to constitute a serious brain drain. Partly this was due to the intrinsic quality of American research, and to the much greater supply of money on the other side of the Atlantic. But there were preventable deficiencies on the European side—the somewhat rigid departmental systems still prevailing in many European universities prevented adequate support and funding from being channeled to a new interdisciplinary field which crossed so many traditional boundaries between the classical fields of biology; research groups were too small to allow the necessary multiple approaches to be concentrated on a single problem, and the number of large interdisciplinary laboratories in Europe was extremely small; European scientists were not in the habit of travelling in one another's countries, and if they did meet colleagues from another European country it was more often than not at some congress in the United States; and Europe lacked the focus and meeting place provided in the United States by laboratories like Cold Spring Harbor and Woods Hole.

The EMBO idea was to try to find ways of repairing this situation, of improving the quality of European biology. We thought of two main ways of doing this, the first one to create an international laboratory, the second to finance fellowships and summer courses, and EMBO was founded with these two objects in mind. As I indicated, it was and is a private body so it needed money, and at the beginning it had not one penny. For EMBO the good fairy was the Volkswagen Foundation which gave us a very large grant, three-quarters of a million dollars for a 3-year period, non-recurrent, the idea being that this would give us time to prove ourselves and to seek more permanent support from the governments. Well, as you all know, our efforts were in fact successful and the Central Programme of the Conference is now supported by thirteen governments and by the Fonds National de la Recherche Scientifique of Belgium. The funds thus provided are used for fellowships, for short-term travel grants, and for financing summer courses and workshops. We now have a continuous record of running a programme along these lines since the beginning of the Volkswagen grant in 1966.

But meanwhile, EMBO had not forgotten about the project of founding a laboratory. Almost from the beginning the European Molecular Biology Conference was used as a forum for discussing this proposal. Of course it was a much more contentious question than the General Programme of the Conference because it had a certain all-or-nothing quality; either you spend a great deal of money, or you have no laboratory at all. Whereas a programme of fellowships and courses can be small or large according to the availability of funds. As a matter of fact, the first EMBO proposal for a laboratory made in 1968 was for a much larger institute than is now authorised. The first proposal did not meet with the approval of the governments, not only, I think, because it was too large, but also because I believe that EMBO itself had not thought out sufficiently fully what were the purposes of the laboratory and what functions it was supposed to perform. The turning point in our discussions came when EMBO invited a representative group of molecular biologists to a meeting at Konstanz in 1969 at which the whole concept of the laboratory was re-discussed. Out of this meeting came a new proposal which seemed, at least to me, to be much more sharply focused on the real needs than the earlier one had been, and it formed the basis of the Laboratory as it is now to be established. As now conceived it was not just a place for scientists to amuse themselves in, though of course any laboratory having a good scientific programme would in fact be fun to work in, rather, the Laboratory was now intended to provide services of a number of kinds to the national institutes. These services were to include the development of advanced instrumentation now more and more needed in biology on a scale and of a degree of sophistication which make it impracticable to carry out in national laboratories, which, in this field, normally have very limited workshop and engineering facilities. Then it would provide advanced training, accepting postdoctoral fellows and even pre-doctoral fellows from all the European universities. And finally it would be a centre for workshops and advanced courses, as well as for meetings of a more informal kind. Those of you who are familiar with the American biological scene will be aware of the function performed there by the famous laboratory at Cold Spring Harbor. What we have in mind is that the EMB Laboratory should have a similar place in Europe. In general the aim is to establish a laboratory which will not be a rival to national institutes, but which will rather complement them and which will explicitly organise itself as a service institution supporting to the best of its ability the national programmes in the field.

Of course a proposition like this was bound to lead to many hard discussions. In the first place it could not be cheap, indeed its budget would be a number of time greater that the whole of the rest of the EMBC programme. Then there must certainly be political problems about the site of the laboratory and about the number of countries ready to support it. Inevitably it took time to sort all this out, but here today we see the successful conclusion of that difficult process of negotiation. Ten governments are ready to sign the Agreement establishing the Laboratory. We are sorry that there were several other members of the EMBC who did not feel able to sign, but I am happy to say that all of them have expressed the hope that they may be able to do so later. So this is certainly a very pleasant occasion for those of us who have been involved in these negotiations now for a number of years. Today's result is due to many factors, and especially to the work and spirit of cooperation of many people and many organizations. I should, however, like to single out several particular debts we ow. First of all to the Swiss government. This Government is our host today and has been responsible for all the excellent arrangements for this meeting, but it is not the first time that it has acted in this capacity. The Swiss Government has been our host on many previous occasions and has taken many initiatives to call the participants together. One can think of it as an essential catalyst, or perhaps as a mid-wife, I leave it to you to choose either the chemical or the biological metaphor. Then we owe a very great debt to the Government of the Federal Republic of Germany—how great that debt is, you can see for yourselves if you visit the beautiful site that the German Government has provided at Heidelberg for the Laboratory, deep in the forest next to the Max Planck Institut für Kernphysik. Some of you will have visited it already and I hope that the rest may do so at the next meeting of the Conference and at the first meeting of the Laboratory Council, which we are planning to hold in Heidelberg. Now the Government of the Federal Republic has been especially helpful not only in providing this extremely nice site, but in many other ways, indirect and direct, including a large and special finan-cial contribution to the capital cost of the Laboratory. Third, I think we, as scientists, owe a very great debt to all of the delegates of the countries of the EMBC, who quite apart from their official instructions from their governments, have come to demonstrate so much personal enthusiasm for our various projects and have worked so hard to break through the political and financial difficulties, and so have made a major contribution to the final result. In this connection I should most especially like to mention the enormous debt we owe to our President, Monsieur Voirier, who, to our sorrow, must retire from that office during the next weeks. Those of you who have sat under his wise chairmanship will know how big a contribution he has made and how his tact and ability have enabled us to find a way through all kinds of intractable problems.

These are some of those who have helped us along the road which ends at today's meeting. But of course today's meeting is not only an end, an end, that is, of politics and discussion, but it is also a beginning of practical action and of the actual creation of the new Laboratory. After all the delays which have occurred we are doing our best not to allow grass to grow under our feet. You probably know that we plan to have a limited competition among a few architects for the design of the building. You may be interested to hear that we are meeting the architects who will take part in the competition in only four days' time from now, in Heidelberg, to discuss the problem with them. So we hope that active architectural planning can begin within the next week or two. Of course it will take two or three years before our building can concretely exist, but this does not mean that the work of the Laboratory cannot begin much sooner. We have received most gen-erous offers of temporary accommodation from the Deutsches Krebsforschungszentrum and the Max Planck Institut für Kernphysik in Heidelberg, for both offices and labora-tories, so that scientific work can begin quite soon. And some of you will know, we have active plans for establishing outstations at the electron synchrotron in Hamburg and the high flux reactor in Grenoble, the first of these to make possible biological experiments with extremely intense X-ray beams, and the second with neutron beams. At the first of these outstations in Hamburg, we already have some scientific research going on under the auspices of the Laboratory Project, and in Grenoble we are actively negotiating about arrangements with the Institut Laue-Langevin for the outstation there. Finally we hope to open an office for the Laboratory Project in Heidelberg within the next month. So for those of us associated with the Project, today's meeting represents the starting signal for activity on a number of different fronts, activity which we hope will have the result that the build up of the Laboratory can begin almost immediately.

We are deeply grateful to all those governments and individuals who have made it possible for those ideas first discussed in Professor Weisskopf's office here in CERN at the end of 1962, and later developed by EMBO, to become real. We look forward to the future collaboration between governments on the one hand and biologists on the other in the new forum of the Laboratory Council which will be holding its first meeting during the next month or two. And finally we look forward to the practical realisation of an enterprise which we hope will be scientifically successful as well as being of material assistance to the biological institutes of the member countries, and which will make its own distinctive contribution to the creation of the more unified Europe of the future.

Notes

This biography of Sir John Cowdery Kendrew is based in part on material found in the Kendrew Archives in the Department of Western Manuscripts, Special Collections Reading Rooms, at the New Bodleian/Weston Library, Oxford. These archives are described in Catalogues I and II compiled in 1989 and in a Supplementary Catalogue compiled in 1998. The author is grateful to John A. Montgomery, executor of John Kendrew's final will, for granting permission to quote in this biography from letters and other documents in the Kendrew Archives (permission letter dated October 11, 2018).

The biography is also based on interviews, e-mails, letters, and/or telephone calls with many friends, colleagues, and associates of John Kendrew. Particularly valuable to the author was the biography of John Kendrew written for the *Biographical Memoirs of Fellows of the Royal Society* by K.C. Holmes in 2001, as well as K.C. Holmes's interview of John Kendrew at the MRC LMB, Cambridge, in June 1997; obituaries of John Kendrew by several authors following his death in August 1997 (Appendix A.2); remembrances by several friends and colleagues at memorials for John Kendrew held in Oxford in October 1997 and Cambridge in November 1997; and Wikipedia and Google Web Search.

The Kendrew Archives references are provided here using the NCUACS designations as compiled in November 1989 by the late Jeannine Alton, the former executive director of the Contemporary Scientific Archives Center, and described in Appendix A.1, along with the more recent Collection Level Description of the archives compiled in 2017. For full details of books, reviews, obituaries, and other publications used in part as sources for the biography see sources S.2.

Abbreviations

JCK, John C. Kendrew; KA, Kendrew Archives

Preface

1. Perutz, M.F., Cambridge Memorial for JCK, November 1997.
2. Chargaff, E., *Heraclitean Fire: Sketches from a Life Before Nature,* 1978, Rockefeller University Press, NY.
3. Medawar, P., *Memoir of a Thinking Radish,* 1986, Oxford University Press, Oxford, UK.
4. Hager, T., *www.thomashager.net.*
5. Holmes, K.C., Sir John Cowdery Kendrew. *Biographical Memoirs of Fellows of the Royal Society* 47, 311–332, 2001.
6. Ackermann, W., interview by the author, Heidelberg, March 2010.

Chapter 1

This chapter is based on information in the KA, particularly Section R (R.1–R.51) of Catalogue II and Section R (R.52–R.367) of the Supplementary Catalogue; books and reviews listed in sources S.2; Wikipedia and Google Web Search.

1. Dickerson, R.E., A little ancient history, *Protein Science* 1, 182–186, 1992.
2. Dickerson, R.E., Myoglobin: A whale of a structure. *Journal Molecular Biology* 392, 10–23, 2009.
3. JCK interview, *New Scientist* 63, 594–596, 1974.
4. Petsko, G.A., The father of us all. *Genome Biology* 3, 1004.1–1004.2, 2002.
5. JCK interview by Holmes, K.C., Cambridge, June 1997, MRC LMB Archives, Cambridge, UK.
6. Harris-Adams, G., e-mails to the author, January 2011.
7. Watson, J.D., Cambridge Memorial for JCK, November 1997.
8. JCK interview, *New Scientist* 63, 594–596, 1974.

Chapter 2

This chapter is based on information obtained from several sources: interviews by and/ or e-mails to the author; the KA, particularly items listed in Sections B (B.1–B.60) of Catalogue I and R (R.52–R.367) of the Supplementary Catalogue; Wikipedia and Google Web Search.

1. Thomas, J.M., Peterhouse, the Royal Society, and molecular biology. *Notes and Records of the Royal Society London* 54, 369–385, 2000.
2. Picknett, T., Journal of molecular biology: A publishers perspective. *Journal Molecular Biology* 293, 165–171, 1999.
3. KA/R.178.
4. Huxley, H.E., e-mail to the author, January 2011.
5. Marton-Lefèvre, J., interview by the author, New York, February 2010.
6. Alton, J., Description of the KA, Catalogue 1, April 1989.
7. Ackermann, W., interview by the author, Heidelberg, March 2010.
8. Holmes, K.C., Sir John Cowdery Kendrew. *Biographical Memoirs Fellows Royal Society* 47, 311–332, 2001.
9. Earnshaw-Held, C., e-mail to the author, June 2013.
10. McKibbin, R., Oxford Memorial for JCK, October 1997.
11. Montgomery, J.A., interview by the author, Oxford, November 2010.
12. Klug, A., Cambridge Memorial for JCK, November 1997.
13. Earnshaw-Held, C., e-mail to the author, June 2013.
14. KA/R.52.
15. KA/B.17.
16. JCK interview, Pursuit in three dimensions, *Chemistry in Britain* 10, 439–443, 1974.
17. McKibbin, R., Oxford Memorial for JCK, October 1997.

18. KA/B.17.
19. KA/R.113.

Chapter 3

The section on JCK's great-grandparents, grandparents, paternal uncles, paternal aunts, and their offspring is based on information obtained from several sources including *ancestry.com, ancestry.co.uk, direct.gov.uk, familysearch.org, findmypast.uk*, and *freebmd. org.uk*. Additional information was obtained from the 1861, 1891, and 1901 England and Wales Census; 1901 Ireland Census; 1861 and 1881 Scotland Census; 1900, 1910, and 1920 US Census. Other information was obtained from the Cambridgeshire and Oxfordshire City Councils, General Register Office, Oxford City Library, British phonebooks, Wikipedia, and Google Web Search.

The section on Wilfrid George Kendrew is based on information obtained from several sources: J. Kenworthy's Meteorologist's Profile—Wilfrid George Kendrew (1884–1962), *Weather* **62**, 49–52, 2007; C. Smith's Wilfrid George Kendrew, *Geographers: Biobibliographical Studies* 43–51, 1997; R. Beckinsale's Mr. W.G. Kendrew, *Nature* **194**, 817, 1962; J. Baker's Wilfrid George Kendrew, *Geographical Journal* **129**, 127–128, 1963; V. Brook's Wilfrid George Kendrew, *St. Catherine's Association Chronicle*, 7–8, 1962. Additional information came from the KA, British Army WW1 Medal Rolls, UK Incoming Passenger Lists, Oxford City Library, Wikipedia, and Google Web Search.

The section on Evelyn Sandberg-Vavalà is based on information obtained from several sources: J. Pope-Hennessy's Mrs. Evelyn Sandberg-Vavalà, *Burlington Magazine* **103**, 466–467, November 1961; H. Honour's Evelyn Sandberg Vavalà *London Times*, 13, September 1961; Fondazione Giorgio Cini's Fondi Fotografici Evelyn Sandberg-Vavalà (1888–1961); e-mails to the author from D. Smith, St. Anne's College, Oxford; L. Standen, Herefordshire Record Office; and P. Bufton, Herefordshire. Additional information was obtained from the KA, Wikipedia, and Google Web Search; the latter two sources were used in part for the section on marital separation.

1. Oral history interview with Eugene V. Thaw. *Archives American Art*, October 1–2, 2007.
2. Clifford, N., A backward glance at the Villa Capponi. *The Florentine*, no. 38, 2006.
3. University of Michigan Museum of Art, The Divine Comedy, 2006/1.97.

Chapter 4

This chapter is based on information in the KA, particularly items listed in Sections A (A.1–A.120) and E (E.1–E16) of Catalogue I and Section R (R.52–R.367) of the Supplementary Catalogue; e-mails to the author from G. Sturt archivist at the Dragon School, Oxford; e-mails to the author from C. Knighton archivist at Clifton College, Bristol; Science at Clifton, 1977; *Cliftonian; Draconian*; N. Ingram's All That You Can't Leave Behind, 2003; Oxford City Library; Wikipedia and Google Web Search.

1. KA/R.2.
2. Mitchell, P., Memoirs Clifton College, *davidmitchell.co.uk*.
3. Clifton College School of Natural Science, *Nature* 4, August 24, 329, 1871.
4. JCK speech at Clifton College jubilee of the Science School, 1977. Gover, T., e-mail to the author, 2010.
5. KA/R.192.
6. KA/R.53.
7. Farago, P. Interview with Sir John Kendrew. *Impact* 51, 701–704, 1974.
8. Pinkerton, J. Recollections of John Kendrew at Clifton College 1932–1937 and later. 1997. Gover, T., e-mails to the author, 2010.
9. KA/A.30.
10. JCK interview by Holmes, K.C., June 1997, MRC LMB archives, Cambridge, UK.
11. Ibid.
12. Ibid.
13. KA/R.54.
14. Alton, J., Description of the KA, Catalogue 1, 1989.

Chapter 5

This chapter is based on information obtained from the KA, particularly items listed in Section B (B.1–B.60) of Catalogue I and Section R (R.52–R.367) of the Supplementary Catalogue; J. McCloskey's British Operational Research in World War II, *History Workshop Journal Issue 63*; Wikipedia and Google Web Search.

1. Churchill, W.L., *Their Finest Hour: The Second World War*, 11, 1949, Riverside Press, Cambridge, UK.
2. Crowther, J.G., and Whiddington, R., *Science at War*, 1948, Philosophical Library, NY.
3. KA/B.24.
4. KA/R.60.
5. Ibid.
6. KA/B.3.
7. KA/B.2.
8. Eban, S., *A Sense of Purpose: Recollections*, 2008, Halban Publishers, London.
9. Ibid.
10. Ibid.
11. Ibid.
12. Ibid.
13. Ibid.
14. Ibid.
15. Ibid.
16. Ibid.
17. KA/R.117.
18. KA/B.2.
19. KA/B.17.

20. Goldsmith, M., *Sage: Life of J.D. Bernal*, 1980, Hutchinson and Co., London.
21. KA/B.45.
22. KA/R.136.
23. KA/R.7.
24. KA/B.23.
25. KA/B.53.
26. KA/R.7.
27. KA/B.53.
28. Ibid.
29. KA/L.27.

Chapter 6

This chapter is based on information obtained from several sources: The KA, particularly items listed in Section C (C.1–C.308)) of Catalogue I and Section R (R.52–R.367) of the Supplementary Catalogue; *https://www.phy.cam.ac.uk/history*; *https://www.phy.cam.ac.uk/history/old*; Wikipedia and Google Web Search.

1. McKenzie, A.E.E., *The Major Achievements of Science*, 1960, Cambridge University, Press, Cambridge, UK.
2. Perutz, M.F., *Is Science Necessary: Essays on Science and Scientists*, 1991, Oxford University Press, Oxford, UK.
3. The New Physical Laboratory of the University of Cambridge, *Nature* 10, 139–142, June 1874.
4. Perutz, M.F., *Is Science Necessary: Essays on Science and Scientists*, 1991, Oxford University Press, Oxford, UK.
5. Perutz, M.F., *I Wish I'd Made You Angry Earlier*, 2003, Cold Spring Harbor Press, Cold Spring Harbor, NY.
6. Wilson, D., *Rutherford: Simple Genius*, 1984, MIT Press, Cambridge, MA.
7. Longair, M., *Maxwell's Enduring Legacy: A Scientific History of the Cavendish Laboratory*, 2016, Cambridge University Press, UK.
8. Pippard, B., in Thomas, J.M. and Phillips, D. eds. *Selections and Reflections: The Legacy of Sir Lawrence Bragg*, 97–100, 1990, Science Reviews Ltd., Middlesex, UK.
9. Cochran, W., and Devons, S. Norman Feather, 16 November 1904–14 August 1978. *Biographical Memoirs of Fellows of the Royal Society* 27, 255–282, 1981.
10. Hunter, G.K., *Light is a Messenger: The Life and Science of William Lawrence Bragg*, 2004, Oxford University Press, Oxford.
11. KA/R.192.
12. Finch, J. *A Nobel Fellow on Every Floor: A History of the Medical Research Council Laboratory of Molecular Biology*, 2008, MRC LMB, Cambridge, UK.
13. Huxley, H.E. The Cavendish Laboratory and structural biology. *Physics World*, March, 29–33, 2003.
14. Ibid.

15. Edsall, J.T., Memories of early days in protein science. *Protein Science* **1**, 1526–1530, 1992.
16. Watson, J.D., *The Double Helix: A Personal Account of the Discovery of the Structure of DNA*, 1968, Atheneum Press, New York, NY.
17. Perutz, M.F., A sagacious scientist, *New Scientist*, April 1981, 39.
18. Perutz, M.F. *Scientific American* **239**, 92–105, 1978.
19. Ibid.
20. Perutz, V., ed., *What a Time I Am Having: Selected Letters of Max Perutz*, 2009, Cold Spring Harbor Press, Cold Spring Harbor, NY.
21. JCK interview by Holmes, K.C., June 1997, MRC LMB archives, Cambridge, UK.
22. Kendrew, J.C., Myoglobin and the structure of proteins. *Science* **139**, 1259–1266, 1963.
23. Rossmann, M.G., The beginnings of structural biology. *Protein Science* **3**, 1731–1733, 1994.
24. Perutz, M.F., Sir John Kendrew. *MRC News,* Autumn, 1997.
25. Klug, A., Cambridge Memorial for JCK, November 1997.
26. Perutz, V., ed., *What a Time I Am Having: Selected Letters of Max Perutz*, 2009, Cold Spring Harbor Press, Cold Spring Harbor, NY.
27. Thomas, J.M., Cambridge Memorial for JCK, November 1997.
28. KA/C.27.
29. KA/C.248.
30. Ibid.
31. KA/C.27.
32. KA/C.2.
33. KA/C.3.
34. Ibid.
35. R.48.
36. KA/O.27.
37. KA/C.27.
38. Ibid.
39. Ibid.
40. *Manchester Guardian*, 1959.
41. Dintzis, H.M. The wandering pathway to determining N to C synthesis of proteins. *Biochemistry and Molecular Biology Education* **34**, 241–246, 2006.
42. *Manchester Guardian*, 1959.
43. JCK interview by Holmes, K.C., June 1997, MRC LMB Archives, Cambridge UK.
44. Kendrew, J.C., Myoglobin and the structure of proteins. *Les Prix Nobel En 1962*, 103–125, 1963.
45. KA/C.27.
46. KA/C.199.
47. Ibid.
48. KA/C.27.
49. KA/C.228.
50. Watson, J.D. *The Double Helix; A Personal Account of the Discovery of the Structure of DNA*, 1968, Atheneum Press, New York, NY.

51. KA/C.17.
52. Ibid.
53. KA/C.21.
54. KA/O.7.
55. Rossmann, M.G. Chapter 3: Recollection of the events leading to the discovery of the structure of hemoglobin. *Journal of Molecular Biology* **392**, 23–32, 2009.
56. Longair, M., *Maxwell's Enduring Legacy: A Scientific History of the Cavendish Laboratory*, 2016, Cambridge University Press, Cambridge, UK.
57. JCK interview by B. Dixon. *Scientist* **1**, 182–184, 1987.

Chapter 7

The section on John Kendrew's marriage is based on information obtained from several sources: e-mails to the author from A. Green and A. Hunt in Cambridge, D. Beaumont at the Fettesian Association, L. Sands at the British Medical Association House, London; England and Wales Marriage Index and Death Index; London Gazette; Wikipedia and Google Web Search. The section on John Kendrew's divorce was written using information obtained from the KA Section R (R.52–R.367) of the Supplementary Catalogue (see R.1); H.E. Huxley interview by the author in Woods Hole in 2009, and e-mails to the author from H.E. Huxley in July 2009 and January 2011; F. Gutfreund interview by the author in Didcot in 2009; Wikipedia and Google Web Search.

1. KA/R.73.
2. Watson, J.D., *Genes, Girls, and Gamow*, 2001, Vintage Books, New York.
3. Perutz, M.F., MRC LMB Archives, Cambridge, UK.
4. Watson, J.D., *The Double Helix: A Personal Account of the Discovery of the Structure of DNA*, 1968, Athenium Press, New York.
5. KA/C.17.
6. Perutz, M.F., MRC LMB Archives, Cambridge, UK.
7. Watson, J.D., *Genes, Girls, and Gamow*, 2001, Vintage Books, New York.
8. KA/R.242.
9. Ibid.

Chapter 8

This chapter is based on information obtained from several sources: Articles by JCK in *Nature* **181** (1958) and **185** (1960), *Scientific American* (December 1961), and *Science* **139** (1963); *Journal of Molecular Biology* **392** (2009) and *Protein Science* **1** (1992) and **27** (2018); the KA, particularly items listed in Section C (C.1–C.308), Section N (N.1–N.71) and Section R (R.52–R.367) of the Supplementary Catalogue; Wikipedia and Google Web Search.

1. Kendrew, J.C., The three-dimensional structure of a protein molecule. *Scientific American*, 1–16, December, 1961.

2. KA/C.288.
3. Perutz, V., ed., *What a Time I Am Having: Selected Letters of Max Perutz*, 2009, Cold Spring Harbor Press, Cold Spring Harbor.
4. KA/C.288.
5. Kendrew, J.C., et al., A three-dimensional model of the myoglobin molecule obtained by X-ray analysis. *Nature* 181, 662–666, 1958.
6. KA/C.217.
7. KA/C.288.
8. Dickerson, R.E., A little ancient history. *Protein Science* 1, 182–186, 1992.
9. KA/C.217.
10. Kendrew, J.C., Citation Classic. *Current Contents* July 21, 1986, 12.
11. KA/R.48.
12. KA/R.283.
13. Kendrew, J.C. Myoglobin and the structure of proteins. *Science* 139, 1259–1266, 1963.
14. KA/C.217.
15. KA/N.12.
16. KA/N.13.
17. Ibid.
18. KA/L.34.

Chapter 9

This chapter is based on information obtained from the KA, particularly items in Section M (M.1–M.37) of Catalogue II and Section R (R.52–R.367) of the Supplementary Catalogue; Wikipedia and Google Web Search.

1. KA/M.1.
2. Ibid.
3. KA/M.20.
4. Ibid.
5. KA/M.27.
6. KA/M.32.
7. KA/M.29.
8. KA/M.33.
9. Ibid.
10. KA/M.34.
11. KA/M.35.

Chapter 10

This chapter is based on information obtained from several sources: E. Norrby's *Nobel Prizes and Life Sciences*, 2010, World Scientific Publishers, Singapore; *NobelPrize. org*; Nobel Lectures, Chemistry 1942–1962, Elsevier; the KA Section R (R.1–R.51) of

Catalogue II and Section R (R.52–R.367) of the Supplementary Catalogue; Wikipedia and Google Web Search.

1. *NobelPrize.org.*
2. KA/R.20.
3. KA/R.63
4. Perutz, V., ed., *What a Time I Am Having: Selected Letters of Max Perutz,* 2009, Cold Spring Harbor Press, Cold Spring Harbor, NY.
5. KA/R.101.
6. Ibid.

Chapter 11

This chapter is based on information obtained from several sources: an advertisement from Gordon Long and Company that appeared in *Country Life,* June 1964; descriptions of The Old Guildhall by Gordon Long and Company; Ward Howard Rowlett Chartered Surveyors; English Heritage; documents viewed during a visit to The Old Guildhall in Linton by the author in 2010; F. Bickerstaff interview by the author in Linton in 2010; the KA, Wikipedia, and Google Web Search.

Chapter 12

This chapter is based on information obtained from several sources: interviews by the author with and/or e-mails to the author from several past or present members of the MRC LMB; the MRC LMB Archives in Cambridge; J. Finch, *A Nobel Fellow on Every Floor,* 2008, MRC LMB, Cambridge, UK; Huxley, H.E., ed., *Memories and Consequences: Visiting Scientists at the MRC Laboratory of Molecular Biology,* 2013, MRC LMB, Cambridge, UK; the KA Section D (D.1–D.39) of Catalogue I, Sections J (J.1–J.176) and L (L.1–L.149) of Catalogue II, and Section R (R.52–R.367) of the Supplementary Catalogue; Wikipedia and Google Web Search.

1. Huxley, H.E., *Memories and Consequences: Visiting Scientists at the MRC Laboratory of Molecular Biology,* 2013, MRC LMB, Cambridge, UK.
2. KA/J.99.
3. Ibid.
4. Ibid.
5. Ibid.
6. de Chadarevian, S., *Designs for Life: Molecular Biology after World War II,* 2002, Cambridge University Press, Cambridge, UK.
7. KA/L.46.
8. KA/R.347.

Chapter 13

This chapter is based on information obtained from several sources: the KA Sections F (F.1–F.232), G (G.1–G.150), and H (H.1–H.417) of Catalogue I and Section R (R.52–R.367) of the Supplementary Catalogue; interviews with and/or e-mails from W. Ackermann, J. Dubochet, K. Holmes, M. Holmes, F. Lennert-Glöckner, K. Leonard, I. Mattaj, M. Mlodzik, K. Möller, C. Nüsslein-Volhard, K. Simons, N. Straussfeld, J. Tooze, and E. Wieschaus; *https://www.embl.de/* and *https://www.embl.org/*; Wikipedia and Google Web Search.

1. KA/R.136.
2. KA/F.231.
3. KA/F.132.
4. KA/F.147.
5. Holmes, K.C., Sir John Cowdery Kendrew. *Biographical Memoirs Fellows Royal Society* 47, 311–332, 2001.
6. KA/H.13.
7. KA/F.133.
8. KA/F.221.
9. KA/H.346.
10. KA/H.397.
11. Ibid.
12. Holmes, K.C., Sir John Cowdery Kendrew. *Biographical Memoirs Fellows Royal Society* 47, 311–332, 2001.
13. KA/H.365.
14. Ibid.
15. KA/H.383.
16. Holmes, K.C., Sir John Cowdery Kendrew. *Biographical Memoirs Fellows Royal Society* 47, 311–332, 2001.
17. EMBL 40 Years, 1974–2014. EMBL Alumni Relations Office, 2014, Heidelberg, Germany.

Chapter 14

This chapter is based on information obtained from several sources: interviews with J. Couling, M. Freedland, W. Hayes, C. Llewellyn-Smith, R. McKibbin, S. Petersen, R. Tourneau, and M. Yee; the KA Section R (R.52–R.367) of the Supplementary Catalogue; Wikipedia and Google Web Search.

1. McKibbin, R., Oxford Memorial for JCK, October 1997.
2. Harris, P.L., Oxford Memorial for JCK, October 1997.
3. KA/R.311.
4. KA/R.266.

5. Keynes, R.D., Cambridge Memorial for JCK, November 1997.
6. Marton-Lefèvre, J., *Science International*, December 1997.
7. Black-Wilsmore, S., e-mail to the author, November 2010.
8. McKibbin, R., Oxford Memorial for JCK, October 1997.
9. Thomas, J.M., Cambridge Memorial for JCK, November 1997.

Sources

S.1 Books on Crystallography

Blow, D. (2002). *Outline of Crystallography for Biologists*. Oxford University Press, Oxford, UK.

Clegg, W. (2015). *X-Ray Crystallography*. Oxford Chemistry Primers. Oxford University Press, Oxford, UK.

Drenth, J. (2010). *Principles of Protein X-Ray Crystallography*. 3rd ed. Springer, New York, NY.

Hammond, C. (2009). *The Basis of Crystallography and Diffraction*. 3rd ed. Oxford University Press, Oxford, UK.

Holmes, K.C., and Blow, D.M. (1965). *The Use of X-Ray Diffraction in the Study of Protein and Nucleic Acid Structure*. Interscience, New York, NY.

McPherson, A. (2002). *Introduction to Macromolecular Crystallography*. John Wiley and Sons, Hoboken, NJ.

McRee, D.E., and David, P.R. (1999). *Practical Protein Crystallography*. 2nd ed. Academic Press, London, UK.

Rhodes, G. (2006). *Crystallography Made Crystal Clear*. 3rd ed. Academic Press, London, UK.

Rupp, B. (2009). *Biomolecular Crystallography: Principles, Practice, and Application to Structural Biology*. Garland Science, New York, NY.

Sherwood, D., and Cooper, J. (2015). *Crystals, X-rays, and Proteins: Comprehensive Protein Crystallography*. Oxford University Press, Oxford, UK.

Wilson, H.R. (1966). *Diffraction of X-rays by Proteins, Nucleic Acids and Viruses*. Edward Arnold, London, UK.

S.2 Bibliography: Books, Reports, and Reviews

Aprahamian, F., and Swann, B., eds. (1999). *J.D. Bernal: A Life in Science and Politics*. Verso Press, London, UK.

Baker, J. (1963). Wilfrid George Kendrew. *Geographical Journal* **129**, 127–128.

Beckinsale, R. (1962). Mr. W.G. Kendrew. *Nature* **194**, 817.

Berol, D.N. (2001). Living materials and the structural ideal: The development of the protein crystallography community in the 20th century. PhD diss., Princeton University, UMI Microform 9993686.

Blow, D.M. (2004). Max Ferdinand Perutz OM CH CBE: 19 May 1914–6 February 2002. *Biographical Memoirs of Fellows of the Royal Society, London* **50**, 227–256.

Bragg, W.L. (1960). British achievements in X-ray crystallography. *Science* **131**, 1870–1874.

310 SOURCES

Branden, C., and Tooze, J. (1999). *Introduction to Protein Structure*. Garland Science, New York, NY.

Brenner, S. (2001). *My Life in Science*. BioMed Central Ltd., London, UK.

Brook, V. (1962). Wilfrid George Kendrew. *St. Catherine's Association Chronicle*, 7–8.

Brown, A. (2005). *J.D. Bernal: The Sage of Science*. Oxford University Press, Oxford, UK.

Brownlee, G.G. (2014). *Fred Sanger—Double Nobel Laureate: A Biography*. Cambridge University Press, Cambridge, UK.

Cochran, W., and Devons, S. (1981). Norman Feather, 16 November 1904–14 August 1978. *Biographical Memoirs of Fellows of The Royal Society* 27, 255–282.

Crick, F. (May 21, 1987). Ruthless research in a cupboard. *New Scientist*, 66–68.

Crowther, J.G., and Whiddington, R. (1948). *Science at War*. Philosophical Library, New York, NY.

Dear, C.B., and Foot, M.R.D, ed. (2001). *Scientists at War*. The Oxford Companion to World War II. Oxford University Press, Oxford, UK.

de Chadarevian, S. (1996). Sequences, conformation, information: Biochemists and molecular biologists in the 1950s. *Journal of the History of Biology* 29, 361–386.

de Chadarevian, S. (2002). *Designs for Life: Molecular Biology After World War II*. Cambridge University Press, Cambridge, UK.

de Chadarevian, S. (2018). John Kendrew and myoglobin: Protein structure determination in the 1950s. *Protein Science* 27, 1136–1143.

Dickerson, R.E. (1992). A little ancient history. *Protein Science* 1, 182–186.

Dickerson, R.E. (2005). *Present At the Flood: How Structural Molecular Biology Came About*. Sinauer Associates, Sunderland, MA.

Dickerson, R.E. (2009). Chapter 2: Myoglobin: A whale of a structure. *Journal Molecular Biology* 392, 10–23.

Dintzis, H.M. (2006). The wandering pathway to determining N to C synthesis of proteins. Some recollections concerning protein structure and biosynthesis. *Biochemistry and Molecular Biology Education* 34, 241–246.

Eban, S. (July 15, 1974). A Cairo girlhood. *New Yorker*.

Eban, S. (2008). *A Sense of Purpose: Recollections*. Halban Press, London, UK.

Edsall, J.T. (1992). Memories of early days in protein science, 1926–1940. *Protein Science* 1, 1526–1530.

Eisenberg, D.S. (1994). Max Perutz's achievements: How did he do it? *Protein Science* 3, 1625–1628.

EMBL. (1978). Annual Report EMBL, Heidelberg, Germany.

EMBL. (1994). 20 Years On, 1974–1994. EMBL, Heidelberg, Germany.

EMBL. (2014). 40 Years, 1974–2014. EMBL Alumni Relations Office, Heidelberg, Germany.

Farago, P. (1974). Interview with Sir John Kendrew. *Impact* 51, 701–704.

Ferry, G. (2007). *Max Perutz and the Secret of Life*. Cold Spring Harbor Press, Cold Spring Harbor, NY.

Ferry, G. (2014). EMBO in perspective: A half-century in the life sciences. EMBO, Germany.

Ferry, G. (2014). Fifty years of EMBO. *Nature* 511, 150–151.

Finch, J. (2008). *A Nobel Fellow on Every Floor: A History of the Medical Research Council Laboratory of Molecular Biology*. MRC LMB, Cambridge, UK.

Friedberg, E.C. (2010). *Sydney Brenner: A Biography*. Cold Spring Harbor Press, Cold Spring Harbor, NY.

Garwin, L., and Lincoln, T., eds. (2003). *A Century of Nature: Twenty-One Discoveries that Changed Science and the World*. Nature Publishing, London, UK.

Goldsmith, M. (1980). *Sage: Life of J.D. Bernal*. Hutchinson and Co., Cambridge University Press, Cambridge, UK; London, UK.

Hager, T. (1995). *Force of Nature: The Life of Linus Pauling*. Simon and Schuster, New York, NY.

Hargittai, I. (2002). *The Road to Stockholm: Nobel Prizes, Science, and Scientists*. Oxford University Press, Oxford, UK.

Hodgkin, D.M.C. (1980). John Desmond Bernal: 10 May 1901–15 September 1971. *Biographical Memoirs of Fellows of the Royal Society, London* **26**, 17–84.

Holmes, K.C. (1997). Pioneer in structural biology and collaborative biological research in Europe. *Nature* **389**, 340.

Holmes, K.C. (2001). Sir John Cowdery Kendrew: 24 March 1917–23 August 1997. *Biographical Memoirs of Fellows of the Royal Society, London* **47**, 311–332.

Holmes, K.C. (2006). The life of a sage. *Nature* **440**, 149–150.

Holmes, K.C., and Weeds, A. (2017). Hugh Esmor Huxley MBE: 25 February 1924–25 July 2013. *Biographical Memoirs of Fellows of the Royal Society, London* **63**, 309–344.

Holmes, K.C. (2017). *Aaron Klug—A Long Way from Durban: A Biography*. Cambridge University Press, Cambridge, UK.

Honour, H. (September 1961). Evelyn Sandberg-Vavala. *London Times*, 13.

Hunter, G.K. (2004). *Light is a Messenger: The Life and Science of William Lawrence Bragg*. Oxford University Press, Oxford, UK.

Huxley, H.E. (March 2003). The Cavendish Laboratory and structural biology. *Physics World*, 29–35.

Huxley, H.E., ed. (2013). *Memories and Consequences: Visiting Scientists at the MRC Laboratory of Molecular Biology, Cambridge*. MRC LMB, Cambridge, UK.

Ingram, N.R. (2003). All that you can't leave behind. *Notes and Records of The Royal Society London* **57**, 177–184.

Jenkin, J. (2008). *William and Lawrence Bragg, Father and Son: The Most Extraordinary Collaboration in Science*. Oxford University Press, Oxford, UK.

Judson, H.F. (1979). *The Eighth Day of Creation: The Makers of the Revolution in Biology*. Simon and Schuster, New York, NY.

Kendrew, J.C. (December 1961). The three-dimensional structure of a protein molecule. *Scientific American* **205**, 96–110.

Kendrew, J.C. (1963). Myoglobin and the structure of proteins. *Les Prix Nobel En 1962*, 103–125.

Kendrew, J.C. (1963). Myoglobin and the structure of proteins. *Science* **139**, 1259–1266.

Kendrew, J.C. (1966). *The Thread of Life: An Introduction to Molecular Biology*. Harvard University Press, Cambridge, UK.

Kendrew, J.C. (1968). EMBO and the idea of a European laboratory. *Nature* **218**, 840–842.

Kendrew, J.C. (1980). The European Molecular Biology Laboratory. *Endeavour, New Series* **4**, 166–170.

Kendrew, J.C., ed. (1994). *The Encyclopedia of Molecular Biology*. Blackwell Science Ltd., Oxford, UK.

Kenworthy, J. (2007). Meteorologist's profile—Wilfrid George Kendrew (1884–1962). *Weather* **62**, 49–52.

Kim, D.-W. (2002). *Leadership and Creativity: A History of the Cavendish Laboratory, 1871–1919*. Springer, Dordrecht, The Netherlands.

Krige, J. (2002). The birth of EMBO and the difficult road to EMBL. *Studies in History and Philosophy of Biological and Biomedical Sciences* 33, 547–564.

Krohn, P.L. (1995). Solly Zuckerman, Baron Zuckerman of Burnham Thorpe, OM, KCB: 30 May 1904–1 April 1993. *Biographical Memoirs of Fellows of the Royal Society, London* 41, 576–598.

Larsen, E. (1962). *The Cavendish Laboratory: Nursery of Genius.* Edmund Ward, London, UK.

Longair, M. (2016). *Maxwell's Enduring Legacy: A Scientific History of the Cavendish Laboratory.* Cambridge University Press, Cambridge, UK.

McCloskey, J.F. (1987). British operational research in World War II. *Operations Research* 35, 453–470

Newmark, P. (1978). European molecular biology. *Nature* 273, 182–183.

Nobel Foundation. (1964). *Nobel Lectures, Chemistry 1942–1962.* Elsevier Publishing, Amsterdam.

Norrby, E. (2010). *Nobel Prizes and Life Sciences.* World Scientific Publishing, Singapore.

Olby, R. (2009). *Francis Crick: Hunter of Life's Secrets.* Cold Spring Harbor Press, Cold Spring Harbor, NY.

Pennisi, E. (2003). A hothouse of molecular biology. *Science* 300, 278–282.

Perutz, M.F. (November 1964). The hemoglobin molecule. *Scientific American* 211, 64–76.

Perutz, M.F. (August 1985). That was the war: Enemy alien. *New Yorker*, 35–54.

Perutz, M.R. (May 21, 1987). The birth of molecular biology. *New Scientist*, 38–42.

Perutz, M.F. (1991). *Is Science Necessary? Essays on Science and Scientists.* Oxford University Press, Oxford, UK.

Perutz, M.F. (1996). The Medical Research Council, Laboratory of Molecular Biology. *Molecular Medicine* 2, 659–662.

Perutz, M.F. (Autumn 1997). Sir John Kendrew. *MRC News.*

Perutz, M.F. (2003). *I Wish I'd Made You Angry Earlier.* Cold Spring Harbor Press, Cold Spring Harbor, NY.

Perutz, V., ed. (2009). *What a Time I Am Having: Selected Letters of Max Perutz.* Cold Spring Harbor Press, Cold Spring Harbor, NY.

Petsko, G.A. (2002). The father of us all. *Genome Biology* 3, 1004.1–1004.2.

Peyton, J. (2002). *Solly Zuckerman: A Scientist Out of the Ordinary.* John Murray, London, UK.

Phillips, D. (1979). William Lawrence Bragg: 31 March 1890–1 July 1971. *Biographical Memoirs of Fellows of the Royal Society, London* 25, 75–136.

Picknett, T. (1999). Journal of Molecular Biology: A publisher's perspective. *Journal of Molecular Biology* 293, 165–171.

Pope-Hennessy, J. (November 1961). Mrs. Evelyn Sandberg-Vavala. *Burlington Magazine* 103, 466–467.

Ridley, M. (2006). *Francis Crick: Discoverer of the Genetic Code.* Harper Collins, New York, NY.

Rosenhead, J. (Winter 1991). Swords into ploughshares: Cecil Gordon's role in the post-war transition of operational research to civilian uses. *Public Administration* 69, 481–501.

Rossmann, M.G. (1994). The beginnings of structural biology. *Protein Science* 3, 1731–1733.

Rossmann, M.G. (2009). Chapter 3: Recollection of the events leading to the discovery of the structure of hemoglobin. *Journal Molecular Biology* 392, 23–32.

Saxon, W. (August 30, 1997). John C. Kendrew dies at 80; biochemist won Nobel in '62. *New York Times*.

Senechal, M. (2012). *I Died for Beauty: Dorothy Wrinch and the Cultures of Science*. Oxford University Press, Oxford, UK.

Shi, Y. (2014). A glimpse of structural biology through X-ray crystallography. *Cell* **159**, 995–1014.

Smith, C.G. (1997). Wilfrid George Kendrew, 1884–1962. *Geographers: Biobibliographical Studies*, 43–51.

Strandberg, B. (1997). Sir John Kendrew. *Structure* **5**, 1713–1715.

Strandberg, B. (2009). Chapter 1: Building the ground for the first two protein structures: Myoglobin and hemoglobin. *Journal Molecular Biology* **392**, 2–10.

Strandberg, B., Dickerson, R.E., and Rossmann, M.G. (2009). 50 years of protein structure analysis. *Journal of Molecular Biology* **392**, 2–32.

Strasser, B. (2003). The transformation of the biological sciences in post-war Europe. *EMBO Reports* **4**, 540–543.

Stryer, L. (1995). *Biochemistry*. W.H. Freeman and Co., New York, NY.

Synge, A. (1992). J.D. Bernal, F.R.S.: Family, school, and university. *Notes and Records of the Royal Society London* **46**, 267–278.

Thomas, J.M. (September 12, 1997). Sir John Kendrew. *Independent*.

Thomas, J.M. (2000). Peterhouse, the Royal Society and molecular biology. *Notes and Records of the Royal Society London* **54**, 369–385.

Thomas, J.M. (2002). The scientific and humane legacy of Max Perutz (1914–2002). *Angewandte Chemie International Edition* **41**, 3155–3166.

Thomas, J.M. (2004). Bragg reflections. *Science* **306**, 2043–2044.

Thomas, J.M. (2012). The birth of X-ray crystallography. *Nature* **491**, 186–187.

Thomas, J.M., and Phillips, D. eds. (1990). *Selections and Reflections: The Legacy of Sir Lawrence Bragg*. Science Reviews Ltd., Middlesex, UK.

Tooze, J. (1986). The role of European Molecular Biology Organization (EMBO) and European Molecular Biology Conference (EMBC) in European molecular biology. *Perspectives in Biology and Medicine* **29**, S38–S46.

Tyack, G. (1998). *Oxford: An Architectural Guide*. Oxford University Press, Oxford, UK.

Wassarman, P.M. (2017). A personal perspective: My four encounters with John Kendrew. *Journal of Molecular Biology* **429**, 2594–2600.

Watson, J.D. (1968). *The Double Helix: A Personal Account of the Discovery of the Structure of DNA*. Atheneum Press, New York, NY.

Watson, J.D. (2001). *Genes, Girls, and Gamow*. Vintage Books, New York, NY.

Wilkins, M. (2003). *Maurice Wilkins: The Third Man of the Double Helix*. Oxford University Press, Oxford, UK.

Wilson, D. (1984). *Rutherford: Simple Genius*. MIT Press, Cambridge, MA.

Index

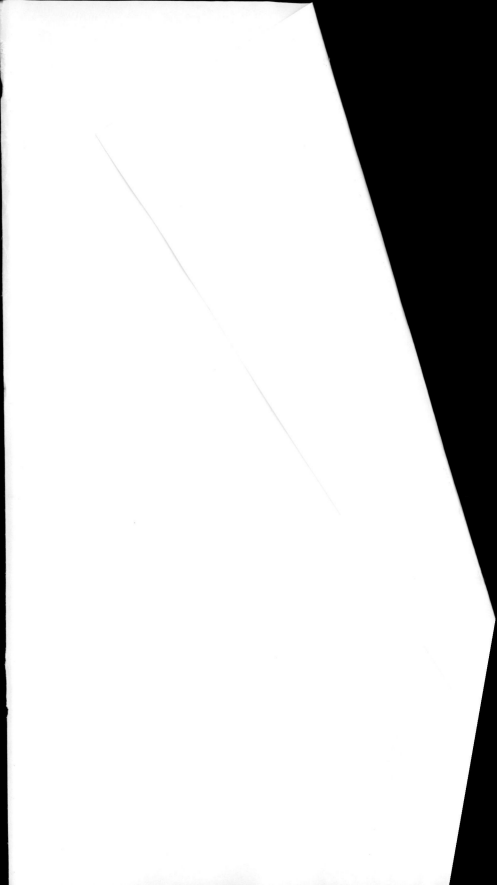